Theory of Solidification

The processes of freezing and melting were present at the beginnings of the Earth and continue to affect the natural and industrial worlds. The solidification of a liquid or the melting of a solid involves a complex interplay of many physical effects. This book systematically presents the field of continuum solidification theory based on instability phenomena. An understanding of the physics is developed by using examples of increasing complexity with the object of creating a deep physical insight applicable to more complex problems.

Applied mathematicians, engineers, physicists and materials scientists will all find this volume of interest.

Stephen H. Davis is McCormick Professor and Walter P. Murphy Professor of Applied Mathematics at Northwestern University.

CAMBRIDGE MONOGRAPHS ON MECHANICS

FOUNDING EDITOR

G. K. Batchelor

GENERAL EDITORS

S. H. Davis

McCormick Professor and Walter P. Murphy Professor
Applied Mathematics
Northwestern University

L. B. Freund

Henry Ledyard Goddard University Professor
Division of Engineering
Brown University

S. Leibovich

Sibley School of Mechanical and Aerospace Engineering
Cornell University

V. Tvergaard

Department of Solid Mechanics
The Technical University of Denmark

THEORY OF SOLIDIFICATION

STEPHEN H. DAVIS
Northwestern University

PUBLISHED BY THE PRESS SYNDICATE OF THE UNIVERSITY OF CAMBRIDGE
The Pitt Building, Trumpington Street, Cambridge, United Kingdom

CAMBRIDGE UNIVERSITY PRESS
The Edinburgh Building, Cambridge CB2 2RU, UK
40 West 20th Street, New York, NY 10011-4211, USA
10 Stamford Road, Oakleigh, VIC 3166, Australia
Ruiz de Alarcón 13, 28014 Madrid, Spain
Dock House, The Waterfront, Cape Town 8001, South Africa

© Cambridge University Press 2001

This book is in copyright. Subject to statutory exception
and to the provisions of relevant collective licensing agreements,
no reproduction of any part may take place without
the written permission of Cambridge University Press.

First published 2001

Printed in the United Kingdom at the University Press, Cambridge

Typeface Times Roman 10/13 pt.   *System* LATEX 2_ε [TB]

A catalog record for this book is available from the British Library.

Library of Congress Cataloging in Publication Data
Davis, Stephen H., 1939–
Theory of solidification / Stephen H. Davis.
p. cm. – (Cambridge monographs on mechanics)
Includes bibliographical references and index.
ISBN 0-521-65080-1
1. Solidification.   I. Title.   II. Series.
QC303 .D38   2001
530.4′14 – dc21   2001018487

ISBN 0 521 65080 1 hardback

I dedicate this book to the wonderful women in my life,
my mother Eva
and
my wife Suellen

Contents

Preface			*page* xiii
1	**Introduction**		1
2	**Pure Substances**		7
	2.1	Planar interfaces	7
		2.1.1 Mathematical model	7
		2.1.2 One-dimensional freezing from a cold boundary	9
		2.1.3 One-dimensional freezing from a cold boundary: Small undercooling	13
		2.1.4 One-dimensional freezing into an undercooled melt	15
		2.1.5 One-dimensional freezing into an undercooled melt: Effect of kinetic undercooling	18
	2.2	Curved interfaces	21
		2.2.1 Boundary conditions	21
		2.2.2 Growth of a nucleus in an undercooled melt	26
		2.2.3 Linearized instability of growing nucleus	32
		2.2.4 Linearized instability of a plane front growing into an undercooled melt	35
		2.2.5 Remarks	39
3	**Binary Substances**		42
	3.1	Mathematical model	42
	3.2	Directional solidification	45

	3.3	Basic state and approximate models	46
	3.4	Linearized instability of a moving front in directional solidification	48
	3.5	Mechanism of morphological instability	56
	3.6	More general models	57
	3.7	Remarks	59
4	**Nonlinear theory for directional solidification**		**62**
	4.1	Bifurcation theory	62
		4.1.1 Two-dimensional theory	62
		4.1.2 Two-dimensional theory for wave number selection	66
		4.1.3 Three-dimensional theory	72
	4.2	Long-scale theories	76
		4.2.1 Small segregation coefficient	77
		4.2.2 Small segregation coefficient and large surface energy	78
		4.2.3 Near absolute stability	80
	4.3	Remarks	82
5	**Anisotropy**		**86**
	5.1	Surface energy and kinetics	86
	5.2	Directional solidification with "small" anisotropy	91
	5.3	Directional solidification with "small" anisotropy: Stepwise growth	97
	5.4	Unconstrained growth with "small" anisotropy	105
		5.4.1 Two-dimensional crystal and one-dimensional front	110
		5.4.2 Three-dimensional crystal and two-dimensional front	111
	5.5	Unconstrained growth with "large" anisotropy – One-dimensional interfaces	121
	5.6	Unconstrained growth with "large" anisotropy – Two-dimensional interfaces	135
	5.7	Faceting with constant driving force	139
	5.8	Coarsening	152
	5.9	Remarks	156
6	**Disequilibrium**		**162**
	6.1	Model of rapid solidification	164
	6.2	Basic state and linear stability theory	167

	6.3	Thermal effects	171
	6.4	Linear-stability theory with thermal effects	172
		6.4.1 Steady mode	173
		6.4.2 Oscillatory mode	173
		6.4.3 The two modes	177
	6.5	Cellular modes in the FTA: Two-dimensional bifurcation theory	181
	6.6	Oscillatory modes in the FTA: Two-dimensional bifurcation theory	183
	6.7	Strongly nonlinear pulsations	189
		6.7.1 Small β	190
		6.7.2 Large β	198
		6.7.3 Numerical simulation	203
	6.8	Mode coupling	204
		6.8.1 Pulsatile–cellular interactions	204
		6.8.2 Oscillatory–cellular interactions	205
		6.8.3 Oscillatory–pulsatile interactions	206
	6.9	Phenomenological models	208
	6.10	Remarks	211
7	**Dendrites**		**215**
	7.1	Isolated needle crystals	217
	7.2	Approximate selection arguments	221
	7.3	Selection theories	229
	7.4	Arrays of needles	237
	7.5	Remarks	251
8	**Eutectics**		**255**
	8.1	Formulation	256
	8.2	Approximate theories for steady growth and selection	261
	8.3	Instabilities	267
	8.4	Remarks	270
9	**Microscale Fluid Flow**		**274**
	9.1	Formulation	276
	9.2	Prototype flows	279
		9.2.1 Free convection	279
		9.2.2 Bénard convection	280
	9.3	Directional solidification and volume-change convection	283

	9.4	Directional solidification and buoyancy-driven convection	287
	9.5	Directional solidification and forced flows	292
	9.6	Directional solidification with imposed cellular convection	304
	9.7	Flows over Ivantsov needles	311
	9.8	Remarks	319
10	**Mesoscale Fluid Flow**	**324**
	10.1	Formulation	325
	10.2	Planar solidification between horizontal planes	326
	10.3	Mushy-zone models	331
	10.4	Mushy zones with volume-change convection	336
	10.5	Mushy zones with buoyancy-driven convective instability	341
	10.6	An oscillatory mode of convective instability	349
	10.7	Weakly nonlinear convection	356
	10.8	Chimneys	357
	10.9	Remarks	363
11	**Phase-Field Models**	**366**
	11.1	Pure materials – A model system	367
	11.2	Pure materials – A deduced system	372
	11.3	Pure materials – Computations	374
	11.4	Remarks	376

Index   379

Preface

Materials Science is an extremely broad field covering metals, semiconductors, ceramics, and polymers, just to mention a few. Its study is dominated by the *fabrication* of specimens and the *characterization* of their properties. A relatively small portion of the field is devoted to *phase transformation*, the dynamic process by which in the present context a liquid is frozen or a solid is melted.

This book is devoted to the study of liquid (melt)-solid transformations of atomically rough materials: metals or semiconductors, including model organics like plastic crystals. The emphasis is on the use of instability behavior as a means of understanding those processes that ultimately determine the microstructure of a crystalline solid. The fundamental building block of this study is the Mullins–Sekerka instability of a front, which gives conditions for the growth of infinitesimal disturbances of a soild–liquid front. This is generalized in many ways: into the nonlinear regime, including thermodynamic disequilibrium, anisotropic material properties, and effects of convection in the liquid. Cellular, eutectic, and dendritic behaviors are discussed. The emphasis is on dynamic phenomena rather than equilibria. In a sense then, it concerns "physiology" rather than "anatomy."

The aim of this book is to present in a systematic way the field of continuum solidification theory. This begins with the primitive field equations for diffusion and the derivation of appropriate jump conditions on the interface between the solid and liquid. It then uses such models to explore morphological instabilities in the linearized range and gives physical explanations for the phenomena uncovered. To this point the discussion is elementary in terms of mathematical sophistication. It then enters into the nonlinear theories of morphological change with the use of bifurcation theory for wave number and pattern selection, long-wave theories in the strongly nonlinear range, and numerical simulation. The reader is assumed to be reasonably sophisticated in the mathematical methods, that is, stability theory and its nonlinear extensions and some asymptotic and perturbation theory, but having little background in materials science. Thus, the book is deliberately nonuniform in its "degree of difficulty." Those with limited mathematical background can skip the nonlinear theories and read about the physical phenomena and the linearized theories in the various chapters. The text should take the reader from the elements of the physics to the latest developments of the theory. It would be hoped that applied mathematicians, engineers, and physicists would profit from the material presented as would theoretically inclined materials scientists who could see how mathematics can generate understanding of relevant physical phenomena. An understanding of the physics is developed by using examples of increasing complexity with the objective of creating a deep physical insight applicable to more complex problems.

My interest in solidification was first stimulated by Jon Dantzig in his Ph.D. thesis of 1977 and permanently triggered by Ulrich Müller in our 1984 work on Bénard convection coupled to a freezing front. When learning a new subject as an "adult," one leans heavily on the expertise of senior colleagues for their wisdom. I thus wish to publicly thank Sam Coriell, Jon Dantzig, Paul Fife, Marty Glicksman, Wilfried Kurz, Jeff McFadden, Uli Müller, Bob Sekerka, Peter Voorhees, and Grae Worster for their contributions to my education.

I have always learned more from my graduate students, post-doctoral fellows, and visiting scientists than they have from me. I wish to thank them for their willingness to try something new. They are V. S. Ajaev, K. Brattkus, R. J. Braun, L. Bühler, D. J. Canright, Y.-J. Chen, J. A. Dantzig, A. A. Golovin, H.-P. Grimm, D. A. Huntley, P.-Q. Luo, G. B. McFadden, G. J. Merchant, P. Metzener, U. Müller, D. S. Riley, T. P. Schulze, B. J. Spencer, A. Umantsev, G. W. Young, and J.-J. Xu.

I am grateful to several people for reading selected chapters of the book and making important suggestions. They are Dan Anderson, Kirk Brattkus, Yi-Ju Chen, Jon Dantzig, Sasha Golovin, Jeff McFadden, Tim Schulze, Peter Voorhees, Grae Worster, and J.-J. Xu.

This book could not have been written without the generous support of the National Aeronautics and Space Administration Microgravity Sciences and Applications Program.

Finally, I would like to thank my secretary, Judy Piehl, not only for her impeccable typing, but for her sense of joy in her work. Her presence in the department makes it possible for all of us to do better what we do.

1

Introduction

The processes of freezing and melting were present at the beginning of the Earth and continue to affect the natural and industrial worlds. These processes created the Earth's crust and affect the dynamics of magmas and ice floes, which in turn affect the circulation of the oceans and the patterns of climate and weather. A huge majority of commercial solid materials were "born" as liquids and frozen into useful configurations. The systems in which solidification is important range in scale from nanometers to kilometers and couple with a vast spectrum of other physics.

The solidification of a liquid or the melting of a solid involves a complex-interplay of many physical effects. The solid–liquid interface is an active free boundary from which latent heat is liberated during phase transformation. This heat is conducted away from the interface through the solid and liquid, resulting in the presence of thermal boundary layers near the interface. Across the interface, the density changes, say, from ρ^ℓ to ρ^s. Thus, if $\rho^s > \rho^\ell$, so that the material shrinks upon solidification, a flow is induced toward the interface from "infinity."

If the liquid is not pure but contains solute, preferential rejection or incorporation of solute occurs at the interface. For example, if a single solute is present and its solubility is smaller in the (crystalline) solid than it is in the liquid, the solute will be rejected at the interface. This rejected material will be diffused away from the interface through the solid, the liquid, or both, resulting in the presence of concentration boundary layers near the interface. The thermal and concentration boundary layer structures determine, in large part, whether morphological instabilities of the interface exist and what the ultimate microstructure of the solid becomes. Many a solidification problem of interest couples the preceding purely diffusive effects with effects of thermodynamic disequilibrium, crystalline anisotropy, and convection in the melt.

On the coarsest level of understanding, freezing is of concern only as a heat or mass transfer process. Thus, one cools a glass of bourbon by inserting ice cubes that extract heat by melting. Likewise, one places salt on icy roads in Evanston to facilitate melting because salt water has a lower melting temperature than pure water.

On a finer level of understanding, freezing can create solids whose microstructures are determined by the process parameters and the intrinsic instabilities of the solid–liquid front. Figure 1.1 shows a longitudinal section of a Zn–Al alloy casting. Notice the dendritic structures that extend inward from the cold boundary and a core region in which no microstructure is visible. At later times, spontaneous nucleation in the core can cause "snowflakes" to grow in the core. The coarseness or fineness of the microstructure helps determine whether mechanical and thermal reprocessing can be accomplished without the appearance of cracks.

Under certain conditions of freezing, the moving solidification front can be susceptible to traveling-wave instabilities, giving structural patterns that can be made visible; see Figure 1.2.

When a eutectic alloy is frozen, the solid can take the form of a lamellar structure, alternate plates of two alloys spatially periodic perpendicular to the freezing direction. Under certain conditions this mode of growth is stable, giving rise to the more complex modes of growth, an example of which is shown in Figure 1.3.

Under conditions of rapid solidification, the microstructure can take on metastable states and patterns inconsistent with equilibrium thermodynamics. Figure 1.4 shows a banded structure in an Al–Cu alloy consisting of alternate layers of structured and unstructured material spatially periodic in the freezing direction. The structured layers may contain cells, dendrites, or eutectic material, whereas the alternate layers seem to have no visible microstructure.

If the solidification process occurs in a gravitational field, the thermal and solutal gradients may induce buoyancy-driven convection, which is known to affect the interfacial patterns greatly and, hence, the solidification microstructures present in the solidified material. The coupling of fluid flow in the melt with phase transformation at the interface can result in changes of microstructure scale and pattern due to alterations of frontal instabilities and the creation of new ones.

When an alloy is frozen at moderate speeds and dendritic arrays are formed, interesting dynamics occur in the dendrite–liquid mixture – the mushy zone. Here, solutal convection can be localized, creating channels parallel to the freezing direction, as shown in Figure 1.5. The channels frozen into the solid are called freckles, and their presence can significantly weaken the structure of the solid.

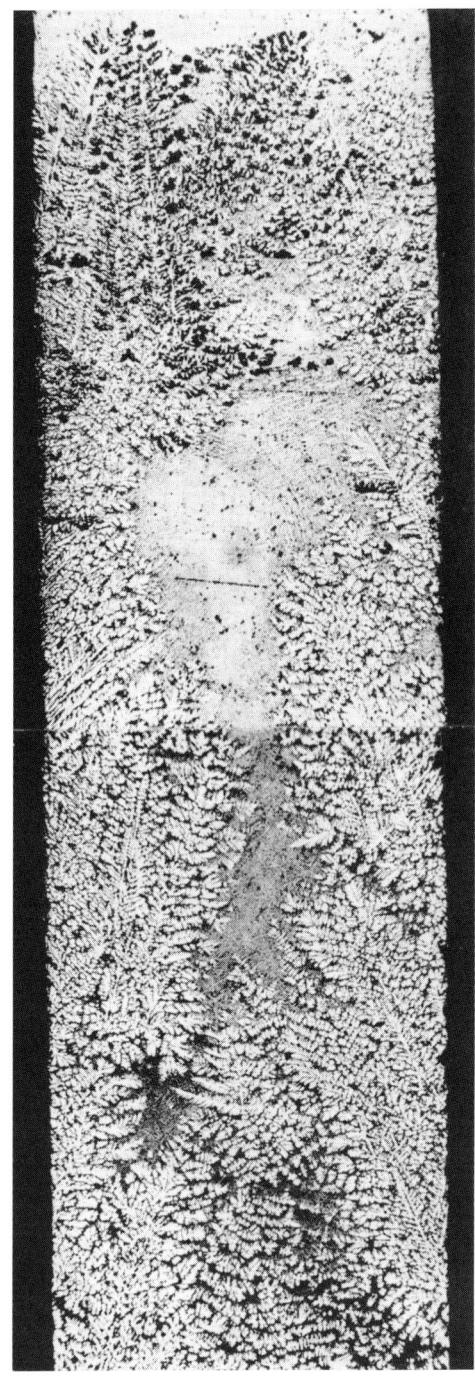

Figure 1.1. Longitudinal section of the quenched interface of the Zn–27%Al alloy. From Ayik et al. (1986).

Figure 1.2. Etched longitudinal section of a Ga-doped Ge single crystal showing traveling waves on the interface. The *arrow* indicates the growth direction. From Singh, Witt, and Gatos (1974).

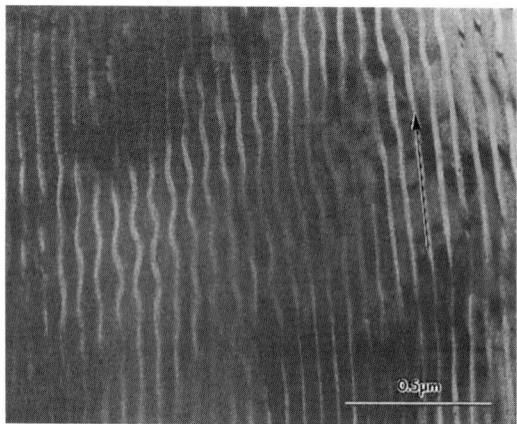

Figure 1.3. TEM micrographs of laser rapidly solidified Al–40 wt % Cu alloy oscillatory instabilities. $V = 0.03$ m/s. From Gill and Kurz (1993).

Given that the solid has crystalline structure, intrinsic symmetries in the material properties help define the continuum material. The surface energy and the kinetic coefficient on the interface as well as the bulk transport properties inherit the directional properties of the crystal, and thus anisotropies are often significant in determining the cellular or dendritic patterns that emerge. If the anisotropy is strong enough, the front can exhibit facets and corners.

1. Introduction

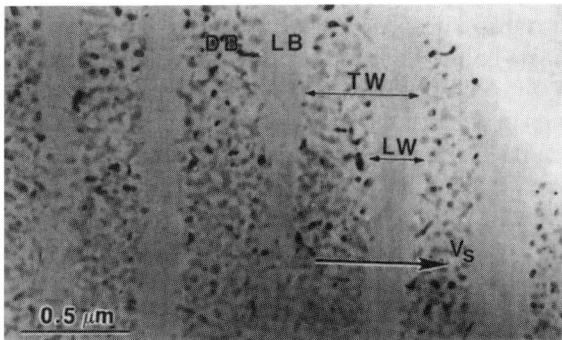

Figure 1.4. Enlarged view of the banded structure in Al–Cu 17 wt %. The dark bands have a dendritic structure, whereas the light bands are microsegregation free. DB = dark band, LB = light band, TW = total bandwidth, LW = light bandwidth, and V_s = growth rate. From Zimmermann et al. (1991).

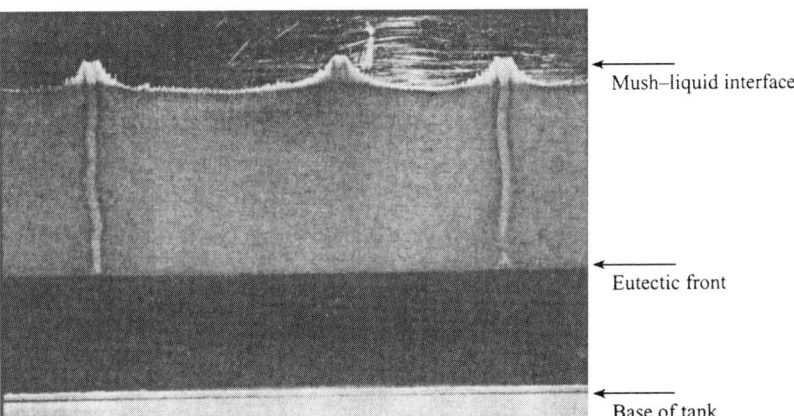

Figure 1.5. A photograph of mushy layer chimneys during an experiment with an ammonium chloride solution. In this system, pure ammonium chloride crystals are formed when the solution is cooled below its freezing temperature, leaving behind a diluted solution with a density lower than that of the bulk fluid. In the present case, the mushy layer is growing away from a fixed cold base that is at a temperature below the eutectic point, and thus both the solid–mush and mush–liquid interfaces are advancing at a decreasing rate. At the time the photograph was taken the distance between the base of the tank and the eutectic front was about 3 cm. Notice that the chimney walls and the mush–liquid interface are flat to a good first approximation. From Schulze and Worster (1998).

Finally, single crystals can be grown having, one would hope, uniform properties as long as the growth rate is very small. However, even in such cases the structure can be interrupted by defects or striations. In Figure 1.6, thermal fluctuations have created solute variations in the form of concentric rings, making the crystal inhomogeneous. If the crystal were rotated to remove azimuthal

1. Introduction

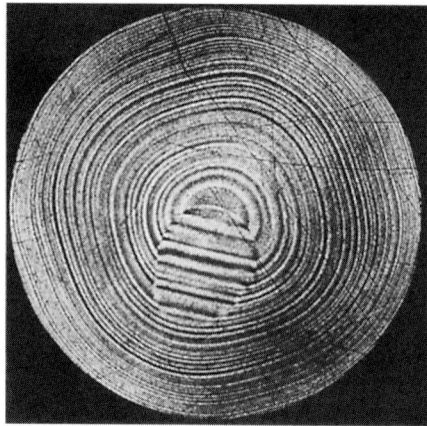

Figure 1.6. Transverse section of a $Ba_2NaNb_5O_{15}$ crystal whose rotational striations form concentric closed loops. The striations are caused by temperature fluctuations in the melt. From Hurle (1993).

thermal variations, rotational striations could occur having the form of spirals emanating from the center of rotation.

The challenge to the scientist is to understand the sources of such inhomogeneities, quantify the phenomena at work, and learn to control the processes so as to create desired microstructures in situ on demand. Significant progress has been made in these directions, though the end point is not at hand. Clearly, this is a huge field, and inevitably an author must make subjective choices of what material to include. The view taken here is that one should delve into a "core" of the field. A grasp of the physics is developed by using examples of increasing complexity intended to create a deep physical insight applicable to more complex problems.

References

Ayik, O., Ghoreshy, M., Sahoo, M., and Smith, R. W. (1986). Solidification and foundry studies of Zn/Al alloys, *J. Crystal Growth* **79**, 594–603.

Gill, S. C., and Kurz, W. (1993). Rapidly solidified Al–Cu alloys – I. Experimental determination of the microstructure selection map, *Acta Metall. Mater.* **41**, 3563–3573.

Hurle, D. T. J. (1993). *Crystal Pulling from the Melt*, Springer-Verlag, Berlin.

Schulze, T. P., and Worster, M. G. (1998). A numerical investigation of steady convection in mushy layers during directional solidification of binary alloys, *J. Fluid Mech.* **356**, 199–220.

Singh, R., Witt, A. F., and Gatos, H. C. (1974). Oscillatory interface instability during Czochralski growth of a heavily doped germanium, *J. Electrochem. Soc.* **121**, 380–385.

Zimmermann, M., Carrard, M., Gremaud, M., and Kurz, W. (1991). Characterization of a banded structure in rapidly solidified Al–Cu alloys, *Mater. Sci. Eng.* **A134**, 1279–1282.

2

Pure substances

2.1 Planar Interfaces

2.1.1 Mathematical Model

Consider a system in thermal equilibrium so that the temperature T is uniform. Part of the system is liquid and part is solid. For the two phases to coexist, the solid–liquid interfaces must be *planar*, and the temperature must be T_m, the *melting temperature*; T_m may depend on pressure and is here taken to be constant.

The amount of heat required to change a unit *mass* of solid into liquid at $T = T_m$ is the *latent heat* L; if ρ^s is the density of the solid, then the latent heat per unit volume is L_v, $L_v = \rho^s L$. The amount of heat required to raise, without change of phase, the temperature of a unit mass of solid or liquid by 1°C is the specific heat c_p.

Consider now a system in which temperature gradients are present so that there are heat fluxes. The bulk heat balance in either phase alone can be obtained by considering a material volume $\mathcal{V}(t)$, as shown in Figure 2.1, and is given by

$$\frac{d}{dt} \int_{\mathcal{V}(t)} \rho c_p T \, dV = - \int_{\partial \mathcal{V}(t)} \mathbf{q} \cdot \mathbf{n} \, dS, \quad (2.1)$$

where ρ is the density, $\mathbf{q}$ is the heat flux, and $\mathbf{n}$ is the unit outward–normal vector to $\mathcal{V}$ on its (closed) boundary $\partial \mathcal{V}$.

The transport theorem for any smooth field F passing through $\mathcal{V}$ states that

$$\frac{d}{dt} \int_{\mathcal{V}(t)} F \, dV = \int_{\mathcal{V}(t)} \left[\frac{\partial F}{\partial t} + \nabla \cdot (F \mathbf{v}) \right] dV, \quad (2.2)$$

where $\mathbf{v}$ is the velocity field of the material (see, e.g., Serrin 1959).

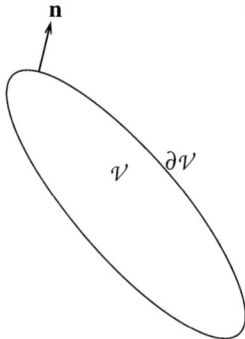

Figure 2.1. A control volume $\mathcal{V}$ entirely with a bulk phase; $\partial\mathcal{V}$ is its boundary and $\mathbf{n}$ is the unit outward normal.

If Gauss's theorem and identity (2.2) are used on relation (2.1), then

$$\int_{\mathcal{V}(t)} \left\{ \frac{\partial}{\partial t}(\rho c_p T) + \nabla \cdot (\rho c_p T \mathbf{v}) \right\} dV = -\int_{\mathcal{V}(t)} \nabla \cdot \mathbf{q} \, dV,$$

and since $\mathcal{V}$ is arbitrary and the integrands are supposed smooth, the point form of the bulk mass balance is obtained as

$$\frac{d}{dt}(\rho c_p T) + \rho c_p T \nabla \cdot \mathbf{v} = -\nabla \cdot \mathbf{q}, \tag{2.3}$$

where the material derivative is given by

$$\frac{d}{dt} = \frac{\partial}{\partial t} + \mathbf{v} \cdot \nabla. \tag{2.4}$$

To complete the specification of the heat balance, a constitutive law is required that relates $\mathbf{q}$ to the temperature field. It is assumed here that the Fourier law of heat conduction holds, that is

$$\mathbf{q} = -k_T \nabla T, \tag{2.5}$$

where k_T is the thermal conductivity of the phase. Thus, the final form of the bulk heat balance is given by

$$\frac{d}{dt}(\rho c_p T) + \rho c_p T \nabla \cdot \mathbf{v} = \nabla \cdot k_T \nabla T. \tag{2.6}$$

In the absence of bulk flow, $\mathbf{v} = \mathbf{0}$, and for ρ, c_p, k_T constant, Eq. (2.6) reduces to the standard heat-conduction equation

$$\frac{\partial T}{\partial t} = \kappa \nabla^2 T, \tag{2.7}$$

2.1 Planar interfaces

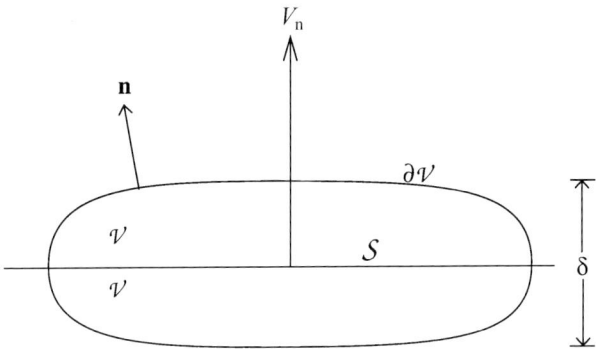

Figure 2.2. A control voume $\mathcal{V}$ spanning the interface $\mathcal{S}$ that moves at speed V_n normal to itself: $\partial\mathcal{V}$ is the boundary of $\mathcal{V}$ and **n** is the unit outward normal.

where

$$\kappa = k_T/\rho c_p$$

is the thermal diffusivity of the phase.

On a moving (planar) interface, there is a heat balance. Consider a (two-dimensional) volume of height δ spanning the interface, as shown in Figure 2.2. If V_n is the speed of the interface (normal to itself), then in a time δt and for $\delta \to 0$,

$$\rho^s L V_n \delta t = (\mathbf{q}_\ell - \mathbf{q}_s) \cdot \mathbf{n} \delta t$$

because the (smooth) heat accumulation vanishes as $\delta \to 0$. Thus, if Fourier heat conduction is applied, Eq. (2.5), the interfacial heat balance is

$$\rho^s L V_n = (k_T^s \nabla T^s - k_T^\ell \nabla T^\ell) \cdot \mathbf{n}. \qquad (2.8)$$

One sees that the net heat entering the interface, the right-hand side, determines the speed V_n of the front.

In addition, the temperature is continuous across the interface and is known to be the equilibrium melting temperature T_m,

$$T^s = T^\ell = T_m. \qquad (2.9)$$

2.1.2 One-Dimensional Freezing from a Cold Boundary

Consider a plane boundary at $z = 0$, which is adjacent to a liquid at initial temperature $T = T_m$, as shown in Figure 2.3. At $t = 0$, the boundary is impulsively

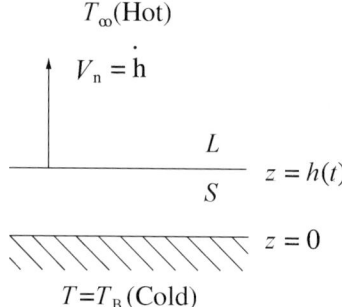

Figure 2.3. Planar solidification from a cold boundary at $z = 0$ with temperature T_B into a warmer melt at temperature T_∞. The interface between solid and liquid is at $z = h(t)$.

cooled to a temperature T_B, such that the undercooling ΔT is

$$\Delta T = T_m - T_B > 0, \qquad (2.10)$$

creating a solid–liquid interface at $z = h(t)$ and it will be supposed that $\rho^s = \rho^\ell$.

Because for $t < 0$, $T = T_m$, the temperature in the liquid will not fall below T_m, hence, the temperature in the liquid is constant for all time,

$$T^\ell = T_m \qquad z > h(t). \qquad (2.11a)$$

In the solid there is heat conduction

$$T^s_t = \kappa^s T^s_{zz} \qquad 0 < z < h(t). \qquad (2.11b)$$

For $t > 0$

$$T^s = T_B \qquad z = 0 \qquad (2.11c)$$

$$T^\ell = T^s = T_m \qquad z = h(t) \qquad (2.11d)$$

$$\rho^s L \dot{a} = k^s_T T^s_z \qquad z = h(t). \qquad (2.11e)$$

For $t = 0$,

$$T^s = T_m, \quad h = 0. \qquad (2.11f)$$

Note that the heat flux in the liquid is zero because the temperature there is constant.

There are no natural time and space scales here, and therefore a similarity solution can be sought. Let the new independent variable be η,

$$\eta = \frac{z}{2\sqrt{\kappa^s t}}, \qquad (2.12a)$$

2.1 Planar interfaces

define the nondimensional temperature by θ,

$$T^s = T_b + (\Delta T)\theta(\eta), \tag{2.12b}$$

and thus $\theta = 0$ at the base and $\theta = 1$ at the front. Finally, consistent with the preceding equations, the interface position is written as

$$h(t) = 2\Lambda\sqrt{\kappa^s t}, \tag{2.12c}$$

where the value of Λ, as yet unknown, determines the speed and position of the front. Through the use of these forms, system (2.11a) becomes

$$\theta'' + 2\eta\theta' = 0 \qquad 0 < \eta < \Lambda \tag{2.13a}$$

$$\theta = 0 \qquad \eta = 0 \tag{2.13b}$$

$$\theta = 1 \qquad \eta = \Lambda \tag{2.13c}$$

$$\theta' = 2S\Lambda \qquad \eta = \Lambda \tag{2.13d}$$

where the Stefan number S is

$$S = \frac{L}{c_p^s \Delta T}. \tag{2.14}$$

Notice that the initial conditions (2.11f) applied at $t = 0$ corresponds to $\eta \to \infty$ and that the temperature is constant in (h, ∞). Thus, the thermal condition can be applied at $\eta = \Lambda$, as shown in Eq. (2.13c). One integral of Eq. (2.13a) gives

$$\theta' = Ae^{-\eta^2}, \tag{2.15a}$$

where the integration constant A satisfies

$$2S\Lambda = Ae^{-\Lambda^2}. \tag{2.15b}$$

A second integral that satisfies Eqs. (2.13b,c) is

$$\theta = \frac{\int_0^\eta e^{-s^2} ds}{\int_0^\Lambda e^{-s^2} ds} = \frac{\operatorname{erf}(\eta)}{\operatorname{erf}(\Lambda)}, \tag{2.15c}$$

where the error function is defined by

$$\operatorname{erf}(z) \equiv \frac{2}{\sqrt{\pi}} \int_0^z e^{-s^2} ds. \tag{2.16}$$

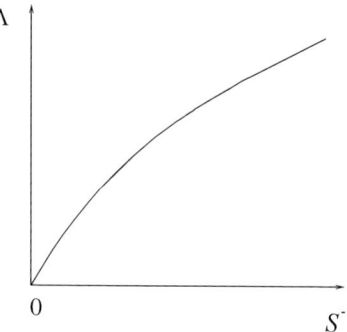

Figure 2.4. A sketch of Eq. (2.17) for the interface speed Λ versus the undercooling S^{-1}. Solutions exist for all S^{-1}.

When form (2.15c) is substituted into the flux condition (2.13d), one finds that

$$\sqrt{\pi}\, \Lambda e^{\Lambda^2} \operatorname{erf}(\Lambda) = S^{-1}, \tag{2.17}$$

which gives $\Lambda = \Lambda(S)$, the speed of the front as a function of the undercooling, as shown in Figure 2.4.

Notes

1. Solutions exist for all values of the nondimensional undercooling S^{-1}.
2. Notice that $h \sim t^{1/2}$, and hence $\dot{h} \sim t^{-1/2}$. The solution fails at $t = 0$, where the front speed is infinite (owing to the assumption of impulsive heating) and decreases with time.
3. The temperature gradient G_T at the interface is

$$\begin{aligned} G_T = T_z^s \mid_{z=h} &= \frac{\Delta T}{2\sqrt{\kappa^s t}} \theta'(\eta) \mid_{\eta=\Lambda} \\ &= \frac{\Delta T}{\sqrt{\kappa^s t}} \Lambda S > 0, \end{aligned} \tag{2.18}$$

and thus the heat flows downward through the solid. Consequently, the front speed depends on κ^s, and not κ^ℓ. As will be seen, $G_T > 0$ indicates that the front is stable to disturbances periodic along the front.
4. If, initially, one sets the temperature of the liquid at $T^\ell = T_\infty > T_m$, then there would also be heat flow in the liquid and the profiles would look as shown in Figure 2.5; the front speed would then depend on both diffusivities. In both cases T is continuous at $z = h(t)$, but the gradient T_z is not.
5. The similarity solution posed is a "preferred" solution in the sense that, under rather weak conditions, all solutions of the initial-value problem (2.11) approach the similarity solution as $t \to \infty$.
6. When S^{-1} is small, ΔT is small or L is large. Figure 2.4 or Eq. (2.17) shows that Λ is small so that freezing takes place very slowly. It is useful to analyze the limit of small ΔT separately.

2.1 Planar interfaces

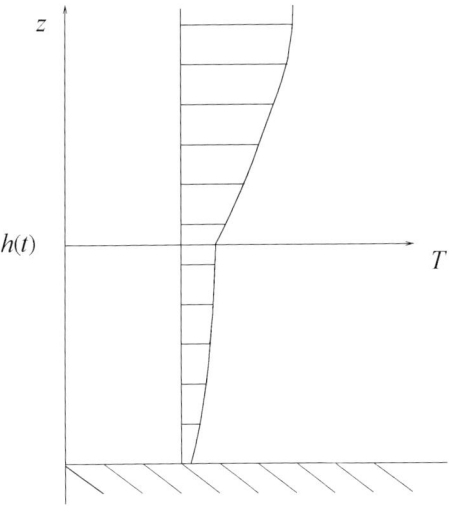

Figure 2.5. A sketch of the temperature profiles in the solid and liquid for the two-phase conduction problem.

2.1.3 One-Dimensional Freezing from a Cold Boundary: Small Undercooling

When $S^{-1} \equiv \epsilon \ll 1$, then the growth rate is small. Scale the original system Eqs. (2.11) as follows:

$$\zeta = z/\ell, \quad \tau = t/t_o \tag{2.19a}$$

$$A(\tau) = h(t)/\ell, \quad T^s(z,t) = T_B + (\Delta T)\theta(\zeta, \tau), \tag{2.19b}$$

where ℓ and t_o are scales undefined for the moment. From the previous solution it is seen that heat conduction in the solid is important; therefore let $t_o = \ell^2/\kappa^s$. System (2.11) then becomes

$$\theta_\tau = \theta_{\zeta\zeta} \quad 0 < \zeta < A(\tau) \tag{2.20a}$$

$$\theta = 0 \quad \zeta = 0 \tag{2.20b}$$

$$\theta = 1 \quad \zeta = A(\tau) \tag{2.20c}$$

$$A_\tau = \epsilon \theta_\zeta \quad \zeta = A(\tau) \tag{2.20d}$$

$$\left.\begin{array}{l} A = A_0 \\ \theta(\zeta, 0) = \theta_0(0) \end{array}\right\}, \quad \tau = 0. \tag{2.20e}$$

In the preceding, the problem has been generalized to allow nonzero initial temperature distributions $\theta_0(\zeta)$ and initial-front positions A_0.

If $\partial/\partial\tau = O(1)$ and $\epsilon \to 0$, the resulting system remains second order in time and hence is capable of satisfying both initial conditions. However, at first approximation $A_\tau \sim 0$, and thus from time zero to $\tau = O(1)$ the interface is stationary at its initial position and solidification does not occur. In this time interval one then has a standard heat-conduction problem for θ on a fixed domain, $0 < \zeta < A$. This represents the *inner solution in time*. The *outer solution in time*, valid for long periods, requires a rescaling of time

$$\hat{\tau} = \epsilon\tau, \tag{2.21}$$

which represents a time scale based on latent heat and undercooling, namely $\rho^s L \ell^2 / k_T \Delta T$, and so describes the solidification process. In this case, system (2.11) becomes

$$\epsilon \theta_{\hat{\tau}} = \theta_{\zeta\zeta} \qquad 0 < \zeta < A(\hat{\tau}) \tag{2.22a}$$

$$\theta = 0 \qquad \zeta = 0 \tag{2.22b}$$

$$\theta = 1 \qquad \zeta = A(\hat{\tau}) \tag{2.22c}$$

$$A_{\hat{\tau}} = \theta_\zeta \qquad \zeta = A(\hat{\tau}) \tag{2.22d}$$

$$A = A_0, \ \theta = \theta_0 \qquad \hat{\tau} = 0. \tag{2.22e}$$

The limit $\partial/\partial\hat{\tau} = O(1)$ and $\epsilon \to 0$ is a singular perturbation; it is seen that at first approximation the temperature is quasi-steady,

$$\theta_{\zeta\zeta} = 0. \tag{2.23}$$

The solution that satisfies Eqs. (2.22b,c) is

$$\theta = \zeta/A. \tag{2.24}$$

Now the flux condition (2.22d) gives

$$AA_{\hat{\tau}} = 1, \tag{2.25}$$

which is a nonlinear evolution equation for A (Young 1994). Thus, with the first of condition (2.22e),

$$A^2(\hat{\tau}) - A_0^2 = 2\hat{\tau}. \tag{2.26}$$

In dimensional terms,

$$h^2 - h_0^2 = 2\epsilon\kappa^s t = \frac{2k_T^s(\Delta T)t}{\rho^s L}. \tag{2.27}$$

The inner and outer solutions automatically match asymptotically.

2.1 Planar interfaces

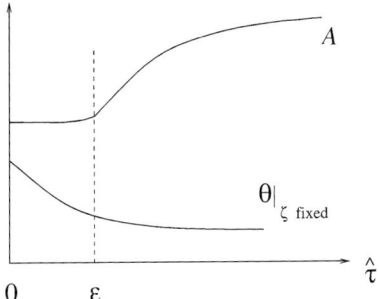

Figure 2.6. Sketches of the temperature and interface position as functions for the outer time $\hat{t}$ for small ϵ.

Solution (2.27) coincides with the similarity solution for $S \to \infty$ and $h_0 = 0$ where $h = 2\Lambda\sqrt{\kappa^s t}$ and $\Lambda \sim (2S)^{-1/2}$.

Note: The length scale ℓ was never defined, and because the original problem has no intrinsic length scale, ℓ cancels from the results, as seen in Eq. (2.27).

The solutions can be sketched symbolically, as shown in Figure 2.6. For $\hat{t} \sim \epsilon, \tau \sim 1$, A is constant, and θ develops. For $\hat{t} \sim 1$, θ is quasi-steady, and its time evolution is determined by that of A, as shown.

2.1.4 One-Dimensional Freezing into an Undercooled Melt

Consider the semi-infinite body of fluid shown in Figure 2.7 that is cooled below T_m to T_∞, $\Delta T = T_m - T_\infty$.

At $t = 0$, a plate is inserted at $z = 0$ at temperature T_m. For $\rho^s = \rho^\ell$, one wishes to determine how the system evolves. The temperature profile at a fixed time is shown in Figure 2.8; in the solid, $T = T_m$ always, and thus heat conduction is absent there.

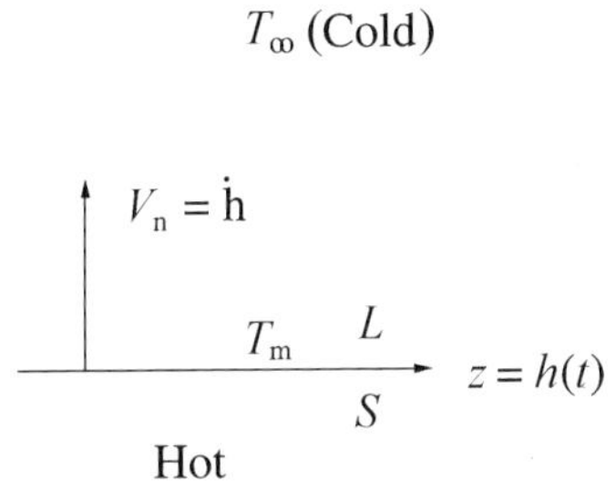

T_∞ (Cold)

$V_n = \dot{h}$

T_m L

  S $z = h(t)$

Hot

Figure 2.7. Planar solidification into an undercooled melt, where the interface has position $z = h(t)$.

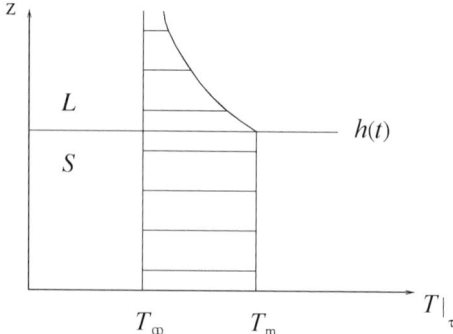

Figure 2.8. A sketch of the temperature profiles for planar solidification into an undercooled melt.

The governing system is as follows:
In solid, $z < h(t)$,

$$T^s = T_m \tag{2.28a}$$

In liquid, $z > h(t)$,

$$T^\ell_t = \kappa^\ell T^\ell_{zz} \tag{2.28b}$$

As $z \to \infty$,

$$T^\ell \to T_\infty \tag{2.28c}$$

On $z = h(t)$

$$T^\ell = T^s = T_m. \tag{2.28d}$$

$$\rho^s L \dot{h} = -k^\ell_T T^\ell_z. \tag{2.28e}$$

Again, because there are no natural spatial and time scales in the problem, a similarity solution can be sought. Let the new independent variable be

$$\eta = \frac{z}{2\sqrt{\kappa^\ell t}}, \tag{2.29a}$$

and let the scaled temperature be θ,

$$T^\ell = T_\infty + (\Delta T)\theta(\eta). \tag{2.29b}$$

For consistency, let

$$h = 2\Lambda\sqrt{\kappa^\ell t}, \tag{2.29c}$$

where the speed coefficient Λ is to be determined.

2.1 Planar interfaces

The solution for the temperature is

$$\theta(\eta) = \frac{\text{erfc}(\eta)}{\text{erfc}(\Lambda)}, \tag{2.30}$$

where the complementary error function is defined by

$$\text{erfc}(z) = \frac{2}{\sqrt{\pi}} \int_z^\infty e^{-s^2} ds. \tag{2.31}$$

The flux condition (2.28e) then gives the characteristic equation

$$\sqrt{\pi} \Lambda e^{\Lambda^2} \text{erfc}(\Lambda) = S^{-1}. \tag{2.32}$$

Note: The temperature gradient G_T in the liquid at the interface

$$G_T = T_z[h(t), t] < 0$$

always, and thus the heat flows through the liquid. Consequently, the front speed depends on κ^ℓ and not κ^s. As will be seen, $G_T < 0$ indicates that the front is unstable to disturbances periodic along the front.

If one plots Eq. (2.32), Figure 2.9 is obtained. The curve approaches $S^{-1} = 1$, which is called *unit undercooling*. There exist *no* solutions for $S^{-1} \geq 1$. Note that as $S^{-1} \to 1^-$, $\Lambda \to \infty$, and thus the front speed approaches infinity. This suggests the breakdown of the validity of the thermodynamic equilibrium assumption appropriate to relatively small front speeds.

Solidification is a surface reaction whose rate depends upon the degree of undercooling that drives it. The argument of Worster (private communication 1993) will be followed. At $T^1 = T_m$ a solid–liquid interface is in a dynamic equilibrium with molecules attaching and detaching continually and at equal rates. When $T^1 < T_m$, molecules become more strongly bound to the interface

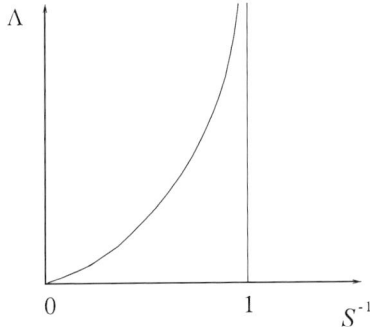

Figure 2.9. A sketch of Eq. (2.32) for the interface speed Λ versus the undercooling S^{-1}. Solutions exist only for $S^{-1} < 1$.

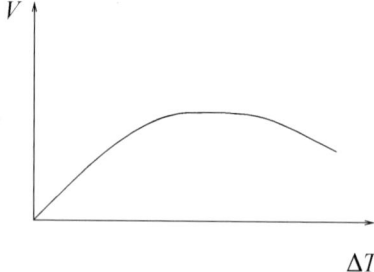

Figure 2.10. A sketch of the kinetic undercooling versus the speeds.

and thus the number detaching per unit time decreases and the interface advances at speed V_n; V_n increases with $T_m - T^l$. However, as T^l decreases further, the molecules in the liquid become sluggish and the rate of attachment decreases. Hence, the figure presumably looks like Figure 2.10. This relation may in fact not be a simple function of V directly, but a more complicated functional (Bates, Fife, Gardner, and Jones 1997). Here the model of Figure 2.10 will be used.

For ΔT small, the graph depends on the mode of attachment (e.g., by adding molecular planes, screw dislocations, or random attachments). However, for substances (e.g., metals) having low latent heats, V_n can be approximated as a linear function of ΔT,

$$V_n = \mu(T_m - T^l), \tag{2.33}$$

where the positive constant μ is called the *kinetic coefficient*. Equation (2.33) can be rewritten as

$$T^l = T_m - \mu^{-1} V_n \tag{2.34}$$

and represents the effect of *kinetic undercooling*.

Equation (2.34) determines the interfacial temperature for each front speed V_n. The presence of kinetic undercooling, $\mu^{-1} \neq 0$, lowers the interfacial temperature below that in equilibrium.

As will be seen in a moment, replacing Eqn. (2.28d) by (2.34), makes solutions possible for all ΔT.

2.1.5 One-Dimensional Freezing into an Undercooled Melt: Effect of Kinetic Undercooling

The generalized governing system now takes the form
  In solid, $z < h(t)$,

$$T^s = T^l \tag{2.35a}$$

2.1 Planar interfaces

In liquid, $z > h(t)$

$$T_t^\ell = \kappa^\ell T_{zz}^\ell \tag{2.35b}$$

As $z \to \infty$

$$T^\ell \to T_\infty \tag{2.35c}$$

On $z = h(t)$,

$$T^\ell = T^{\mathrm{I}}, \tag{2.35d}$$

$$\rho^s L \dot{h} = -k^\ell T_z^\ell, \tag{2.35e}$$

$$\dot{h} = \mu(T_{\mathrm{m}} - T^{\mathrm{I}}). \tag{2.35f}$$

Notice here that T^{I} is now unknown and must be determined as part of the solution.

An important difference between the equilibrium and nonequilibrium formulations is seen by dimensional analysis. Previously, the scales of length ℓ and time t_o were arbitrary, and a relation between them associated with heat conduction, $t_o = \ell^2/\kappa^\ell$, was used; however, ℓ was still arbitrary. Now, however, ℓ is determined by the kinetic undercooling, namely

$$\ell \left(\frac{\kappa^\ell}{\ell^2} \right) = \mu \Delta T,$$

and thus

$$\ell = \frac{\kappa^\ell}{\mu \Delta T}; \tag{2.36a}$$

hence, using $t_o = \ell^2/\kappa^\ell$, we obtain

$$t_o = \frac{\kappa^\ell}{\mu^2 (\Delta T)^2}. \tag{2.36b}$$

Write again

$$T^\ell = T_\infty + (\Delta T)\theta, \tag{2.37}$$

and the scaled system becomes

$$\theta_t = \theta_{zz} \qquad z > A(t) \tag{2.38a}$$

$$\theta \to 0 \qquad z \to \infty \tag{2.38b}$$

$$\left. \begin{array}{l} \theta = \theta^{\mathrm{I}} \\ S\dot{A} = -\theta_z \\ \dot{A} = 1 - \theta^{\mathrm{I}} \end{array} \right\}, \quad z = A(t) \tag{2.38c}$$

The kinetic condition suggests seeking a solution with θ^I constant; hence, $\dot{A}(t) = V$. Let us seek a traveling-wave solution

$$\theta = \theta(\zeta), \qquad (2.39a)$$

where

$$\zeta = z - Vt, \qquad (2.39b)$$

and so the interface lies at $\zeta = 0$. Here V is a constant and ζ is measured in a moving frame of reference. The diffusion equation (2.38a) then becomes

$$-V\theta' = \theta'', \qquad (2.40)$$

where a prime denotes $d/d\zeta$.

The solution of Eq. (2.40), subject to conditions (2.38b) and the first of (2.38c), is

$$\theta = \theta^I e^{-V\zeta}. \qquad (2.41)$$

The Stefan and kinetic conditions, the second and third of Eqs. (2.38c) then give

$$\theta^I = S \qquad (2.42)$$

and

$$V = 1 - S, \ S < 1. \qquad (2.43)$$

Thus, for all $S < 1$,

$$\theta = Se^{-(1-S)(z-Vt)}, \qquad (2.44)$$

and solutions for $S^{-1} > 1$ have been found (Glicksman and Schaefer 1967).

Note: Again, $G_T < 0$.

When there is unit undercooling, $S = 1$, yet a different solution exists (Umantsev 1985) with $h(t) \sim t^{2/3}$. With kinetic undercooling present, there are solutions for all S:

$$\left.\begin{array}{ll} S^{-1} < 1, & h \sim t^{\frac{1}{2}} \\ S^{-1} = 1, & h \sim t^{\frac{2}{3}} \\ S^{-1} > 1, & h \sim t \end{array}\right\}. \qquad (2.45)$$

2.2 Curved Interfaces

2.2.1 Boundary Conditions

Consider a two-dimensional solid "drop" on a substrate, as shown in Figure 2.11; no phase transformation is present. Thermodynamic equilibrium implies that the Helmholtz free energy E of the system, the sum of the surface energies, must be at a minimum for an equilibrium state to exist. Note that other energies, such as the elastic energy of "drop" and substrate, are ignored here, as is usual. The analysis follows Mullins (1963).

Let the system be uniform in the direction normal to the page and let w be a unit of length in that direction. The Helmholtz free energy is then

$$E = w \left\{ \int_{\ell_1}^{\ell_2} \gamma \left(1 + h_x^2\right)^{1/2} dx + \gamma_1(\ell_2 - \ell_1) + \gamma_2[\ell - (\ell_2 - \ell_1)] \right\}, \quad (2.46a)$$

where $z = h(x)$ is the height of the interface, γ is the energy per unit area on the drop–liquid interface, and γ_1 and γ_2 are the corresponding surface energies per unit area on the solid–substrate and liquid–substrate interfaces, respectively; all of these are taken to be constants, ℓ is a fixed length, always larger than drop width, which is introduced to keep the energies finite; $x = \ell_1$ and ℓ_2 are the endpoints of the drop, and θ, measured within the drop, is called the contact angle. Consider variations in h, ℓ_1, and ℓ_2 such that the volume $\mathcal{V}$ of the drop is preserved, where

$$\mathcal{V} = w \int_{\ell_1}^{\ell_2} h \, dx. \quad (2.46b)$$

For a discussion of the variational calculus needed for this section, see Courant and Hilbert (1953), Chap. 4.

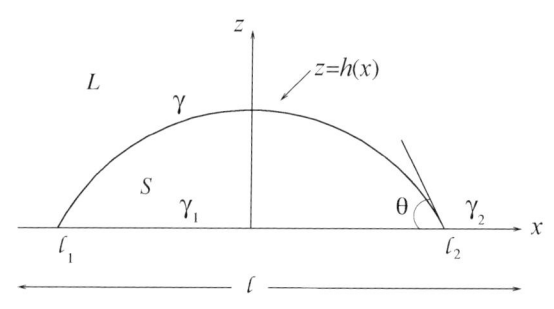

Figure 2.11. A solid, two-dimensional drop on a substrate; $z = h(x)$ is the drop shape, $x = \ell_1$ and ℓ_2 are the locations of the contact lines, ℓ indicates an expanse larger than the drop, and θ is the contact angle.

The constrained problem defined above can be written as an unconstrained variational problem by introducing the Lagrange multiplier λ and writing

$$E' = E + \lambda w \int_{\ell_1}^{\ell_2} h \, dx$$

$$= w \left\{ \int_{\ell_1}^{\ell_2} \left[\gamma \left(1 + h_x^2\right)^{1/2} + \lambda h \right] dx + \gamma_1(\ell_2 - \ell_1) + \gamma_2 \left[\ell - (\ell_2 - \ell_1)\right] \right\}. \tag{2.47}$$

In order for E' to be minimum, it is necessary for the first variation, $\delta E'$, to be zero,

$$w^{-1}\delta E' = \int_{\ell_1}^{\ell_2} \left\{ \frac{\gamma h_x \delta(h_x)}{\left(1 + h_x^2\right)^{1/2}} + \lambda \delta h \right\} dx$$

$$+ \left\{ \gamma \left[\left(1 + h_x^2\right)^{1/2}\right] + \lambda h \Big|_{\ell_2} + \gamma_1 - \gamma_2 \right\} \delta \ell_2$$

$$+ \left\{ -\gamma \left[\left(1 + h_x^2\right)^{1/2}\right] - \lambda h \Big|_{\ell_1} - \gamma_1 + \gamma_2 \right\} \delta \ell_1 = 0 \tag{2.48}$$

Formally, one writes that

$$\delta(h_x) = (\delta h)_x \tag{2.49}$$

and uses integration by parts to obtain

$$\int_{\ell_1}^{\ell} \frac{h_x \delta h_x}{\left(1 + h_x^2\right)^{1/2}} dx = -\int_{\ell_1}^{\ell_2} \left\{ \frac{h_x}{\left(1 + h_x^2\right)^{1/2}} \right\}_x \delta h \, dx + \frac{h_x \delta h}{\left(1 + h_x^2\right)^{1/2}} \Big|_{\ell_1}^{\ell_2}. \tag{2.50}$$

Figure 2.12 shows a neighborhood of the right-hand contact line at $x = \ell_2$. One can then write that

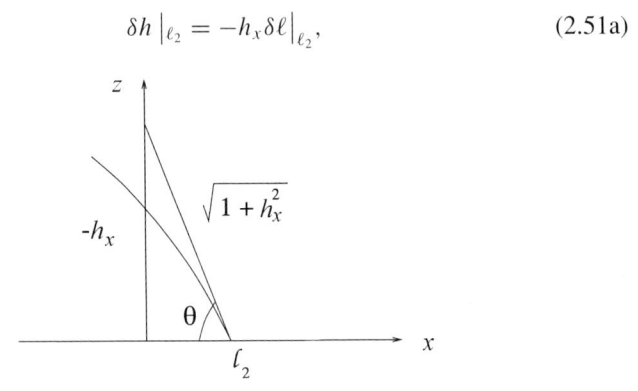

$$\delta h \big|_{\ell_2} = -h_x \delta \ell \big|_{\ell_2}, \tag{2.51a}$$

Figure 2.12. A close-up of the local geometry of the contact-line region near $x = \ell_2$ of Figure 2.11.

and similarly at the other contact line

$$\delta h \,|_{\ell_1} = -h_x \delta \ell \,|_{\ell_1} \qquad (2.51b)$$

Thus,

$$\frac{h_x \delta h}{\left(1+h_x^2\right)^{1/2}}\bigg|_{\ell_1}^{\ell_2} = -\frac{h_x^2 \delta \ell}{\left(1+h_x^2\right)^{1/2}}\bigg|_{\ell_1}^{\ell_2}. \qquad (2.52)$$

Finally,

$$w^{-1}\delta E' = \int_{\ell_1}^{\ell_2} \left\{ -\gamma \left[\frac{h_x}{\left(1+h_x^2\right)^{1/2}} \right]_x + \lambda \right\} \delta h \, dx$$

$$+ \left\{ \frac{\gamma}{\left(1+h_x^2\right)^{1/2}} \bigg|_{\ell_2} + \lambda h + \gamma_1 - \gamma_2 \right\} \delta \ell_2$$

$$- \left\{ \frac{\gamma}{\left(1+h_x^2\right)^{1/2}} \bigg|_{\ell_1} + \lambda h + \gamma_1 - \gamma_2 \right\} \delta \ell_1 = 0, \qquad (2.53)$$

where the boundary terms have been combined.

Consider first variations in which the endpoints are fixed, that is, $\delta\ell_1 = \delta\ell_2 = 0$. Given that the integrand of Eq. (2.53) is smooth and δh is otherwise arbitrary, that integrand must vanish. This gives the Euler–Lagrange equation

$$-\gamma \left[\frac{h_x}{\left(1+h_x^2\right)^{1/2}} \right]_x + \lambda = 0$$

or

$$2H\gamma = \lambda \qquad (2.54)$$

Here, for a one-dimensional interface the mean curvature H is defined by

$$2H = \frac{h_{xx}}{\left(1+h_x^2\right)^{3/2}}. \qquad (2.55)$$

The interfacial shape h has constant curvature, and thus in two dimensions is an arc of a circle. By this definition a solid finger extending into the liquid has $H < 0$. The Lagrange multiplier is a constant that cannot be determined by variational calculus but requires some additional physical statement (Hills and Roberts 1993).

Let the Gibbs free energy Φ of the system depend on the pressure p and the temperature T. Let Φ on the solid side of the interface be equal to that on the liquid side. Expand this equation about $p = p^\ell$, the pressure on the liquid side, let $T = T_m$ and identify $\partial \Phi / \partial p$ by $1/p^s$, and let $\partial \Phi / \partial T = -s$, the entropy, and $\Delta s = L_v/T_m$; then one can write

$$\lambda = L_v \Delta T / T_m, \tag{2.56a}$$

where

$$\Delta T = T^1 - T_m. \tag{2.56b}$$

Here T^1 is the drop–liquid interface temperature and T_m is the equilibrium melting temperature of the material in the drop.

The results above can be obtained by using the total grand potential (Gibbs 1948, p. 229; Wettlaufer and Worster 1995), and this approach has the advantage of easy generalization to multicomponent systems.

Equation (2.54) is the Laplace relation for the interface, and the following is the Gibbs–Thomson relation giving *capillary undercooling*:

$$T^1 = T_m \left[1 + 2H \frac{\gamma}{L_v} \right] \tag{2.57}$$

If H is replaced by its generalization to a two-dimensional surface, then Eq. (2.57) holds for three-dimensional systems. See Chapter 5 for generalizations to systems where γ depends on the orientation of the surface.

Consider next variations in which the endpoints may move. Because Eq. (2.54) holds already, one has at each endpoint $x = \ell_i$, $i = 1, 2$, that

$$\frac{\gamma}{\left(1 + h_x^2\right)^{1/2}} + \gamma_1 - \gamma_2 = 0, \tag{2.58}$$

because h is zero at the endpoints. If Figure 2.12 is used to evaluate this, $\left(1 + h_x^2\right)^{-1/2} = \cos \theta$, and thus the Young–Laplace relation emerges,

$$\gamma \cos \theta = \gamma_2 - \gamma_1. \tag{2.59}$$

At equilibrium at each contact line the contact angle θ adjusts itself to give this surface–energy balance. See Chapter 5 for generalizations to systems where γ depends on the orientation of the surface.

On a moving interface, there is a heat balance, applied to the domain shown in Figure 2.13. Call V_n the speed of the interface normal to itself and s the arc length. In Figure 2.13 as $\delta \to 0$ in a time δt,

$$\rho^s L V_n A \delta t - \gamma A |_{s+\delta s} + \gamma A|_s = (\mathbf{q}_\ell - \mathbf{q}_s) \cdot \mathbf{n} A \delta t$$

or

$$\rho^s L V_n - \frac{1}{A} \frac{\partial}{\partial t} (\gamma A) = k_T^s T_n^s - k_T^\ell T_n^\ell. \tag{2.60}$$

2.2 Curved interfaces

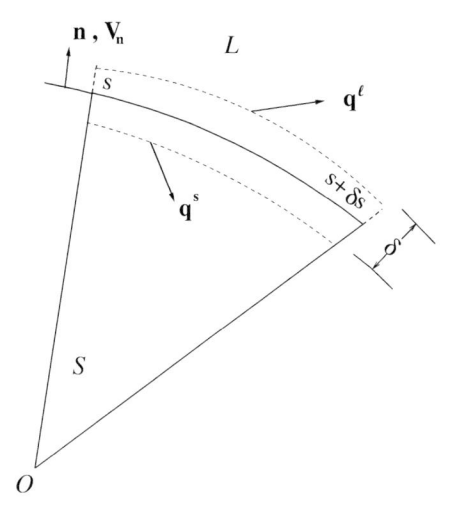

Figure 2.13. A sketch of a sector of interface from s to $s + \Delta s$ and a control volume spanning it.

One can obtain an identity from differential geometry (e.g., see Aris 1989),

$$\frac{1}{A}\frac{\partial A}{\partial t} = -2HV_n, \qquad (2.61)$$

where $\partial A/\partial t$ represents the stretching of the area A. Hence,

$$(\rho^s L + 2H\gamma)V_n = (k_T^s \nabla T^s - k_T^\ell \nabla T^\ell) \cdot \mathbf{n}. \qquad (2.62)$$

The usual form of this heat balance ignores the second term on the left-hand side, which was first derived by Wollkind (1979). It represents the energy expended by interfacial stretching. For a solid finger extending into the liquid $H < 0$, and thus a portion of the heat liberated by phase transformation goes into the creation of interface. For a given difference in heat fluxes, L is effectively decreased, and so V_n is increased.

In sum, the required conditions on a curved interface are as follows:
There is continuity of temperature

$$T^I = T^s = T^\ell \qquad (2.63a)$$

and the generalized Gibbs–Thomson equation

$$T^I = T_m\left(1 + 2H\frac{\gamma}{L_v}\right) - \mu^{-1}V_n \qquad (2.63b)$$

where we have included the kinetic undercooling discussed earlier with V replaced by V_n for a curved front. Thus, the interface temperature is reduced for "fingers" of solid growing into melt ($H < 0$) and for increased speed V_n of the interface. In addition, there is the heat balance

$$k_T^s \nabla T^s - k_T^\ell \nabla T^\ell = L_v V_n \left(1 + 2H \frac{\gamma}{L_v}\right). \tag{2.63c}$$

2.2.2 Growth of a Nucleus in an Undercooled Melt

Consider at time $t = 0$, a small spherical nucleus of solid surrounded by a large volume of its melt. Under what conditions does the nucleus melt back and disappear? Under what conditions does it grow, and then what are the characteristics of this growth? The analysis follows from Mullins and Sekerka (1963).

Consider at time $t = 0$ a small solid sphere of radius R_0, as shown in Figure 2.14 in an undercooled melt at temperature $T^\ell = T_\infty < T_m$. Assume that the two phases have equal densities, $\rho^\ell = \rho^s$, and equal specific heats, $c_p^\ell = c_p^s$. Call $R(t)$ the radius of the nucleus at time t and let the temperature field be $T = T(r, t)$, where the spherically symmetric growth is measured by radial coordinate r.

In the solid, $r < R(t)$, and in the liquid $r > R(t)$ there is thermal conduction

$$T_t = \kappa \nabla^2 T, \tag{2.64a}$$

where

$$\nabla^2 = \frac{1}{r^2} \frac{\partial}{\partial r} \left(r^2 \frac{\partial}{\partial r}\right). \tag{2.64b}$$

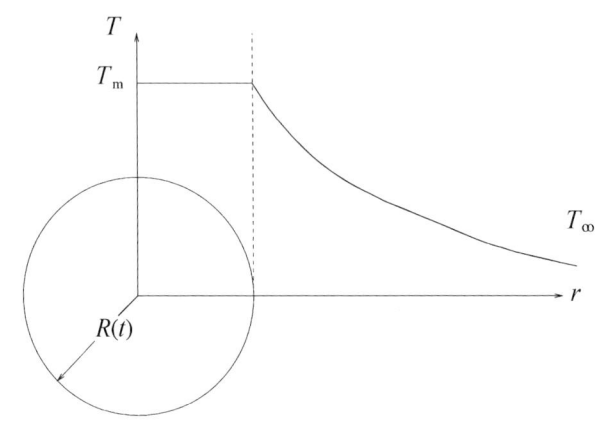

Figure 2.14. A sketch of a cross section of a growing spherical nucleus of radius $R(t)$ into an undercooled melt and the corresponding temperature profile.

Far from the sphere, $r \to \infty$,

$$T^\ell \to T_\infty, \tag{2.64c}$$

whereas at the origin, $r = 0$, the temperature is bounded,

$$|T^s| < \infty. \tag{2.64d}$$

On the sphere $r = R(t)$, there is the continuity of temperature, the Gibbs–Thomson equation, and the heat balance,

$$T^\ell = T^s = T^{\text{I}} = T_{\text{m}}\left[1 + 2H\frac{\gamma}{L_\nu}\right] \tag{2.64e}$$

and

$$L_\nu \frac{dR}{dt} = \left[k_{\text{T}}^s \nabla T^s - k_{\text{T}}^\ell \nabla T^\ell\right] \cdot \mathbf{n}, \tag{2.64f}$$

where $\mathbf{n}$ is the unit vector pointing out of the nucleus. The curvature-dependent heat rise in the heat balance, usually small, and the kinetic undercooling, have both been ignored for simplicity. Here the mean curvature is

$$H = -\frac{1}{R} \tag{2.64g}$$

and

$$\mathbf{n} \cdot \nabla = \frac{\partial}{\partial r}. \tag{2.64h}$$

Finally, there are initial conditions,

$$\left.\begin{array}{c} R(0) = R_0 \\ T^\ell(r, 0) = T_o^\ell(r) \\ T^s(r, 0) = T_o^s(r) \end{array}\right\}, \quad t = 0. \tag{2.64i}$$

Define scales as follows:

$$\left.\begin{array}{rl} \text{length} & \to R_0 \\ \text{temperature} & \to \Delta T = T_{\text{m}} - T_\infty \\ \text{speed} & \to v_p = \dfrac{k_{\text{T}}^\ell \Delta T}{R_0 L_\nu} \\ \text{time} & \to t_p = R_0/v_p = \dfrac{R_0^2 L_\nu}{k_{\text{T}}^\ell \Delta T}. \end{array}\right\} \tag{2.65}$$

Here ΔT is the undercooling and v_p is determined by the heat balance at the interface.

Define now the following nondimensional variables:

$$\hat{r} = r/R_0, \quad \hat{T} = (T - T_m)/\Delta T, \quad \hat{t} = t/t_p. \tag{2.66}$$

The governing system (2.64) can be written as follows:

$$\epsilon T_t^\ell = \nabla^2 T^\ell \qquad r > R(t) \tag{2.67a}$$

$$\epsilon T_t^s = \kappa \nabla^2 T^s \qquad r < R(t) \tag{2.67b}$$

$$T^\ell = -1 \qquad r \to \infty \tag{2.67c}$$

$$|T^s| < \infty \qquad r \to 0 \tag{2.67d}$$

$$\left. \begin{array}{l} T^\ell = T^s = -\dfrac{\Gamma}{R} \\[1em] \dfrac{dR}{dt} = k_T T_r^s - T_r^\ell \end{array} \right\} \qquad r = R(t) \tag{2.67e}$$

$$\left. \begin{array}{l} R = 1 \\[0.5em] T^\ell = T_o^\ell \\[0.5em] T^s = T_o^s \end{array} \right\} \quad t = 0, \tag{2.67f}$$

where the carets have been dropped. The nondimensional parameters present are the modified Stefan number S,

$$S^{-1} = \frac{k_T^\ell \Delta T}{L_v \kappa^\ell} = \frac{c_p^\ell \Delta T}{L} \equiv \epsilon, \tag{2.68a}$$

the surface energy parameter Γ,

$$\Gamma = \frac{2\gamma}{R_0 L_v} \frac{T_m}{\Delta T} \tag{2.68b}$$

and the thermal parameters κ and k_T,

$$\left. \begin{array}{l} \kappa = \dfrac{\kappa^s}{\kappa^\ell} \\[1em] k_T = \dfrac{k_T^s}{k_T^\ell} \end{array} \right\}. \tag{2.68c}$$

Let us estimate the values of the physical constants for typical materials. These are shown in Table 2.1.

2.2 Curved interfaces

Table 2.1. *Typical values of S and T_m*

Material	$S\Delta T(°C)$	$T_m(°C)$
Water	80	0
Al	316	660
Cu	446	1084
SCN	23	58

For Cu, say, $k_T \approx 1.47$, $\kappa \approx 1.60$,

$$\Gamma \approx \frac{1.85 \times 10^{-14}}{R_0(\text{cm})} \frac{T_m}{\Delta T} \tag{2.69a}$$

and

$$\epsilon = \frac{\Delta T °C}{409 °C}. \tag{2.69b}$$

Notice that Γ is small except for very small nuclei and for small undercoolings, whereas ϵ is always small.

Let us fix all parameters except ϵ, and let $\epsilon \to 0$. As long as $\partial/\partial t$ is of unit order, this limit makes the bulk temperatures quasi-static as expressed by

$$\nabla^2 T = 0. \tag{2.70}$$

The solution of this approximate system cannot satisfy all of the initial conditions because the system is first order in time; the limit $\epsilon \to 0$ is a singular perturbation. The initial conditions on T are dropped for the moment, and the limiting system is solved by the *outer solution* valid for $\partial/\partial t = O(1)$.

It is easy to solve Eq. (2.70) in the liquid,

$$T^\ell = a_0(t) + \frac{b_0(t)}{r}.$$

The condition (2.67c) (at infinity) makes $a_0(t) = -1$, and the Gibbs–Thomson condition, Eq. (2.67e), makes $b_0 = R(1 - \Gamma/R)$. Thus,

$$T^\ell = -1 + \frac{R}{r}\left(1 - \frac{\Gamma}{R}\right). \tag{2.71a}$$

Similarly, in the solid, the Gibbs–Thomson condition and Eq. (2.67d) (at $r = 0$) give

$$T^s = -\frac{\Gamma}{R}. \tag{2.71b}$$

Now, use these temperatures in the heat-flux condition, Eq. (2.67e), to obtain

$$\frac{dR}{dt} = \frac{1}{R}\left(1 - \frac{\Gamma}{R}\right) \tag{2.72a}$$

with

$$R(0) = 1. \tag{2.72b}$$

This is analogous to a "bubble equation" in fluid mechanics.

If the surface energy $\Gamma = 0$, then $R = (1 + 2t)^{1/2}$, which represents unconditional growth, that is, growth to infinity for all size particles. If $\Delta T \to 0$, $\Gamma \to \infty$; if one reverts to the dimensional version of Eq. (2.72a),

$$L_v \frac{dR}{dt} = \frac{k_T^\ell}{R}\left(\Delta T - \frac{2\gamma T_m}{L_v}\frac{1}{R}\right),$$

one sees that in the zero undercooling limit, $\Delta T \to 0$, (back in nondimensional form)

$$R = (1 - 3\Gamma t)^{1/3}, \tag{2.73}$$

which represents unconditional shrinkage because there is no driving force.

The sphere grows only if at $t = 0$, R_0 is such that

$$1 - \frac{\Gamma}{R(0)} > 0,$$

and then it will continue to grow. Because $R(0) = 1$, the condition for growth is

$$\Gamma < 1 \quad \text{(growth)}. \tag{2.74}$$

The critical nucleation radius R_* is the radius that makes $\Gamma = 1$,

$$R_* = \frac{2\gamma}{L_v}\frac{T_m}{\Delta T}. \tag{2.75}$$

When $R_0 > R_*$, the sphere will grow to $r = \infty$. For metals, $R_* \approx 2.0 \times 10^{-7}\frac{T_m}{\Delta T} \approx 0.02\,\mu\text{m}$ for $\Delta T/T_m = 0.1$.

Note: Growth is independent of the thermal properties of the solid because the limit $\epsilon \to 0$ was taken, and therefore the solid is isothermal.

The thermal gradient G_T in the liquid at the interface is given by

$$G_T = -\frac{1}{R}\left(1 - \frac{\Gamma}{R}\right) < 0, \tag{2.76a}$$

and thus the bubble equation (2.72a) has the form

$$\frac{dR}{dt} = -G_T(t); \tag{2.76b}$$

therefore, growth requires a negative gradient.

The preceding analysis is valid for $\epsilon \to 0$ and $\partial/\partial t = O(1)$, a time interval from $O(\epsilon)$ to ∞. It does not satisfy the initial conditions on T. In order to solve

2.2 Curved interfaces

for this early-time behavior, one must rescale time and hence seek the *inner solution*.

Let t be the dimensional time and $\hat{t}$ be the "outer" time. Then

$$t_* = \frac{\hat{t}}{\epsilon} = \frac{t}{\epsilon t_p} = \frac{t}{\left(R_0^2/\kappa^\ell\right)}. \tag{2.77a}$$

Thus, $\hat{t}$, the outer time, derives its scale from the interfacial balance of heat, and t_*, the inner time, is determined by thermal conduction. In fact it is seen that

$$\frac{\hat{t}}{t_*} = \epsilon. \tag{2.77b}$$

When the change of variable (2.77a) is made, the bulk diffusion equations become free of ϵ, but now the interfacial heat balance is as follows:

$$\frac{dR}{dt_*} = \epsilon \left(k_\mathrm{T} T_r^s - T_r^\ell\right) \tag{2.78}$$

with all of the original initial conditions (2.67f) applying. Clearly, the inner system remains second order in time when $\epsilon \to 0$ with $\partial/\partial t_* = O(1)$. In this limit, $dR/dt_* = 0$ so that $R(0) = 1$ in the whole time interval. The interface remains at its initial position, and heat conduction takes place in a fixed spatial domain. When $t_* \sim 1/\epsilon$, this solution automatically matches the outer solution for $\hat{t} \sim \epsilon$.

In Figure 2.15 are shown sketches of the full solution drawn versus the outer time $\hat{t}$. It is seen that in an $O(\epsilon)$ neighborhood, $R \sim 1$, whereas the temperatures develop from their initial values toward the values consistent with the outer solution.

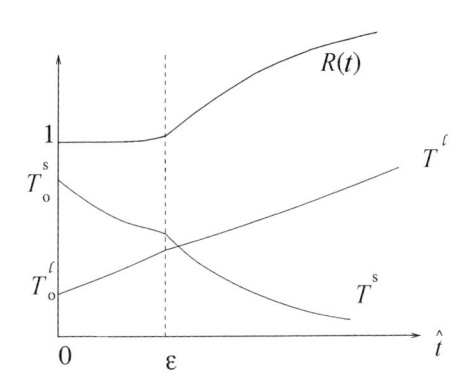

Figure 2.15. Sketches for $\Gamma < 1$ and r fixed of the temperatures and interface position as functions of the outer time $\hat{t}$ for small ϵ.

2.2.3 Linearized Instability of Growing Nucleus

As the spherical nucleus grows in time, as given by Eq. (2.72a), there is the possibility that at some time (or equivalently at some radius) an instability will occur that destroys the spherical symmetry of the particle. In order to determine whether instabilities occur, begin with the full time-dependent system (2.67), the *exact* solution given asymptotically for $\epsilon \to 0$ in the last section, namely

$$\bar{T} = \bar{T}(r, t), \quad R = \bar{R}(t),$$

and disturb the system by small disturbances T' and R' as follows:

$$T = \bar{T}(r, t) + T'(r, \theta, \phi, t), \quad R = \bar{R}(t) + R'(\theta, \phi, t). \quad (2.79)$$

Substitute these into the system, and linearize in primed quantities. The resulting linear system as follows:

$$\epsilon T_t^{\ell'} = \nabla^2 T^{\ell'}, \quad r > \bar{R} \quad (2.80\text{a})$$

$$\epsilon T_t^{s'} = \kappa \nabla^2 T^{s'}, \quad r < \bar{R} \quad (2.80\text{b})$$

$$T' \to 0, \quad r \to \infty \quad (2.80\text{c})$$

$$|T'| < \infty, \quad r \to 0 \quad (2.80\text{d})$$

$$\left. \begin{array}{c} R'_t = k_T T_r^{s'} - T_r^{\ell'} + \left(k_T \bar{T}_{rr}^s - \bar{T}_{rr}^\ell \right) R' \\ T^{\ell'} + \bar{T}_r^\ell R' = T^{s'} + \bar{T}_r^s R' = H' \Gamma \end{array} \right\}, \quad r = \bar{R}, \quad (2.80\text{e})$$

where

$$\nabla^2 = \frac{\partial^2}{\partial r^2} + \frac{2}{r} \frac{\partial}{\partial r} + \frac{1}{r^2} \mathcal{L} \quad (2.80\text{f})$$

$$\mathcal{L} = \frac{\partial^2}{\partial \theta^2} + \cot \theta \frac{\partial}{\partial \theta} + \frac{1}{\sin^2 \theta} \frac{\partial^2}{\partial \varphi^2} \quad (2.80\text{g})$$

$$\mathbf{n} = \left(1, -\frac{1}{R} R_\theta, -\frac{1}{R \sin \theta} R \right) \left(1 + \frac{R_\theta^2}{R^2} + \frac{R_\varphi^2}{R^2 \sin^2 \theta} \right)^{-1/2} \quad (2.80\text{h})$$

$$2H = -\frac{2}{\bar{R}} + \frac{2R'}{\bar{R}^2} + \mathcal{L} \frac{R'}{\bar{R}^2} \sim -\frac{2}{\bar{R}} + \frac{(\mathcal{L} + 2) R'}{\bar{R}^2} + \cdots. \quad (2.80\text{i})$$

Notice that all interfacial conditions apply on $r = \bar{R}$ rather than on $r = \bar{R} + R'$ because Taylor series have been used to transfer the site of their application to the undisturbed interface. In the second of Eqs. (2.80e), H' is the disturbed curvature that is defined in Eq. (2.80i).

2.2 Curved interfaces

Notice, as well, that the initial conditions have been dropped because the interest lies in the long-time behavior of disturbances, that is, if (T', R') grows with time as $t \to \infty$, then the interface is unstable. If (T', R') decays always, then the interface is stable.

Given that the long-time behavior is of interest, consider only the outer solution and set $\epsilon = 0$. Thus, the temperatures in each phase are quasi-steady. For $r > \bar{R}(t)$ then,

$$\nabla^2 T^{\ell'} = 0. \tag{2.81a}$$

Let us consider normal-mode solutions (separation of variables) and write

$$T^{\ell'} = f(r) Y_{\ell m}, \tag{2.81b}$$

where $Y_{\ell m}$ are spherical harmonics that satisfy

$$\mathcal{L} Y_{\ell m} = -\ell(\ell + 1) Y_{\ell m} \tag{2.81c}$$

(see Morse and Feshbach 1953, 1264ff.). Then, the function f satisfies

$$f'' + \frac{2}{r} f' - \frac{\ell(\ell+1)}{r^2} f = 0. \tag{2.81d}$$

This is an equidimensional equation whose solutions have the form $f \propto r^k$, where $k = \ell, -\ell - 1$. Thus, here

$$T'^{\ell} = A_{\mathrm{O}}(t) r^{-\ell-1} Y_{\ell m}(\theta, \phi), \tag{2.81e}$$

which satisfies the condition (2.80c) at infinity.

For $r < \bar{R}(t)$,

$$\nabla T'^s = 0 \tag{2.82a}$$

and

$$T'^s = A_{\mathrm{I}}(t) r^{\ell} Y_{\ell m}(\theta, \varphi), \tag{2.82b}$$

which satisfies the condition (2.80d) as $r \to 0$.

Let $R' = \hat{R}(t) Y_{\ell m}$ and substitute forms (2.81e) and (2.82b) into the Gibbs–Thomson equation (2.80e) to obtain

$$A_{\mathrm{O}}(t) = \left[\bar{R}^{\ell} - \frac{1}{2} \Gamma \ell(\ell+1) \bar{R}^{\ell-1} \right] \hat{R}. \tag{2.82c}$$

Finally, use the flux condition, the first part of Eq. (2.80e), to find after some algebra that

$$\frac{d\hat{R}}{dt} = \left\{ 1 - \frac{\Gamma}{2\bar{R}(t)} [(\ell+1)(\ell+2) + 2 + k_{\mathrm{T}} \ell(\ell+2)] \right\} \frac{(\ell-1)}{\bar{R}^2} \hat{R}. \tag{2.83}$$

Notice that for large ℓ the bracketed term is always negative; spatially rapid oscillations are damped by surface energy. The most dangerous modes have the smallest ℓ.

The mode $\ell = 1$ is neutrally stable and corresponds to a translation of the sphere. It is then the mode $\ell = 2$ that determines instability. For $\ell = 2$

$$\frac{d\hat{R}}{dt} = \frac{1}{\bar{R}^2}\left\{1 - \frac{1}{\bar{R}}(4k_T + 7)\Gamma\right\}\hat{R}. \qquad (2.84)$$

Notice that for small enough surface energy Γ or large enough $\bar{R}$, $\hat{R}$ grows to infinity. The mode $\ell = 2$ grows if

$$\bar{R} > R_c \equiv (4k_T + 7)\Gamma, \qquad (2.85)$$

which is the instability condition at $t = 0$, where $\bar{R} = 1$. As time progresses, surface energy becomes less important.

Notice that R_c, the critical radius for *instability*, can be related to R_*, the critical *nucleation* radius that guarantees a growing nucleus, by

$$\frac{R_c}{R_*} = 4k_T + 7, \qquad (2.86)$$

and thus for a typical metal $R_c \approx 0.30\,\mu\text{m}$, whereas $R_* \approx 0.02\,\mu\text{m}$.

Three interesting cases are

$$\frac{R_c}{R_*} = \begin{cases} 7 & \text{if } k_T = 0 \\ 11 & \text{if } k_T = 1 \\ 15 & \text{if } k_T = 2. \end{cases} \qquad (2.87)$$

Conduction in the solid stabilizes the particle so it grows to larger radius before it becomes unstable.

The mechanism for instability depends on the thermal boundary layer on the particle (on the liquid side). Figure 2.14 shows a spherical particle at time t with the basic-state temperature profile superposed. Then $\bar{T}^s$ is constant (for $\epsilon \to 0$), and $\bar{T}^\ell$ decreases with radius. Thus, always $d\bar{T}^\ell/dr = -G^\ell(t) < 0$.

Consider now a shape perturbation on the sphere as shown. A solid bump pushes further into the liquid, steepens the magnitude of the gradient G^ℓ, which, by the flux balance, causes the bump to grow faster. The surface energy is available to oppose this growth. It is able to do this for $R_* < R < R_c$, but it cannot for $R > R_c$. The bump would be further warmed if heat conduction were present in the solid, and thus its presence is also stabilizing.

When $R > R_c$, bumps on the particle grow and grow without bound, leading to the development of "snow flakes." When $|R'| \sim \bar{R}$, and $|T'| \sim \bar{T}$, linear stability theory has broken down, and the description of the further evolution must involve nonlinear effects. This discussion is postponed to Chapter 4.

2.2.4 Linearized Instability of a Plane Front Growing into an Undercooled Melt

In Section 2.1.4 the solution is given for a moving planar front. If and when the front becomes unstable to cellular modes, periodic along the interface, the stability problem becomes one in which curved interfaces are important. Only the case of larger undercooling, $S^{-1} > 1$, (called hypercooled) will be given here, and hence kinetic undercooling is retained.

Consider a front moving in the z-direction, $z = h(x, t)$ and only the two-dimensional problem.

The governing system is as follows:
In liquid, $z > h$,

$$T_t^\ell = \kappa^\ell \nabla^2 T^\ell. \tag{2.88a}$$

In solid, $z < h$,

$$T^s = T^1. \tag{2.88b}$$

where $T^1(x, t)$ is unknown a priori. As $z \to \infty$

$$T^\ell \to T_\infty. \tag{2.88c}$$

On $z = h(x, t)$

$$T^s = T^\ell = T^1 = T_m \left(1 + 2H \frac{\gamma}{L_v}\right) - \mu^{-1} V_n \tag{2.88d}$$

$$(L_v + 2H\gamma)V_n = -k_T^\ell \nabla T^\ell \cdot \mathbf{n}, \tag{2.88e}$$

where

$$V_n = \frac{h_t}{\left(1 + h_x^2\right)^{1/2}} \tag{2.88f}$$

is the speed of the front normal to itself, and the effects of kinetic undercooling are included in the Gibbs–Thomson relation (2.88e) through coefficient μ.

Introduce scales as follows: length $\sim \ell$, time $\sim t_o$, temperature $\sim \Delta T = T_m - T_\infty > 0$, and define θ through $T = T_\infty + (\Delta T)\theta$. Here

$$\ell = \frac{\kappa^\ell}{\mu \Delta T}, \text{ and } t_o = \frac{\kappa^\ell}{\mu^2 (\Delta T)^2}. \tag{2.89}$$

The scaled equations now have the form:
In liquid $z > h$,

$$\theta_t = \nabla^2 \theta. \tag{2.90a}$$

In solid $z < h$,
$$\theta = \theta^1, \qquad (2.90b)$$
where
$$\theta^1 = (T^1 - T_\infty)/\Delta T \qquad (2.90c)$$
As $z \to \infty$
$$\theta \to 0. \qquad (2.90d)$$
On the interface $z = h(x, t)$
$$(S + 2HC^{-1}) V_n = -\nabla\theta \cdot \mathbf{n} \qquad (2.90e)$$
$$1 + 2H\Gamma_1 - V_n = \theta, \qquad (2.90f)$$
where
$$C = \frac{k_T^\ell}{\mu\gamma}, \quad S = \frac{L}{c_p^\ell \Delta T}, \quad \Gamma_1 = \frac{\gamma}{L} \frac{T_m c_p^\ell \mu}{k_T^\ell}. \qquad (2.91)$$

The first step is defining a basic state, the solution that was obtained in Section 2.1.4, as follows:
$$\bar{\theta} = S e^{-V\hat{z}} \qquad (2.92a)$$
$$\bar{h} = Vt, \qquad (2.92b)$$
where
$$\hat{z} = z - Vt \qquad (2.92c)$$
and
$$S = 1 - V. \qquad (2.92d)$$

One changes coordinates $(x, z, t) \to (\hat{x}, \hat{z}, \hat{t})$, where $\hat{x} = x$, $\hat{z} = z - Vt$, $\hat{t} = t$, and Eq. (2.90b) becomes
$$\theta_t - V\theta_z = \nabla^2\theta; \qquad (2.93)$$
the carats have been dropped.

Disturb this exact solution as follows:
$$\theta = \bar{\theta} + \theta', \quad h = h'. \qquad (2.94)$$

If one substitutes forms (2.94) into system (2.90) and linearizes in primed quantities, the linearized stability problem results. In liquid, $z > 0$
$$\theta'_t - V\theta'_z = \nabla^2 \theta' \qquad (2.95a)$$

2.2 Curved interfaces

At $z \to \infty$,
$$\theta' \to 0 \tag{2.95b}$$

On $z = 0$,
$$Sh'_t + C^{-1}Vh'_{xx} + V^2 Sh' = -\theta'_z \tag{2.95c}$$
$$\Gamma_1 h'_{xx} - h'_t + V Sh' = 0, \tag{2.95d}$$

where the interfacial-boundary conditions have been linearized, as discussed earlier. Note that because $\theta^s = \theta^l$, the temperature in the solid decouples from this system and hence can be determined after (2.95) is solved.

To solve system (2.95), use normal modes
$$[\theta'(x,t), h'(t)] = [\hat{\theta}(z), \hat{h}]e^{\sigma t + i a_1 x}. \tag{2.96}$$

Note that because time does not appear explicitly in the coefficients of system (2.95), exponential variations in time occur in the disturbances; this contrasts to the case of the nucleus that grows like $t^{1/2}$.

The heat equation (2.95a) becomes
$$\left(\sigma - V \frac{d}{dz}\right)\hat{\theta} = \left(\frac{d^2}{dz^2} - a_1^2\right)\hat{\theta}, \tag{2.97a}$$

which has solutions $\hat{\theta} = e^{-pz}$, $\text{Re } p > \infty$ that satisfy the condition at $z \to \infty$. Here
$$p^2 - pV - (a_1^2 + \sigma) = 0$$

so that
$$p = \frac{1}{2}\left\{V + [V^2 + 4(a_1^2 + \sigma)]^{1/2}\right\}. \tag{2.97b}$$

The interfacial conditions are
$$(\sigma S - C^{-1}V a_1^2 + V^2 S)\hat{h} = p\hat{\theta}(0) \tag{2.97c}$$

and
$$(-\Gamma_1 a_1^2 - \sigma + V S)\hat{h} = \hat{\theta}(0). \tag{2.97d}$$

This linear, homogeneous system has nontrivial solutions only when the determinant of the matrix of coefficients is zero. This gives the characteristic equation
$$\sigma(S + p) = p(VS - \Gamma a_1^2) + C^{-1}a_1^2 - V^2 S \tag{2.97e}$$

with
$$V = 1 - S. \tag{2.97f}$$

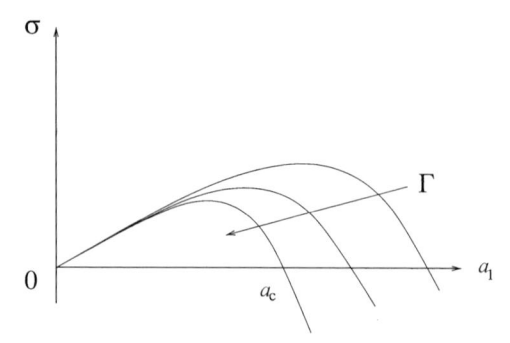

Figure 2.16. The growth rate σ versus the wave number a_1 for various surface tensions Γ.

Note that when $S = 0$, $V = 1$; then

$$\sigma = \left(\frac{C^{-1}}{p} - \Gamma_1\right) a_1^2$$

and there is linearized instability when

$$\Gamma_1 < \frac{C^{-1}}{p} \qquad (2.98)$$

(Note that $p = p(\sigma)$). The same result holds for $S = 1$, $V = 0$.

Figure 2.16 shows σ versus a_1, for various values of Γ. When $\Gamma \to 0$, the basic state is always unstable, and as Γ increases, the instability ($\sigma > 0$) becomes confined to an interval $(0, a_c)$ of a_1, where a_c is small, which suggests looking at a small wave number expansion of Eq. (2.97e). Let

$$\sigma = \sigma_0 + a_1^2 \sigma_2 + \cdots$$
$$p = p_0 + a_1^2 p_2 + \cdots \qquad (2.99)$$

because a_1 only appears in the combination a_1^2.

If forms (2.99) are substituted into the characteristic equation and coefficients of like powers of a_1 are equated, one obtains a sequence of simplified problems.
At $O(1)$

$$\sigma_0(S + p_0) = (p_0 - V)VS \qquad (2.100a)$$

with

$$p_0 = V + \frac{\sigma_0}{V}. \qquad (2.100b)$$

2.2 Curved interfaces

The leading-order solutions are $\sigma_0 = 0, -V^2$. A small perturbation will not change the sign of $\sigma \sim -V^2$, which will always correspond to a decaying mode. Thus, consider $\sigma_0 = 0$. At $O(a_1^2)$, one can find σ_2,

$$\sigma_2 = -\Gamma_1 + \frac{S + C^{-1}}{1 - S}. \tag{2.101}$$

The neutral stability curve is obtained for $\sigma \sim \sigma_0 + a^2\sigma_2 = 0$, giving $S = S_c$,

$$S_c^{-1} = \frac{\Gamma_{1c} + 1}{\Gamma_{1c} - C^{-1}}. \tag{2.102}$$

Equation (2.102) gives a critical value Γ for neutral stability. One can then write it as

$$\sigma_2 = \frac{S_c^{-1} - S^{-1}}{S^{-1} - 1}(1 - C^{-1}) + O[(S_c - S)^2]$$

so that as the undercooling is decreased from S_c^{-1}, the planar front becomes unstable; there is instability for

$$S^{-1} < S_c^{-1}. \tag{2.103}$$

One can estimate the size of C^{-1} and hence the magnitude of the curvature effect in the flux condition: S^{-1} is large in metals and $C^{-1}/\Gamma_1 = L/T_m c_p \approx 409/T_m \approx 0.3$–$0.5$, even at these large undercoolings.

When $S^{-1} < 1$, the planar growth has $h \sim t^{1/2}$, and thus the linearized disturbance equations have time-dependent coefficients, and normal modes can reduce this system only to a set of partial differential equations in z and t. One must then use numerical methods to solve these as initial value problems. However, there is no convenient representation for the solutions, and somewhat odd initial conditions can give rather large growth rates; moreover, no simple basis exists for judging the criteria for stability. See Davis (1976) for a discussion of this matter in fluid-dynamical systems.

2.2.5 Remarks

The equations governing a solidifying pure melt have been derived. In particular, across the interface between solid and liquid, there is the heat balance, Eq. (2.62),

$$L_v\left(1 + 2H\frac{\gamma}{L_v}\right)V_n = \left(k_T^s \nabla T^s - k_T^\ell \nabla T^\ell\right) \cdot \mathbf{n}, \tag{2.104a}$$

continuity of temperature

$$T^I = T^s = T^\ell, \tag{2.104b}$$

and the Gibbs–Thomson equation

$$T^l = T_m \left(1 + 2H\frac{\gamma}{L_v}\right) - \mu^{-1} V_n. \qquad (2.104c)$$

Condition (2.104a) shows that the magnitudes of the temperature gradients at the interface determine the propagation speed V_n, and their signs determine whether planar or spherical propagation is susceptible to instabilities periodic along the fronts.

Consider the same geometries that have been discussed above, except now a boundary temperature is *raised* above T_m. A block of solid at $T^s = T_m$ is then *melted*, and the interface moves like $h(t) = 2\Lambda\sqrt{\kappa^\ell t}$, where now h depends on κ^ℓ rather than κ^s (see Eq. (2.12c)); the heat conduction is principally in the liquid. Thus, the times it takes to freeze a liquid and melt a solid are intrinsically different because their ratio depends on κ^s/κ^ℓ.

In this chapter, pure liquids and thermal gradients have been studied. There is an analogous problem in which the temperature is uniform and a solid is created by growth from a saturated liquid. The *same* systems govern both if the mass flux is Fickian, that is $\mathbf{q}_c = -D\nabla C$, cross-diffusion is absent, C replaces T, and D replaces κ and k_T.

References

Aris, R. (1989). *Vectors, Tensors, and the Basic Equations of Fluid Mechanics*, Dover, New York.

Bates, P. W., Fife, P. C., Gardner, R. A., and Jones, C. K. R. T. (1997). Phase field models for hypercooled solidification, *Physica D* **104**, 1–31.

Courant, R., and Hilbert, D. (1953). *Methods of Mathematical Physics*, Vol. 1, Interscience, New York.

Davis, S. H. (1976). The stability of time-periodic flows, *Ann. Rev. Fluid Mech.* **8**, 57–74.

Gibbs, J. W. (1948). *The Collected Works of J. Willard Gibbs*, Vol. 1, Yale University Press, New Haven.

Glicksman, M. E., and Schaefer, R. J. (1967). Investigation of solid/liquid interface temperatures via isenthalpic solidification, *J. Crystal Growth* **1**, 297–310.

Hills, R. N., and Roberts, P. H. (1993). A note on the kinetic conditions at a supercooled interface, *Intern. Comm. Heat Mass Transf.* **20**, 407–416.

Morse, P. M., and Feshbach, H. (1953). *Methods of Theoretical Physics*, Part II, McGraw-Hill, New York.

Mullins, W. W. (1963). Solid surface morphologies governed by capillarity, in *Metal Surfaces*, pages 17–66, Chap. 2, American Society for Metals, Metals Park, Ohio.

Mullins, W. W., and Sekerka, R. (1963). Morphological stability of a particle growing by diffusion or heat flow, *J. Appl. Phys.* **34**, 323–329.

Serrin, J. (1959). Principles of classical fluid mechanics, in *Handbuch der Physik* Vol. VIII/1 Mathematical, 125–263, Springer-Verlag, Berlin.

Umantsev, A. (1985). Motion of a plane front during cyrstallization, *Sov. Phys. Crystallogr.* **30**, 87–91.

Wettlaufer, J. S., and Worster, M. G. (1995). Dynamics of premelted films: Frost heave in a capillary, *Phys. Rev. E* **51**, 4679–4689.

Wollkind, D. J. (1979). A deterministic continuum mechanical approach to morphological stability of the solid–liquid interfaces, in W. R. Wilcox, editor, *Preparation and Properties of Solid State Materials* **4**, 111–191, Dekker, New York.

Young, G. W. (1994). Mathematical description of viscous free surface flows, in *Free Boundaries in Viscous Flows*: IMA Volumes in Mathematics and Its Applications **61**, 1–27, Springer-Verlag, New York.

3

Binary substances

3.1 Mathematical Model

When a binary liquid is frozen, it usually rejects some or all of its solute because that solute is more soluble in the liquid than in the crystalline solid. The degree of rejection can be obtained from the phase diagram for systems in thermodynamical equilibrium, an example of which is shown in Figure 3.1. The rejected solute in the liquid is subject to diffusion, and thus, a major difference between the dynamics of pure material and that of alloys is the need to track both the temperature and concentration fields.

In the bulk liquid or solid there is again the heat conduction

$$T_t = \kappa \nabla^2 T. \tag{3.1}$$

Consider that the solute is dilute; thus the solute diffusion equation valid in both phases is

$$C_t = D \nabla^2 C, \tag{3.2}$$

where D is the solute diffusivity and C is the solute concentration. Equation (3.2) can be derived exactly as was Eq. (3.1) if one makes the following replacements: $T \to C$, $\kappa \to D$, assumes that there is no cross diffusion and that the solute flux satisfies the relation $\mathbf{q}_s = -D\nabla C$, a Fickian model. In the far fields there are appropriate conditions on both T and C.

Figure 3.2 shows a phase diagram near its left-hand side, where C is small. The curved liquidus and solidus have been represented by straight lines. Above the liquidus all material is liquid, and below the solidus, all is solid. At a fixed interface temperature T_0, as shown, the interface has concentration C^s on the solid side equal to C_∞, and C^ℓ, the concentration at the interface on the liquid side is given by C_∞/k; k is the segregation or distribution coefficient

3.1 Mathematical model

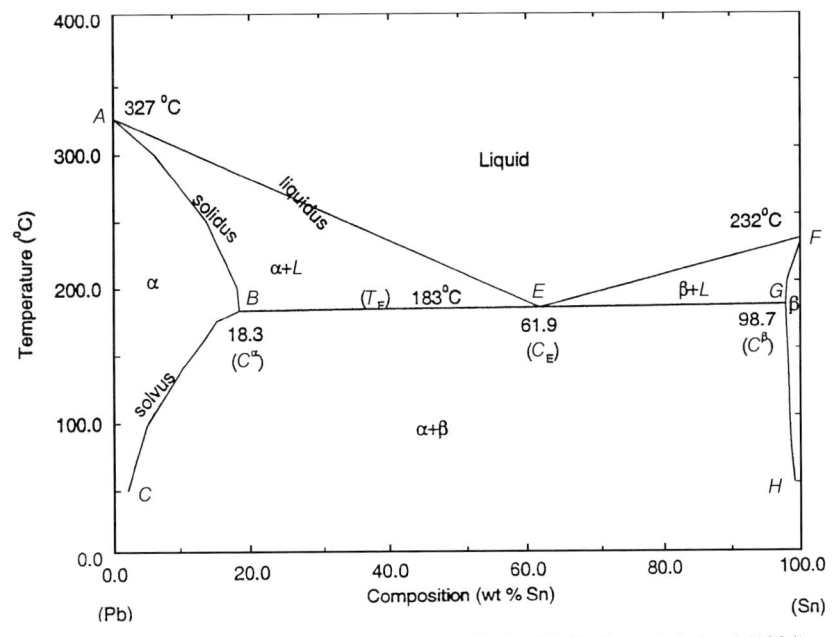

Figure 3.1. A typical phase diagram, T versus C, for Pb–Sn from Askeland (1984).

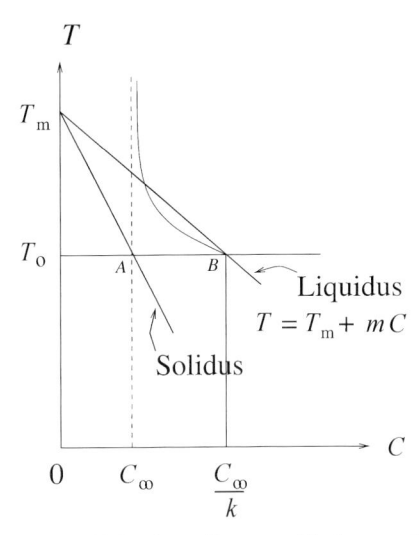

Figure 3.2. A linearized, small C, phase diagram with the constitutional undercooling profile superimposed.

defined by
$$k = C^s/C^\ell \tag{3.3}$$
at constant temperature and pressure. Because the interface is considered to have zero thickness, points A and B of Figure 3.2 coincide physically, and then there is a jump in concentration across the interface, due to solute rejection, that is given by
$$\Delta C_0 = (C^\ell - C^s)^{\mathrm{I}} = \frac{1-k}{k} C_\infty. \tag{3.4}$$
The equation of the liquidus is
$$T^{\mathrm{I}} = T_{\mathrm{m}} + mC, \tag{3.5}$$
where m (here $m < 0$) is the slope showing that the melting point is reduced by the presence of solute. In Eq. (3.5) the superscript on C^ℓ has been dropped; in all further references, quantities with no superscripts represent values in the liquid.

Corresponding to ΔC_0 there is a temperature difference ΔT_0, or undercooling,
$$\Delta T_0 = -m \Delta C_0. \tag{3.6}$$

Form (3.6) describes the shift in interface temperature (downward) due to solute rejection. This *constitutional undercooling* can be combined with the capillary and kinetic undercoolings to generalize the Gibbs–Thomson equation on the interface
$$T^s = T^\ell = T^{\mathrm{I}} = T_{\mathrm{m}}\left(1 + 2H\frac{\gamma}{L_v}\right) + mC - \mu^{-1}V_n. \tag{3.7}$$

Had the curvature of the liquidus and solidus been retained, the linear form mC would be replaced by a more general function of C representing the equation of the liquidus. Form (3.7) leaves the undercooling due to surface energy unchanged from its value for pure materials. This can be proven to be the case in the dilute limit (Flemmings 1974, p. 273).

The heat balance is as before
$$(L_v + 2H\gamma)V_n = \left(k_T^s \nabla T^s - k_T^\ell \nabla T^\ell\right) \cdot \mathbf{n}. \tag{3.8}$$

The liquid sees the interface as a source of solute whose balance is given by
$$(C - C^s)V_n = (D^s \nabla C^s - D^\ell \nabla C) \cdot \mathbf{n}. \tag{3.9}$$

Finally, there is the definition of the segregation coefficient that links C^s and C:

$$C^s = kC. \qquad (3.10)$$

This completes the specification of the model except for initial conditions.

3.2 Directional Solidification

Consider a binary liquid encased between two parallel plates, as shown in Figure 3.3. This sandwich is laid across two fixed-temperature blocks, with $T_T > T_B$. If the temperatures are chosen so that T_m lies between them, $T_B < T_m < T_T$, part of the material will freeze and form a planar, stationary freezing front as shown; this is a static state. Now, the blocks are kept fixed in place, and the plates and the enclosed material are pulled "downward" toward the solid at speed V. After transients have vanished, there is a new steady consisting of a planar front (in a slightly different position due to constitutional undercooling effects) and exponential profiles $\bar{T}$ and $\bar{C}$, as shown.

This system is useful for experiment and theory. First, this system represents a continuous process in which new fluid moves "downward," cools, freezes, and

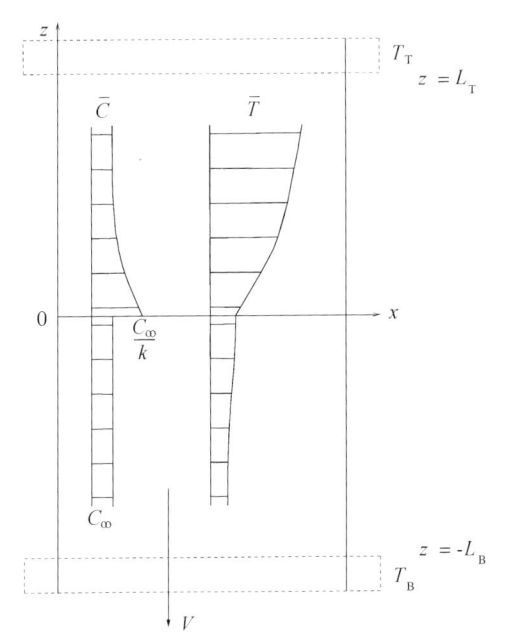

Figure 3.3. A sketch of a directional solidification device with the basic state $\bar{C}(z)$ and $\bar{T}(z)$ indicated.

moves onward. Second, it turns out that the C-field can cause morphological instabilities of the front while the thermal field stabilizes the front. These opposing effects lead to frontal instabilities that equilibrate in amplitude, making this system an excellent vehicle for the careful study of phase transformation dynamics. Third, the thinness of the Hele–Shaw cell and the use of transparent model fluids make for the easy viewing of the interface morphology through a microscope.

A model describing this uses a coordinate system, $z \to z - Vt$, that is moving with the interface at speed V, and thus the diffusion fields satisfy the following equations:

$$T_t - VT_z = \kappa \nabla^2 T \tag{3.11a}$$

$$C_t - VC_z = D\nabla^2 C. \tag{3.11b}$$

In the two phases Eqs. (3.11a,b) apply on the two sides of the interface, which is located at $z = h(x, y, t)$ and $+z$ points into the liquid. At $z = L_T$, say,

$$T^\ell = T_T \tag{3.11c}$$

$$C^\ell_z = 0 \tag{3.11d}$$

and at $z = -L_B$,

$$T^s = T_B \tag{3.11e}$$

$$C^s_z = 0. \tag{3.11f}$$

On the interface, Eqs. (3.7)–(3.10) hold.

3.3 Basic State and Approximate Models

In the laboratory frame one seeks a solution independent of x, y, and t and, hence, dependent on z only.

As seen from Eq. (3.11a),

$$-VT_z = \kappa T_{zz}, \tag{3.12}$$

whose solutions are a constant and $\exp(-Vz/\kappa)$. With conditions at $z = L_T$, $-L_B$, one can obtain appropriate solutions for T (and C) with $h = h_0$ (constant).

For a well-designed experiment, L_T, L_B are very large, and the numerical values of these are expected to have very little effect on instabilities of unit-order wavelength at the interface, and thus it would be convenient to replace these by infinity. This is justified in the liquid phase, for as $z \to \infty$, the exponential decays, and only exponentially small errors are made by taking

3.3 Basic state and approximate models

$L_T \to \infty$. In the solid phase, however, as $z \to -\infty$, the exponentials grow. If κ^s (and D^s) are small in some sense, it is possible to show (see, e.g., Huntley 1993) for $\kappa^s \to 0$, that T^s develops nonuniformities not at the interface but at the lower boundary $z = -L_B$. This "boundary layer" is thin and far removed from the interface, and thus it can be permissible to ignore these boundary-layer behaviors and use as approximate solutions in the solid phase that are unbounded at $z \to -\infty$. A similar argument holds for concentration and $D^s \to 0$.

If $D^s = D^\ell$, one has the *symmetric model*, which is appropriate for certain phase transformation studies of liquid crystals, for example. If $D^s \ll D^\ell$, one has the *one-sided model*. Here D^s can be set to zero and L_T can be set to infinity, as discussed above. The one-sided model is appropriate for solidification in metals, semiconductors, and many transparent organics. The one-sided and extended model will be discussed here.

If the Stefan number S (latent heat) is set to zero in the heat-flux condition and $\Gamma = O(1)$ elsewhere, the thermal properties (e.g., k_T, κ, c_p) in the two phases are assumed to be equal, and, as is usual, $D \ll \kappa$ everywhere, then one can replace T and T^s by a single linear variation valid in both phases,

$$T = T_0 + G_T z, \tag{3.13}$$

where G_T is the constant temperature gradient. Conditions in the far field are deemed not to influence the dynamics of the front strongly. This combination of limits gives the FTA, the *frozen temperature approximation* (Langer 1980). It comes about by scaling lengths on the concentration-boundary-layer thickness δ_c,

$$\delta_c = \frac{D}{V}, \tag{3.14a}$$

which is much smaller than the thermal-boundary-layer thickness δ_T,

$$\delta_T = \frac{\kappa}{V}. \tag{3.14b}$$

The scaled transport equations in nondimensional form are then

$$\frac{D}{\kappa}(T_t - T_z) = \nabla^2 T \tag{3.15a}$$

and

$$C_t - C_z = \nabla^2 C, \tag{3.15b}$$

and when $D/\kappa \to 0$, the thermal-steady-state approximation (see Coriell and McFadden 1993) is enforced, a linear temperature distribution, the "small" z approximation to the exponential behavior, is obtained. As will be seen, such approximations are very good in many systems.

The basic state that will be considered next applies to the one-sided model with the FTA.

Here $\bar{T}$ is given by Eq. (3.13), and

$$\bar{C} = C_\infty \left[1 + \frac{1-k}{k} e^{-\frac{Vz}{D}} \right], \tag{3.16a}$$

$$\bar{C}^s = C_\infty, \tag{3.16b}$$

$$h = 0, \tag{3.16c}$$

where

$$T_0 = T_m + \frac{mC_\infty}{k} - \mu^{-1} V, \tag{3.16d}$$

and

$$\bar{C} \to C_\infty \text{ as } z \to \infty. \tag{3.16e}$$

The concentration profile balances rejection of solute at the front, a source, with diffusion in the liquid. Equation (3.16d) defines T_0, the temperature at the interface in terms of the concentration and the speed. Both the presence of the constitutional undercooling (recall that $m < 0$) and the kinetic undercooling tend to reduce T_0 from T_m. Note that at the interface (on the liquid side) the concentration gradient G_c is given from Eq. (3.16a) by

$$G_c = \frac{(k-1)}{k} C_\infty \frac{V}{D}, \tag{3.17}$$

which is negative for $k < 1$.

3.4 Linearized Instability of a Moving Front in Directional Solidification

Consider directional solidification in two dimensions via the one-sided model with the FTA with kinetics ignored; all quantities are independent of y.

Let us introduce nondimensional variables. Scale all lengths on δ_c

$$(\hat{x}, \hat{z}) = \frac{1}{\delta_c}(x, z), \quad \hat{h} = \frac{h}{\delta_c}$$

$$\hat{t} = \frac{V^2}{D} t, \quad \hat{C} = \frac{C - C_\infty/k}{DG_c/V} \tag{3.18}$$

$$\hat{T} = \frac{(T - T_0 - mC_\infty/k)}{\delta_c G_T},$$

3.4 Linearized instability of a moving front in directional solidification

and then drop the carets. The temperature field is permanent at

$$T = z. \tag{3.19a}$$

For $z > h(x, t)$,

$$C_t - C_z = \nabla^2 C. \tag{3.19b}$$

As $z \to \infty$,

$$C \to 1. \tag{3.19c}$$

On $z = h(x, t)$,

$$M^{-1}T = C + 2H\Gamma \tag{3.19d}$$

and

$$[C(k-1) + 1](1 + h_t) = C_z - h_x C_x, \tag{3.19e}$$

where

$$M = \frac{mG_c}{G_T} = \frac{m(k-1)C_\infty V}{kG_T D} \tag{3.20a}$$

is the morphological number and

$$\Gamma = \frac{\gamma T_m}{L_v m G_c} \frac{V^2}{D^2} = \frac{\gamma T_m V}{C_\infty L_v D} \frac{k}{m(k-1)} \tag{3.20b}$$

is a measure of the surface energy. As can be seen, the problem reduces only to the tracking of solute in the liquid and the shape of the interface.

The basic state, as before, consists of a planar interface and an exponential solute field

$$\bar{h} = 0 \tag{3.21a}$$

$$\bar{C} = 1 - e^{-z} \tag{3.21b}$$

with

$$T = z \tag{3.21c}$$

everywhere.

Let the system be disturbed as follows:

$$h = \bar{h} + h', \ C = \bar{C} + C'. \tag{3.22}$$

There is no need to perturb the thermal field because, in the present approximation, thermal disturbances always decay to zero.

If forms (3.22) are substituted into system (3.19) and linearized in primed quantities, the linearized disturbance equations are obtained.

For $z > 0$,
$$C'_t - C'_z = \nabla^2 C'. \tag{3.23a}$$

As $z \to \infty$,
$$C' \to 0. \tag{3.23b}$$

On $z = 0$,
$$C'_z = (k-1)C' + kh' + h'_t \tag{3.23c}$$

$$C' = (M^{-1} - 1 - \Gamma\nabla^2)h'. \tag{3.23d}$$

This linear system has coefficients independent of x and t, and thus one can introduce normal modes
$$(C', h') = (c_1(z), h_1)e^{\sigma t + i a_1 x}, \tag{3.24}$$
where σ is the growth rate and a_1 is the wave number along the interface. When these are substituted into system (3.23a), one obtains
$$\sigma c_1 = \left(\frac{d^2}{dz^2} - a_1^2 + \frac{d}{dz}\right) c_1, \quad z > 0. \tag{3.25a}$$

On the interface $z = 0$ the linearized boundary conditions are
$$\frac{dc_1}{dz} = (k-1)c_1 + (k+\sigma)h_1, \tag{3.25b}$$

$$c_1 = \left(M^{-1} - 1 + a_1^2 \Gamma\right) h_1, \tag{3.25c}$$

and
$$c_1 \to 0 \quad \text{as} \quad z \to \infty. \tag{3.25d}$$

The solution of Eq. (3.25a) that satisfies condition (3.25d) is
$$c_1(z) = A e^{-\frac{1}{2}\left\{1 + \sqrt{1 + 4(\sigma + a_1^2)}\right\} z}. \tag{3.25e}$$

If Eq. (3.25e) is substituted in conditions (3.25b,c), a homogeneous, linear set of algebraic equations is obtained. To guarantee the existence of nontrivial solutions, the determinant of the matrix of coefficients must be set to zero; this yields the characteristic equation
$$\sqrt{1 + 4\left(\sigma + a_1^2\right)} = 1 - 2k - 2(\sigma + k)\left(M^{-1} - 1 + a_1^2 \Gamma\right)^{-1}. \tag{3.26}$$

This equation of the form $F(\sigma, a_1^2; M, \Gamma, k) = 0$ determines the stability

3.4 Linearized instability of a moving front in directional solidification

characteristics of the moving planar front. Note that σ can be shown to be real when it corresponds to a disturbance that grows or is neutral (Wollkind and Segel 1970).

The characteristic equation (3.26) has been derived for the two-dimensional case. If one rederived this for the three-dimensional case using the normal modes $\exp(\sigma t + ia_1 x + ia_2 y)$, Eq. (3.26) again is obtained if a_1 is replaced by $a = \sqrt{a_1^2 + a_2^2}$; the system is isotropic in x and y.

Consider the case of neutral stability, $\sigma = 0$. Then Eq. (3.26) becomes

$$M^{-1} = 1 - a_1^2 \Gamma + \frac{2k}{1 - 2k - \sqrt{1 + 4a_1^2}}. \tag{3.27}$$

Consider for the moment the special case of zero surface energy, $\Gamma = 0$. It is easy to show that $M^{-1}(a_1^2)$ is monotonically increasing, $M^{-1}(0) = 0$, and $M^{-1}(\infty) = 1$; thus, the basic state is always unstable if $M > 1$, and the instability is predominant as $a_1^2 \to \infty$. $M > 1$ is called the *constitutionally undercooled limit*.

When $\Gamma \neq 0$, as shown in Figure 3.4, M^{-1} crosses zero at a value of $a_1 = a_s$, the shortwave cutoff. Wave number a_s is then given by Eq. (3.27) with $a_1 = a_s$, $M^{-1} = 0$. The root occurs for small values of a_s^2, where one obtains

$$\Gamma \equiv \Gamma_s = \frac{1}{k} \tag{3.28a}$$

with

$$M^{-1} \sim \left(\frac{1}{k} - \Gamma\right) a_1^2 \tag{3.28b}$$

for a_1 near a_s.

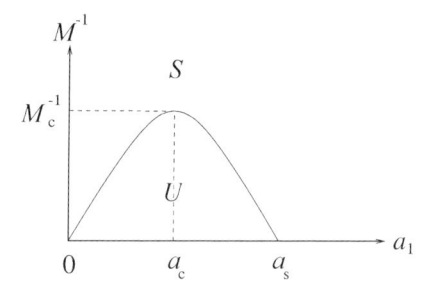

Figure 3.4. The neutral curve, M versus a_1, for directional solidification with k and Γ fixed.

For $\Gamma < \Gamma_s$, Figure 3.4 shows that the existence of surface energy stabilizes the front to shortwave disturbances, $a_1 > a_s$, and that a value (Mullins and Sekerka 1964) $a_1 = a_c$, exists corresponding to the most unstable mode for given Γ and k at $M = M_c = M(a_c^2)$.

As Γ is increased toward Γ_s, $a_s^2, a_c^2 \to 0$, $M_c^{-1} \to 0$, and in this limit no disturbances grow. The limit $\Gamma = \Gamma_s$ is called the *absolute stability boundary* (Mullins and Sekerka 1964).

Figure 3.4 shows $M = M(a_1^2)$ with the physically unacceptable negative branch for $a_1 > a_s$ removed. As M, the measure of solute rejection, is increased from zero, all disturbances decay until $M = M_c$ when the disturbance corresponding to $a_1 = a_c$ becomes neutral. For any $M > M_c$, a whole interval of a_1 corresponds to unstable states.

The stability map can be summarized for k fixed by selecting for each (M_c, Γ) the appropriate a_c and plotting these, as shown in Figure 3.5. Here M_c^{-1} versus Γ for fixed k gives a curve below which the front is morphologically unstable by linear theory. Noted as well are the values of a_c along this curve. When $\Gamma \to 0$, $M \to 1$ and a_c, which is scaled on D/V, is infinite. When $\Gamma \to \Gamma_s = 1/k$, a_c, scaled on D/V, approaches zero.

Note that the three control parameters G_T, C_∞, and V are merged into both M and Γ so that Eq. (3.27) does not show easily the physically relevant

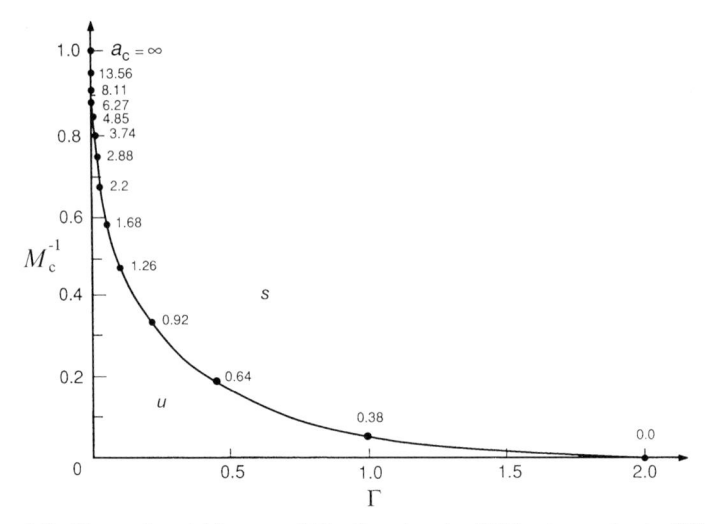

Figure 3.5. The preferred M_c versus Γ for directional solidification under the FTA with $a_1 = a_c$ at each point indicated. From Riley and Davis (1990). (Reprinted by permission of SIAM.)

3.4 Linearized instability of a moving front in directional solidification

situation of G_T fixed as V and C_∞ increase,

$$\Gamma \propto \frac{V}{C_\infty}, \quad M \propto V C_\infty, \quad a \propto \frac{1}{V}. \tag{3.29}$$

This dilemma can be overcome if one defines new measures that separate the control parameters (Merchant and Davis 1990), or, alternatively, one can return to the dimensional system. In either case the new figures have similar shapes.

The previous results can be recast into a set of velocity and concentration curves using nondimensional parameters that isolate V and C_∞ as independently controllable parameters; call these *natural parameters*. The appropriate scalings are described in Merchant and Davis (1990) where, instead of diffusion scales, capillary scales are used. These scales for length and time are

$$\left.\begin{array}{r}\hat{\delta}_L = (\gamma T_M / L G_T)^{1/2} \\ \hat{\delta}_T = \gamma T_M / L G_T D. \end{array}\right\} \tag{3.30}$$

The control parameters M and Γ are replaced by nondimensional speed and concentration parameters $\mathcal{V}$ and $\mathcal{C}$, respectively. These are defined by

$$\left.\begin{array}{r}\mathcal{V}^2 = \dfrac{\gamma T_M V^2}{L G_T D^2} \\ \mathcal{C}^2 = \dfrac{m_E^2 (k_E - 1)^2 L C_\infty^2}{k_E^2 G_T \gamma T_M}. \end{array}\right\} \tag{3.31}$$

For the easy switching from the old parameters M and Γ to the natural ones $\mathcal{V}$ and $\mathcal{C}$, the following transformations can be used:

$$\left.\begin{array}{r}M \to \mathcal{VC}, \quad \Gamma \to \mathcal{V}/\mathcal{C}; \\ a \to a/\mathcal{V}, \quad \sigma \to \sigma/\mathcal{V}^2, \end{array}\right\} \tag{3.32}$$

which can be inserted into the characteristic equation.

Let us plot the neutral stability curve given by Eq. (3.27) when the *dimensional* wave number a_*, and *dimensional* growth rate σ_* are introduced,

$$a = a_* \frac{D}{V}, \quad \sigma = \sigma_* \frac{D}{V^2}. \tag{3.33}$$

Fix G_T and let C_∞ and V be the control variables of interest. For a given set of material constants, the neutral curve is as shown in Figure 3.6, $\ell n V$ versus $\ell n C_\infty$. To the left of the curve the basic state is linearly stable, and to the right it is unstable.

3. Binary substances

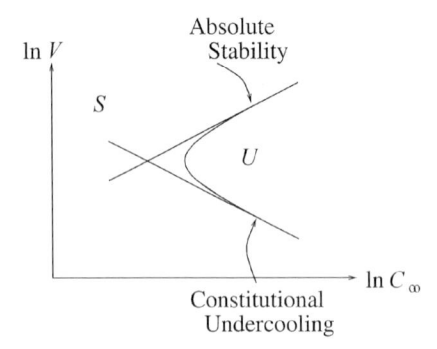

Figure 3.6. The neutral curve, $\ln V$ versus $\ln C_\infty$, G_T fixed, for directional solidification under the FTA.

Recall from Figure 3.4 that the onset of stability occurs at a fixed value of M. Then $M = M_c$ gives from relation (3.20a) that

$$\ln V + \ln C_\infty \propto \ln M_c, \tag{3.34a}$$

and thus a straight line with slope -1 in Figure 3.6 represents one asymptote, the *constitutional undercooling limit*, where surface energy is not very important, though it is necessary for selecting a_c.

Recall, as well, that if $\Gamma > \Gamma_s = 1/k$, then all linear instabilities disappear. Here, using relation (3.20b),

$$\ln V - \ln C_\infty \propto \ln \Gamma_s. \tag{3.34b}$$

In Figure 3.6, a line with slope $+1$ is an asymptote, the *absolute stability limit*.
Likewise,

$$a_c \to \infty \text{ as } \Gamma = 0 \tag{3.35a}$$

and

$$a_c \to 0 \text{ as } \Gamma \to \Gamma_s. \tag{3.35b}$$

In dimensional terms the reverse of Eqs. (3.35) is true

$$a_{*c} \to 0 \text{ as } \Gamma \to 0 \tag{3.36a}$$

$$a_{*c} \to \infty \text{ as } \Gamma \to \Gamma_s. \tag{3.36b}$$

Consider an experiment in which C_∞ is fixed. Figure 3.6 shows that if $C_\infty < C_N$, the value of the concentration at the nose of the neutral curve, then there are no linearized instabilities. For $C_\infty > C_N$, there are two critical values of V, namely, V_c, the lower intersection, and V_A, the upper intersection. As V is increased from zero for $C_\infty > C_N$, the planar front is linearly stable until

3.4 Linearized instability of a moving front in directional solidification

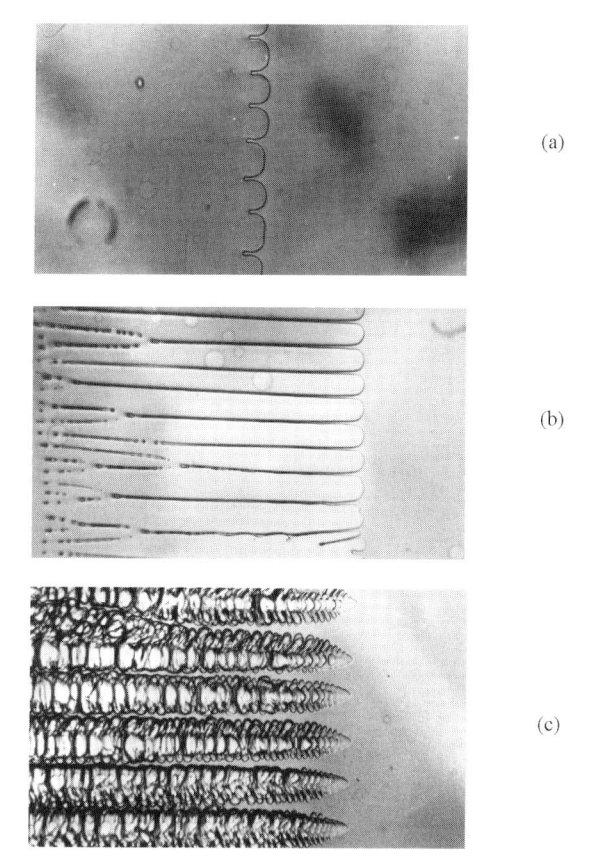

Figure 3.7. Directional solidification for SCN-acetone for (a) V near V_c, (b) for V well above V_c, and (c) still larger V. From Trivedi (1984). (Reprinted with permission from Metall. mater. Trans.)

$V = V_c$. Above V_c, shallow cells appear as a secondary state; see Figure 3.7(a). As V is increased further, the linearized theory loses validity, though one knows from experiment that the cells deepen (see Figure 3.7(b)); then at high V, the front becomes dendritic; see Figure 3.7(c). There is a value of V beyond which the dendrites become deep cells; at higher V, the cells become shallow, and according to this model, as V exceeds V_A, the plane front restabilizes. Note that the critical wave number will change as one goes from V_c to V_A. Figure 3.8(a) shows a cut through the stability curve at fixed C_∞. Inside the closed figure the planar front is unstable. As V increases from zero, the first instability gives a_c at point 1, whereas at the absolute stability boundary the shallow cells that vanish for increasing V have a_c at point 2. The same information is contained in Figure 3.8(b) if one plots it in dimensional variables.

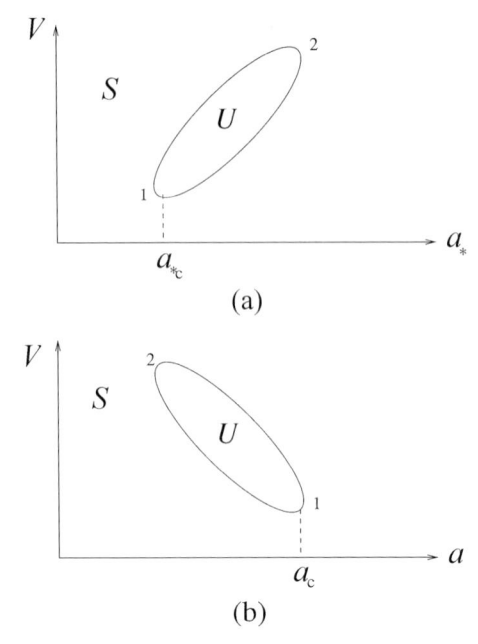

Figure 3.8. Neutral curves for G_T and C_∞ fixed (a) V versus a_* (dimensional) and (b) V versus a (nondimensional).

3.5 Mechanism of Morphological Instability

In nondimensional variables, the basic state is $\bar{T} = z, \bar{C} = 1 - e^{-z}$. In dimensional terms (indicated by asterisks) this is

$$T_* = T_0 + G_T z_* \tag{3.37a}$$

$$C_* = C_\infty \left[1 + \frac{1-k}{k} e^{-\frac{V z_*}{D}} \right], \tag{3.37b}$$

where

$$T_0 = T_m + m C_\infty / k. \tag{3.37c}$$

The basic state can be plotted on the phase diagram T_* versus C_* because T_* is monotone in z_*. Figure 3.2 shows the linearized phase diagram, and at the interfacial temperature T_0 the concentrations on the two sides of the interface are C_∞ (solid) and C_∞/k (liquid). Equations (3.37) define, through parameter z_*, a curve C_* versus T_*. The interface lies at $z_* = 0$ where $(T_*, C_*) = (T_0, C_\infty/k)$, which is the point on the liquidus at $T_* = T_0$. As shown in Figure 3.2, the basic state lies on the exponential that begins at $(T_0, C_\infty/k)$ and approaches C_∞ at $T_* \to \infty$.

In Figure 3.2 the exponential curve dips below the liquidus. This is the case if at the point $(T_0, C_\infty/k)$, $dT_*/dC_* > m$ $(m < 0)$. This slope can be evaluated by

$$\left.\frac{dT_*}{dC_*}\right|_{z_*=0} = \left.\frac{dT_*}{dz_*} \middle/ \frac{dC_*}{dz_*}\right|_{z_*=0} = G_T/G_c.$$

Thus, the operating curve dips below the liquidus if $G_T/G_c > m$ or, because both sides are negative, $mG_L/G_T > 1$. This condition is precisely $M > 1$, which is the definition of *constitutional undercooling*. If $M < 1$, the operating curve would be above the liquidus always.

Consider a perturbed interface in the case $M > 1$. A finger of solid that points into the liquid protrudes into a region where the temperature is higher. Because the solute has lowered the melting point, the tip finds itself in an undercooled environment, causing the finger to grow further. The concentration gradient G_c must be negative when $m < 0$, and it is the solute field that drives the instability. Clearly, the imposed temperature gradient G_T opposes the instability, as does capillary undercooling.

The condition $M > 1$ for instability was first discussed by Rutter and Chalmers (1953), and Mullins and Sekerka (1964) first included surface energy so that a numerical value to M_c could be obtained.

3.6 More General Models

In Section 3.5, the one-sided FTA model was analyzed for instabilities. The characteristic equation (3.26) determines the nondimensional growth rate σ in terms of the nondimensional wave number a.

Consider now a generalization of this model in which latent heat effects are retained, the thermal properties of the two phases may differ, and $D/\kappa \neq 0$. The new characteristic equation, given by Sekerka (1967), Coriell and Sekerka (1983), and Coriell and McFadden (1993), can be expressed in terms of the dimensional growth rate σ_* and wave number a_*, as follows:

$$\sigma_* = \left\{ -\frac{k_T^\ell G_L^l}{2\bar{\alpha}\bar{k}_T}\left(\alpha_L - \frac{V}{\kappa^\ell}\right) - \frac{k_T^s G_s}{2\bar{\alpha}\bar{k}_T}\left(\alpha_s - \frac{V}{\kappa^\ell}\right) - T_m\frac{\gamma}{L_\nu}a_*^2 \right.$$

$$\left. + m'G_c\frac{\alpha - \frac{V}{D^\ell}}{\alpha - p'\frac{V}{D^\ell}} \right\} \middle/ \left\{ \frac{L_\nu}{2\bar{\alpha}\bar{k}_T} + \frac{m'G_c}{V\left(\alpha - p'\frac{V}{D^\ell}\right)} \right\}$$

(3.38a)

where

$$\left.\begin{aligned}
\alpha &= \frac{V}{2D^\ell} + \left[\left(\frac{V}{2D^\ell}\right)^2 + a_*^2 + \frac{\sigma_*}{D^\ell}\right]^{\frac{1}{2}} \\
\alpha_L &= \frac{V}{2\kappa^\ell} + \left[\left(\frac{V}{2\kappa^\ell}\right)^2 + a_*^2 + \frac{\sigma_*}{\kappa^\ell}\right]^{\frac{1}{2}} \\
\alpha_s &= \frac{V}{2\kappa^s} + \left[\left(\frac{V}{2\kappa^s}\right)^2 + a_*^2 + \frac{\sigma_*}{\kappa^s}\right]^{\frac{1}{2}} \\
\bar{\alpha} &= (k_T^s \alpha_s + k_T^\ell \alpha_\ell)/(2\bar{k}_T) \\
\bar{k}_T &= \tfrac{1}{2}(k_T^s + k_T^\ell) \\
m' &= \frac{\partial T_E}{\partial C} \\
p' &= 1 - k' \\
k' &= \frac{\partial C^s}{\partial C} = k + \frac{C_\infty}{k}\frac{\partial k}{\partial C}
\end{aligned}\right\}. \quad (3.38b)$$

The quantities k' and m' allow the solidus and liquidus to be curved.

The heat balance at the interfaces requires that

$$k_T^s G_S - k_T^\ell G_L = V L_v, \quad (3.38c)$$

where G_S and G_L are the temperature gradients at the interface on the solid and liquid sides, respectively.

In the thermal steady-state approximation, $D, D^s \ll \kappa^\ell, \kappa^s$. Let $\kappa^\ell, \kappa^s \to \infty$ as formally in forms (3.38), $\bar{\alpha}, \alpha_s, \alpha_L \to a_*$, and the neutral curve, $\sigma_* = 0$ has the form

$$-\bar{G}_T + T_m \frac{\gamma}{L_v} a_*^2 + m' G_c \frac{\alpha - V/D}{\alpha - p'V/D}, \quad (3.39a)$$

where $\bar{G}$ is the conductivity-weighted temperature gradient

$$\bar{G}_T = \frac{k_T^\ell G_L + k_T^s G_S}{k_T^\ell + k_T^s}. \quad (3.39b)$$

If now, one lets $k_T^s = k_T^\ell$ and $L_v \to 0$ with $\gamma/L_v = O(1)$, then condition (3.38c) gives $G_S = G_L \equiv G_T$, constant, consistent with the FTA.

The results of Section 3.4 hold qualitatively in the more general cases. The addition of latent heat slightly stabilizes the front against morphological changes. Figure 3.9 compares the neutral curves for two one-sided models, one in the FTA and the other generalized to include the full thermal fields. The two asymptotes for constitutional undercooling and absolute instability are unchanged by

3.7 Remarks

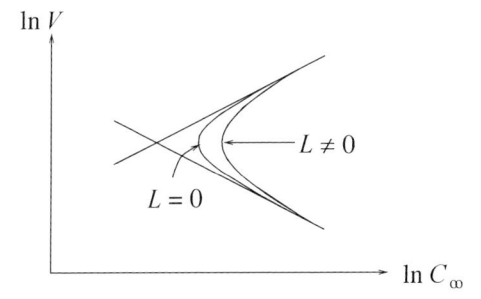

Figure 3.9. Neutral curves, $\ell n V$ versus $\ell n C_\infty$, G_T fixed for $L = 0$ and $L \neq 0$.

the presence of latent heat or the other generalizations of the FTA (Huntley and Davis 1993). The presence of latent heat flattens the nose of the curve and thus stabilizes the front near the nose.

As has been shown, the qualitative behavior in the linear stability theory is well described by the FTA in the one-sided model, and for most of what follows this model will be considered. If quantitative information is needed for comparison with experiment, then the more general models may be required.

3.7 Remarks

The equations governing the solidification of a dilute, binary melt have been derived. In particular across an interface between solid and liquid there are the thermal conditions, the heat balance

$$L_\nu \left(1 + 2H \frac{\gamma}{L_\nu}\right) V_n = (k_T^s \nabla T^s - k_T^\ell \nabla T^\ell) \cdot \mathbf{n}, \tag{3.40a}$$

the continuity of temperature

$$T^I = T^s = T^\ell, \tag{3.40b}$$

and the Gibbs–Thomson equation, with constitutional undercooling,

$$T^I = mC + T_m \left(1 + 2H \frac{\gamma}{L_\nu}\right) - \mu^{-1} V_n. \tag{3.40c}$$

In addition, there are conditions on the solute, the mass balance

$$(1 - k)_C V_n = (D^s \nabla C^s - D^\ell \nabla C^\ell) \cdot \mathbf{n}, \tag{3.40d}$$

and the definition of the segregation coefficient,

$$C^s = kC. \tag{3.40e}$$

Condition (3.40d) shows that the magnitude of the concentration gradients controls the frontal speed.

The mechanism of morphological instability in a dilute binary alloy has been discussed as well as models for the prediction of M_c and a_c. Experiments of de Cheveigné, Guthmann, and Lebrun (1985) on CB_4 with about 0.1% Br as a contaminant show that, on the low-speed branch, theory and experiment agree reasonably well on M_c, but the critical wave numbers can differ by a factor of two or more. Several possible explanations exist for this disagreement, and these are discussed in Chapter 4.

One possibility for the disagreement is that the growth of the instability begins early, when the basic state has not yet approached its final steady form. The analysis of the instabilities of unsteady states is quite difficult; consequently, Warren and Langer (1993) examined a model version of the problem by parameterizing the time dependence through an unsteady solute boundary layer thickness $\ell(t)$ such that $\ell(t) \to \delta_c$ as $t \to \infty$. They then found the critical a_{*_c} smaller than those of the previous theory, which has $\ell = \delta_c$ forever, and obtained predictions qualitatively in agreement with the observations of Trivedi and Somboonsuk (1985) in SCN–acetone. Losert, Shi, and Cummins (1998) used SCN with coumarin 152 as a contaminant and found good agreement with the Warren–Langer predictions (SCN is the transparent organic succinonitrile).

References

Askeland, D. R. (1984). *The Science and Engineering of Materials*, Book-Cole Engineering Division.

Coriell, S. R., and McFadden, G. B. (1993). Morphological stability, in D. T. J. Hurle, editor, *Handbook of Crystal Growth*. **16**, 785–857, North-Holland, Amsterdam.

Coriell, S. R., and Sekerka, R. F. (1983). Oscillatory morphological instabilities due to non-equilibrium segregation, *J. Crystal Growth* **61**, 499–508.

de Cheveigné, S., Guthmann, C., and Lebrun, M.-M. (1985). Nature of the transition of the solidification front of a binary mixture from a plane to a cellular morphology, *J. Crystal Growth* **73**, 242–244.

Flemmings, M. C. (1974). *Solidification Processing*, McGraw–Hill, New York.

Huntley, D. A. (1993). Thermal effects in rapid directional solidification. Ph.D. thesis, Northwestern University.

Huntley, D. A., and Davis, S. H. (1993). Thermal effects in rapid directional solidification: Linear theory, *Acta Metall. Mater.* **41**, 2025–2043.

Langer, J. S. (1980). Instabilities and pattern formation in crystal growth, *Rev. Mod. Phys.* **52**, 1–28.

Losert, W., Shi, B. Q., and Cummins, J. Z. (1998). Evolution of dendritic patterns during alloy solidification: Onset of the initial instability, *Proc. Natl. Acad. Sci. U.S.A.* **95**, 431–438.

Merchant, G. J., and Davis, S. H. (1990). Morphological instability in rapid directional solidification, *Acta Metall. Mater.* **38**, 2683–2693.

Mullins, W. W., and Sekerka, R. F. (1964). Stability of a planar interface during solidification of a dilute binary alloy, *J. Appl. Phys.* **35**, 444–451.

Riley, D. S., and Davis, S. H. (1990). Long-wave morphological instabilities in the directional solidification of a dilute binary mixture, *SIAM J. Appl. Math.* **50**, 420–436.

Rutter, J. W., and Chalmers, B. (1953). A prismatic substructure formed during solidification of metals, *Can. J. Phys.* **31**, 15–49.
Sekerka, R. F. (1967). Time-dependent theory of stability of a planar interface during dilute binary alloy solidification, in H. S. Perser, editor, *Crystal Growth*, pages 691–702, Pergamon Press, Oxford.
Trivedi, R. (1984). Interdendritic spacing: Part II. A comparison of theory and experiment, *Metall. mater. Trans.* **15A**, 977–982.
Trivedi, R., and Somboonsuk, K. (1985). Pattern formation during the solidification of binary systems, *Acta Metall. Mater.* **33**, 1061–1068.
Warren, J. A., and Langer, J. S. (1993). Prediction of dendritic spacings in a directional solidification experiment, *Phys. Rev. E* **47**, 2702–2712.
Wollkind, D. G., and Segel, L. A. (1970). A nonlinear stability analysis of the freezing of a dilute binary alloy, *Phil. Trans. Roy. Soc. London* **A268**, 351–380.

4

Nonlinear theory for directional solidification

In Chapter 2 we saw that a spherical front in a pure material in an undercooled liquid is either unstable or not, depending on the size of the sphere. There is no secondary control that allows one to mediate the growth. In Chapter 3, we saw that, in directional solidification of a binary liquid, the concentration gradient at the interface creates an instability; however, there is a secondary parameter, the temperature gradient, that opposes the instability and thus can be used to control the local growth beyond the linearized stability limit. Thus, much attention has been given to directional solidification both experimentally and theoretically, and it is in this chapter that nonlinear theory will be discussed.

There are two approaches to the nonlinear theory, depending upon whether the critical wave number a_c of linear theory is of unit order, or is small (i.e., asymptotically zero). In the former case one can construct Landau, Ginzburg–Landau, or Newell–Whitehead–Segel equations to study (weakly nonlinear) bifurcation behavior. This gives information regarding the nature of the bifurcation (sub- or supercritical), the question of wave number selection, the preferred pattern of the morphology, and hence the resulting microstructure. If a_c is small one must use a longwave theory that generates evolution equations governing the nonlinear development. Such longwave theories can be weakly or strongly nonlinear, depending on the particular situation. They, too, can then be analyzed to discover the nature of the bifurcation and the selection of preferred wave number and pattern.

4.1 Bifurcation Theory

4.1.1 Two-Dimensional Theory

When directional solidification is performed in a Hele–Shaw cell, a region bounded by two parallel, closely spaced plates, the instability can readily be

4.1 Bifurcation theory

observed if the plates and the fluid are transparent. The instability is sensibly two-dimensional as long as the curvature of the interface across the gap is negligible.

Bifurcation theory in two dimensions in the FTA proceeds as follows. From the linearized theory one has M_c, a_c and the corresponding eigenfunctions. The growth rate σ can be written as

$$\sigma = \sigma_0(M - M_c) + O\left[(M - M_c)^2\right] \quad (4.1)$$

valid near the neutral curve, say for $a_1 = a_c$ $\sigma_0 = O(1)$. Because linearized growth has the form $\exp(\sigma t) \sim \exp[\sigma_0(M - M_c)t]$, one can identify the slow time $\tilde{t}$ by $\tilde{t} = \sigma t$.

If one defines the small parameter $\epsilon \equiv |\sigma|^{1/2}$ as $M \to M_c$, then

$$\tilde{t} = \epsilon^2 t \quad (4.2a)$$

and it follows that

$$M \sim M_c + \delta\epsilon^2, \quad (4.2b)$$

where, for convenience, the factor $\delta = \mathrm{sgn}(M - M_c)$ has been introduced so that both signs of $M - M_c$ can be treated simultaneously.

Forms (4.2) are introduced into Eqs. (3.19) governing directional solidification subject to the FTA and the one-sided model. The dependent variables are written as asymptotic expansions

$$[C(x, z, t), h(x, t)] \sim [\bar{C}(z), \bar{h}]$$
$$+ \epsilon \left\{ A(\tilde{t})[C_1(z), h_1]e^{ia_1 x} + A^*(\tilde{t})[C_1^*(z), h_1^*]e^{-ia_1 x} \right\}, \quad (4.2c)$$

where the amplitude A is for the moment unknown and an asterisk denotes complex conjugate. The forms (4.2) are substituted into the governing system, and coefficients of equal powers of ϵ are equated to zero.

At $O(1)$, $[\bar{C}(z), \bar{h}]$ turns out to be the basic state already calculated. At $O(\epsilon)$, one obtains the linear theory for $M = M_c$. At $O(\epsilon^2)$ an inhomogeneous linear system for $[C_2(z), h_2]$, which can be solved explicitly, is obtained. Finally, at $O(\epsilon^3)$ the inhomogeneous linear system for $[C_3(z), h_3]$ cannot be solved unless the right-hand side satisfies an orthogonality condition. The result of this Fredholm condition is an equation for $A(\tilde{t})$, which until now, was undefined; A satisfies a Landau equation

$$\frac{dA}{d\tilde{t}} = \left[\delta - b_1 |A|^2\right] A. \quad (4.3)$$

Here b_1, the Landau coefficient, is a real, computable quantity that depends on

4. Nonlinear theory for directional solidification

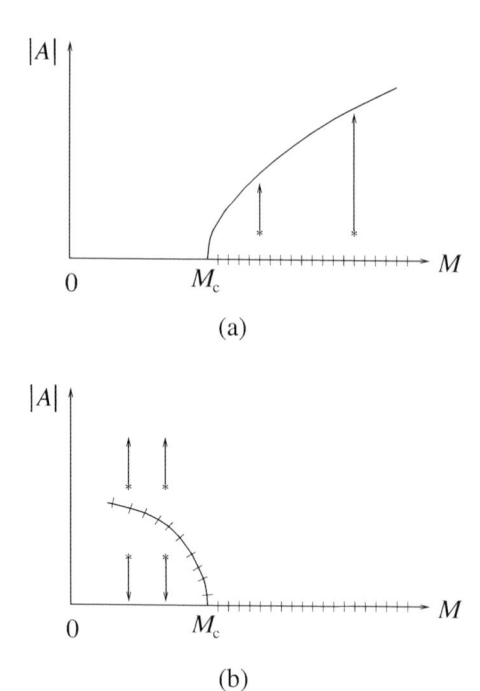

Figure 4.1. Sketches of bifurcation types for the Landau equations (4.3) with cubic nonlinearity. Hatches denote unstable branches. The arrows indicate evolution in time from initial conditions, denoted by asterisks, for (a) supercritical, $b_1 > 0$, and (b) subcritical, $b_1 < 0$, bifurcations.

all of the parameters of the problem. The sign of b_1 determines the type of bifurcation. If $b_1 > 0$, the steady version of Eq. (4.3) requires that $\delta = 1$, and there is supercritical bifurcation to stable, finite-amplitude cells with $|A| = (1/b_1)^{1/2}$; see Figure 4.1(a). If $b_1 < 0$, then necessarily $\delta = -1$, and the bifurcating solution is subcritical and has $|A| = (-1/b_1)^{1/2}$; see Figure 4.1(b).

The first such theory for directional solidification was that of Wollkind and Segel (1970), who extended the model to $k_T \neq 1$. (Thus, the temperature field is piecewise linear.) They showed that the bifurcation is necessarily steady so that b_1 is real and leads to a cellular interface, as shown in Figure 3.7a. They computed b_1 for materials with $k < 1$ and found that the upper branch is generally supercritical, and the lower speed branch is generally subcritical, with the *transition point*, $b_1 = 0$, separating the two near the nose of the lower branch, as shown in Figure 4.2(a). Caroli, Caroli, and Roulet (1982) showed for $k_T = 1$ that the entire lower branch becomes supercritical when the segregation coefficient $k > 0.45$.

4.1 Bifurcation theory

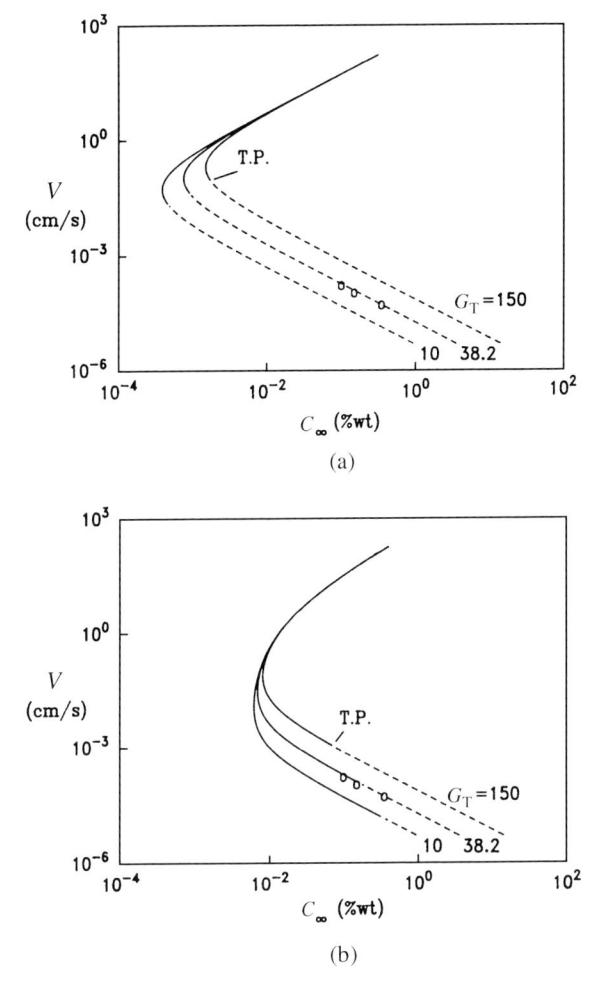

Figure 4.2. Neutral curves, V versus C_∞, for various values of G_T under the FTA for SCN–acetone. T.P. denotes the transition point between supercritical (solid lines) and subcritical (dashed lines) bifurcations for (a) the FTA and (b) the non-FTA. From Merchant and Davis (1989a).

The position of the transition point can be sensitive to thermal parameters of the problem; compare Figures 4.2(a) and (b). Alexander, Wollkind, and Sekerka (1986) included the effects of latent heat and argued that it should be especially important in materials for which $k_T \approx 1$ such as the transparent organics used in experiment. They found in this case that approximately

$$b_1 \sim \frac{1 - k_T + I(k_T + 1)}{(1 + I)(1 + k_T)\Gamma} + O\left(\left(\frac{4k}{\Gamma}\right)^{\frac{2}{3}}\right), \qquad (4.4\text{a})$$

where

$$I = \frac{kL_v D}{m(k-1)(1+k_T)\kappa^\ell C_\infty}. \tag{4.4b}$$

The parameter I is numerically small, and when k_T is not near unity, latent heat provides a small correction to b_1. However, when $k_T \approx 1$, latent heat can determine both the magnitude and the sign of b_1. Merchant and Davis (1989a) presented calculations for SCN–acetone alloys and argued that one could experimentally access the supercritical range on the lower branch and thus test the Mullins–Sekerka theory quantitatively. Using CBr$_4$ with roughly 0.1%Br$_2$, de Cheveigné, Guthmann, and Lebrun (1985) showed that the bifurcation is subcritical to the right of the transition point. Liu, Williams, and Cummins (1994) observed the existence of the transition point using SCN with 0.1% Coumarin 150. Near the nose of the curve, both branches are accessible to analysis, and it was found by Ramprasad, Bennett, and Brown (1988) and Merchant and Davis (1989b) that complicated behavior is possible. Merchant and Davis showed, as k_T departs from unity, that the transition point can move to the upper branch and that isolas, isolated solutions, can exist to the left of the nose.

Finally, de Cheveigné, Guthmann, and Lebrun (1986) showed that the observed wavelengths at the onset of subcritical instability are smaller by a factor of 2–3 from the Mullins–Sekerka predictions. Several possible explanations exist for this discrepency, and these will be discussed below.

4.1.2 Two-Dimensional Theory for Wave Number Selection

When the initial bifurcation is supercritical, one can inquire about wave number selection by deriving a Ginzburg–Landau equation,

$$\frac{\partial A}{\partial \tilde{t}} = \left[\delta - b_1 |A|^2\right] A + d_1 \frac{\partial^2 A}{\partial \tilde{x}^2}, \tag{4.5}$$

where $\tilde{x}$ is a slow space variable,

$$\tilde{x} = \epsilon x, \tag{4.6}$$

where $\tilde{t}$ and b_1 are as before, and $d_1 > 0$. The number d_1 comes from linear theory and is proportional to $\partial^2 M/\partial a_1^2$ at $a_1 = a_c$. From Eqs. (4.5) and (4.6), it can be shown (Eckhaus 1965) that for all wave numbers a_1 with

$$\frac{|a_1 - a_c|}{a_c} > \frac{1}{\sqrt{3}}, \tag{4.7}$$

two-dimensional finite amplitude solutions are unstable to those within the band. See Figure 4.3 for a sketch of the stable band.

4.1 Bifurcation theory

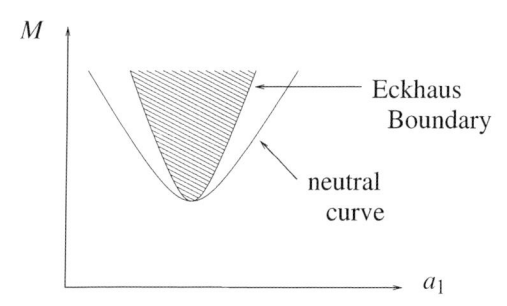

Figure 4.3. The neutral curve, M versus a_1, and the Eckhaus boundary. The unshaded regions correspond to two-dimensional cells that are unstable to the modes in the shaded region.

It has been argued by Bennett and Brown (1989) that the (dimensional) neutral curve, C_∞ versus a_{1_*} for fixed V and G_T is very flat at its nose so that when C_∞ is slightly supercritical there is a wide band of wave numbers correspond to growing modes. As a result, wave number selection is quite weak, or, equivalently, the one-mode bifurcation theory has a very small range of validity.

Such difficulties can be approached in several ways. First, one can "model" a flat neutral curve by allowing higher-order derivatives in the Ginzburg–Landau equation, say

$$\frac{\partial A}{\partial \tilde{t}} = (\delta - b_1 |A|^2)A + d_n \frac{\partial^{2n} A}{\partial \tilde{x}^{2n}}, \quad n = 1, 2, 3 \ldots. \tag{4.7}$$

When $n = 1$, Eq. (4.5) is recovered. When $n > 1$, Riley and Davis (1989) showed under supercritical conditions that the Eckhaus instability becomes successively weaker as n increases in the sense that the wave number interval becomes increasingly Eckhaus stable. The generalized instability condition, corresponding to Eq. (4.7), becomes

$$\frac{|a_1 - a_c|}{a_c} > \left(\frac{2n-1}{4n-1}\right)^{\frac{1}{2n}}, \quad n \geq 1. \tag{4.8}$$

This result in some sense rationalizes the weakening of the wave number selection near $M = M_c$ though it does not provide the selection rules.

Second, one can examine wave number selection for the standard case, $n = 1$, away from the bifurcation point by using phase-equation analysis. The complex amplitude A is expressed in polar form

$$A = Re^{i\varphi},$$

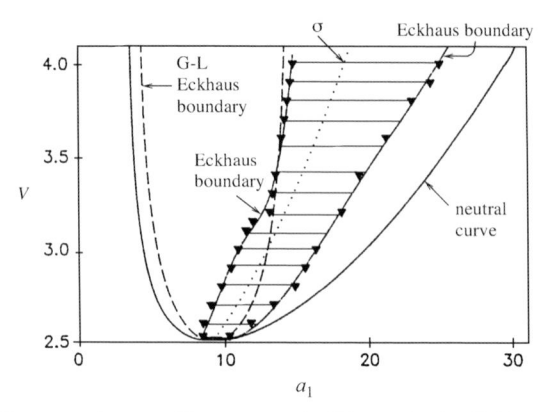

4. Nonlinear theory for directional solidification

Figure 4.4. The neutral curve, V versus a_1 (solid curve), the Eckhaus boundary determined from the Ginzburg–Landau equation (dashed curve), the computed Eckhaus boundary using the phase equation, and the curve of maximum growth rate of linear theory. The unhatched area denotes two-dimensional cells unstable to those in the shaded region. From Brattkus and Misbah (1990).

where R and φ are the amplitude and phase, respectively. As one operates further away from the bifurcation point and assumes that R is constant (or slowly varying), the following phase equation can be derived:

$$\varphi_t = D(\varphi)\varphi_{xx}. \tag{4.9}$$

When $D > 0$, the phase satisfies a forward diffusion equation, and the basic state in question is stable. When $D < 0$, this state is unstable. By tracking D numerically, Brattkus and Misbah (1990) have extended the Eckhaus boundary for the symmetric model as shown in Figure 4.4, which shows that the stable two-dimensional states become greatly restricted as M increases. The upper boundary well approximates the $a_c \to 2a_c$ transition discussed later in this section.

Kopczyński, Karma, and Rappel (1996) have examined the symmetric model of direction solidification in the FTA and extended the analysis by direct numerical simulation and Floquet theory. Figure 4.5(a) shows that the balloon of stable states closes at the top and that, above it, all two-dimensional steady models are unstable. Figure 4.5(b) shows the same type of result except that there is now fourfold anisotropy of the surface energy, which produces a narrow strip of stable two-dimensional cells. Apparently, the anisotropy stabilizes these in some way. See Chapter 5 for a discussion of anisotropy.

Third, as discussed in Chapter 3, the predicted and observed a_c may differ significantly if the instability arises during the *development* of the basic state. Warren and Langer (1993) have examined this phenomenon using a model system, and their predictions seem to represent well some of the observations; see Section 3.7 for a further discussion.

4.1 Bifurcation theory

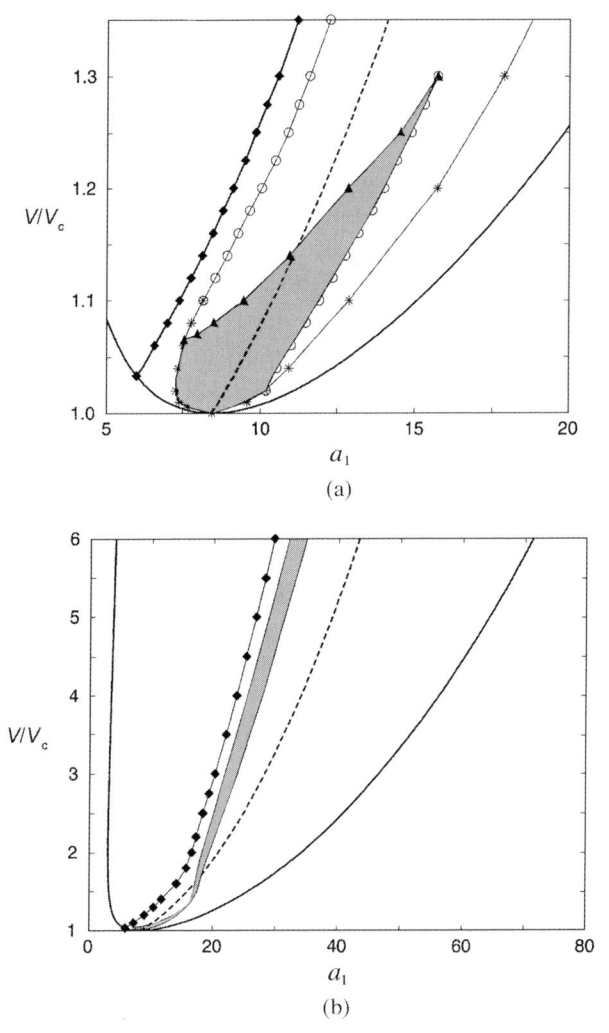

Figure 4.5. The neutral curve, V versus a_1 (solid curve), the Ginzburg–Landau version of the Eckhaus boundary (asterisks), the $(1/2)a_1$ steady mode (open circles), the $1/2a$ oscillatory mode (triangles), and the curve of maximum growth rate of linear theory. The shaded region is the balloon of stable two-dimensional cells for (a) the isotropic case, $\alpha_4 = 0$, and (b) the fourfold anisotropic case, $\alpha_4 = 0.01$. From Kopczyński et al. (1996).

Fourth, one can perform direct numerical simulations. Bennett and Brown (1989) used finite-element numerics on a spatially periodic domain and found in the symmetric model that there is a tendency for tip splitting under supercritical conditions in which wave number $a_c \to 2a_c$. They believe this transition

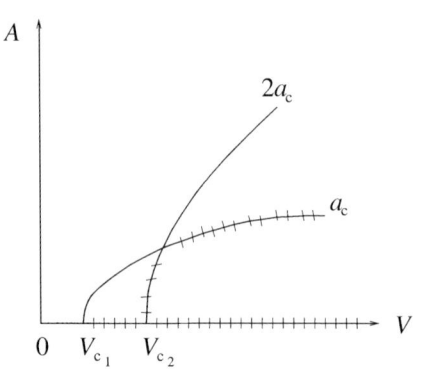

4. Nonlinear theory for directional solidification

Figure 4.6. A simplified sketch of the bifurcation structure in two-dimensional directional solidification on a spatially periodic domain. Hatches denote unstable branches.

occurs because, when the neutral curve is very flat, the mode pair $(a_c, 2a_c)$ is nearly simultaneously excited; see Figure 4.6.

As a result, one can imagine a *model* system in which the 1–2 interaction is dominant and for which one can analyze the set

$$[C(x, z, t), h(x, t)] \sim [\bar{C}(z), \bar{h}] + \epsilon \left\{ A_1(\tilde{t})[C_{01}(z), h_{01}]e^{ia_{1c}x} \right.$$
$$\left. + A_2(\tilde{t})[C_{02}(z), h_{02}]e^{2ia_{1c}x} + c.c. \right\}. \quad (4.10)$$

Such a theory formally proceeds as before. There are now two orthogonality conditions to be enforced because there are two independent eigenfunctions neutral at the same point.

Levine, Rappel, and Riecke (1991) have examined such a pair of disturbances in the symmetric model and obtained the Landau equations

$$\dot{A}_1 = \sigma_1 A_1 - c_1 A_1^* A_2 - \left[b_1 |A_1|^2 + b_2 |A_2|^2 \right] A_1$$
$$\dot{A}_2 = \sigma_2 A_2 - c_2 A_1^2 - \left[a_1 |A_1|^2 + a_2 |A_2|^2 \right] A_2. \quad (4.11)$$

The coefficients were calculated and the results compared with direct numerical calculations by Rappel and Riecke (1992). This analysis correctly predicts the 1–2 interactions, shows how some of the observed behavior may emerge, and predicts a possible secondary state of traveling cells. However, because it ignores disturbances that have larger linear theory growth rates than the modes included, the analysis does not explain *how* this pair is selected.

Two-dimensional morphological instability in a periodic domain (in x) of length $2\pi/a_c$ allows spatial transitions that are superharmonic (length $2\pi n/a_c$) but not subharmonic. Ungar and Brown (1984) used finite element analysis, and McFadden and Coriell (1984) used finite differences to obtain a bifurcation map of the type shown in Figure 4.6. Both theory and numerics show that

4.1 Bifurcation theory

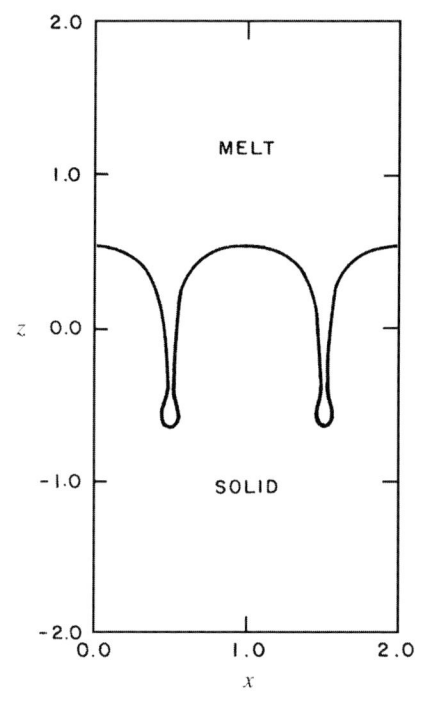

Figure 4.7. A sample interface shape on the stable part of the $2a_c$ branch of Figure 4.5. From Ungar and Brown (1985).

each primary bifurcation is subcritical, though the numerics suggests that each branch reaches a limit point and turns around in such a small neighborhood of the bifurcation point that the subcriticality is not apparent in the figure.

The computations show a secondary bifurcation from mode $a_c \to 2a_c$ at finite amplitude, and the process exhibits the flattening of the cell tips and the creation of new roots at the tip centers. This tip splitting is qualitatively consistent with the observations of Morris and Winegard (1969).

Ungar and Brown (1985) and Ramprasad et al. (1988) have used finite element analysis with multiple coordinate rescalings to predict the shapes of deep cells, as shown in Figure 4.7. The drop-shaped roots are frequently seen in experiment (Kurowski, de Cheveigné, Faivre, and Guthmann 1989), and the prediction of these requires retention of the solute diffusion in the solid, given that some melting is required to evolve to such shapes. Kessler and Levine (1989) used boundary-integral methods to predict similar features.

Bennett and Brown (1989) have observed the onset of time-dependence, and Saito, Misbah, and Müller-Krumbhaar (1989) have seen the onset of secondary side arms of deep cells. Various calculations in two dimensions have been done using the FTA and their generalizations. See Coriell and McFadden (1993) for a detailed summary.

4.1.3 Three-Dimensional Theory

When directional solidification is performed in general applications, the material is not confined to a Hele–Shaw geometry but occupies a three-dimensional space. In this case, the concentration, temperature, and interface position depend on two transverse spatial variables x and y.

One approach to the description of three-dimensional instabilities is the derivation, valid near the bifurcation point, of an amplitude equation that allows long-scale modulations in both x and y to two-dimensional solutions. Such an equation, the Newell–Whitehead–Segel equation, was derived by Segel (1969) and Newell and Whitehead (1969) in the context of Bénard convection.

If the basic two-dimensional solution treated in Section 4.1.1 depends on x, t only, then one lets $\tilde{x} = \epsilon x$, $\tilde{t} = \epsilon^2 t$, as before, but now lets $\tilde{y} = \epsilon^{1/2} y$. The result is the equation for $A(\tilde{x}, \tilde{y}, \tilde{t})$

$$\frac{\partial A}{\partial \tilde{t}} - d_1 \left[\frac{\partial}{\partial \tilde{x}} - i \frac{\partial^2}{\partial \tilde{y}^2} \right]^2 A = (\delta - b_1 |A|^2) A, \qquad (4.12)$$

where d_1, δ, and b_1 are the same constants as before. From this equation the stability of two-dimensional solutions to the Eckhaus instabilities can be examined for $(\partial/\partial \tilde{y} = 0)$, as well as, three-dimensional corrugations.

When in fact three-dimensional states are present, one can directly examine the competition among *regular* planforms by deriving appropriate sets of Landau equations.

In the FTA, the dependent variables are written

$$[C(x, y, z, t), h(x, y, t)] \sim (\bar{C}(z), \bar{h}) + \epsilon f_0 [C_1(z), h_1], \qquad (4.13a)$$

where

$$f_0 = \sum_{j=1}^{J} \{ A_j(\tilde{t}) e^{i a_j \cdot r} + cc \}, \qquad (4.13b)$$

A_j is a complex coefficient and the vector wave number $\mathbf{a}_j = (a_1, a_2)$; in the plane $\mathbf{r} = (x, y)$. The number of modes J depends upon the interaction being considered and determines the number of orthogonality conditions that must be enforced.

When two-dimensional cells are considered, $J = 1$ so that, say $\mathbf{a} = (1, 0) a_c$ and the amplitude A_1 satisfies Eq. (4.3) When one interacts two-dimensional cells and allows squares, $J = 2$, $\mathbf{a} = (1/\sqrt{2}, 1/\sqrt{2})a_c$, and the amplitudes A_1, A_2 satisfy

$$\frac{dA_1}{d\tilde{t}} = \{\delta - b_1 |A_1|^2 - b_2 |A_2|^2\} A_1 \tag{4.14a}$$

$$\frac{dA_2}{d\tilde{t}} = \{\delta - b_2 |A_1|^2 - b_1 |A_2|^2\} A_2. \tag{4.14b}$$

When one interacts two-dimensional cells and allows hexagons, $J = 3$, $\mathbf{a} = (a_1 \cos 1/3 (j-1)\pi, a_2 \sin 1/3 (j-1)\pi)$, $j = 1, 2, 3$, and $|\mathbf{a}_j| = a_c$. In this case

$$\frac{dA_1}{d\tilde{t}} = \delta A_1 - b_0 A_2 A_3^* - A_1 \{b_1 |A_1|^2 + b_3 [|A_2|^2 + |A_3|^2]\} \tag{4.15a}$$

$$\frac{dA_2}{d\tilde{t}} = \delta A_2 - b_0 A_1^* A_3 - A_2 \{b_1 |A_2|^2 + b_3 [|A_3|^2 + |A_1|^2]\} \tag{4.15b}$$

$$\frac{dA_3}{d\tilde{t}} = \delta A_3 - b_0 A_1^* A_2 - A_3 \{b_1 |A_3|^2 + b_3 [|A_1|^2 + |A_2|^2]\}. \tag{4.15c}$$

Equation (4.3) governs the growth of two-dimensional cells. As discussed earlier, b_1, the Landau coefficient, determines whether the bifurcation is supercritical ($b_1 > 0$) or subcritical ($b_1 < 0$). The point where $b_1 = 0$ is the transition point separating the two behaviors.

Equations (4.14) govern the competition between two-dimensional cells ($A_1 = 0, A_2 \neq 0$, or $A_1 \neq 0, A_2 = 0$) and squares ($A_1, A_2 \neq 0$). The coefficient b_2 has not been calculated for directional solidification likely because limited observations indicate that hexagonal arrays are the principal pattern seen. For example if both $b_1, b_2 > 0$, then either rolls or squares, but not both, are stable depending upon whether $b_1/b_2 < 1$ or $b_1/b_2 > 1$, respectively.

Experiments by Cole and Winegard (1963) on tin alloys and Morris and Winegard (1969) show hexagonal structures. As a result, system (4.15) has drawn more attention. The cubic interaction coefficient b_3, is required as well as the quadratic interaction coefficient b_0.

In all of the preceding equations the primitive (unscaled) amplitude is scaled by $\epsilon^{1/2}$, the time by ϵ and the cubic coefficients are of unit order. When quadratic terms are present, as in Eqs. (4.15), and the primitive quadratic coefficient $\bar{b}_0$ is of unit order, the cubic terms are negligible. To retain both the cubic and quadratic terms, $\bar{b}_0$ must be order ϵ in magnitude. Thus, the coefficient b_0 in Eqs. (4.15) is defined by $\bar{b}_0 = b_0 \epsilon$, $b_0 = O(1)$.

When $b_0 \neq 0$, the hexagonal planforms can exist and be stable. Figures 4.8(a) and (b) from McFadden, Boisvert, and Coriell (1987) show physically distinct configurations that depend on sgn b_0 and the relative

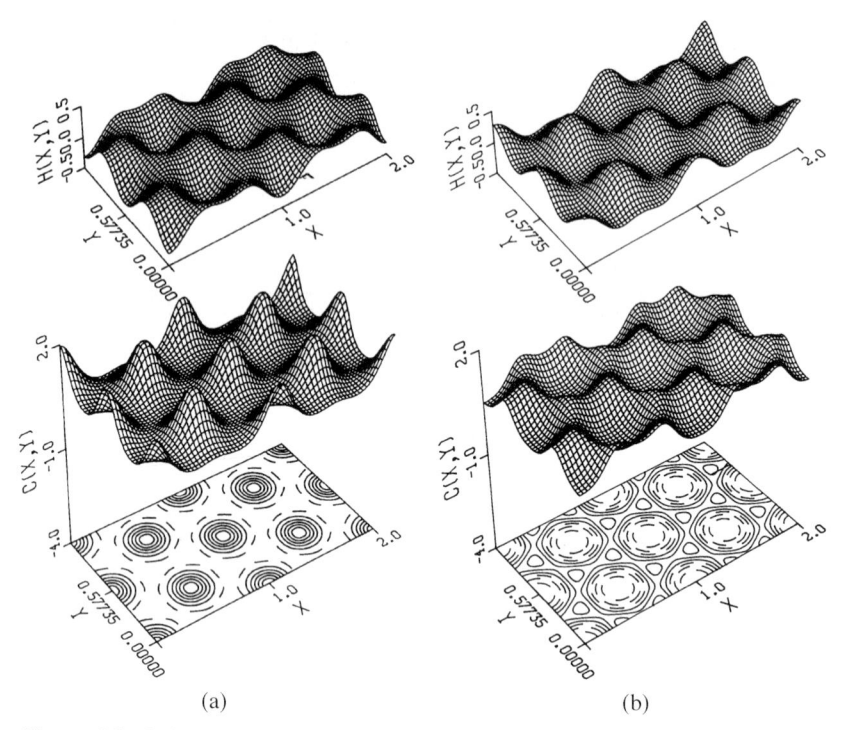

Figure 4.8. Calculated interface shapes and concentration profiles for (a) hexagonal cells and (b) hexagonal nodes. From McFadden et al. (1987).

signs of C and h, which is analogous to the distinction between up and down hexagons in Bénard convection. Hexagonal *cells* in Figure 4.8(a) have dome-shaped centers pointing into the liquid and high concentrations of solute at the edges. Hexagonal *nodes* in Figure 4.8(b) have dome-shaped centers pointing into the solid and high concentrations of solute in the centers; see Wollkind, Sriranganathan, and Oulton (1984). Figures 4.9 show pictures by Morris and Winegard (1969) of quenched interfaces in Pb–Sn for (a) nodes and (b) cells. The darkness of the regions denote solute-concentration level (the darkest region denotes the highest concentration).

Sriranganathan, Wollkind, and Oulton (1983) and Wollkind, Oulton, and Sriranganathan (1984) calculated the coefficients b_0 and b_3 in Eqs. (4.15). Figures 4.10 shows the typical bifurcations diagram for $b_0 > 0$ for cases in which two-dimensional cells branch supercritically ($b_1 > 0$). Wollkind, Oulton, and Sriranganathan found for Pb-Sn alloys that $b_0 > 0$ always. Here, as M increases toward M_c there is a jump transition before M_c to a down hexagon that remains stable up to $M = M^{(2)}$. Meanwhile, two-dimensional cells restabilize

4.1 Bifurcation theory

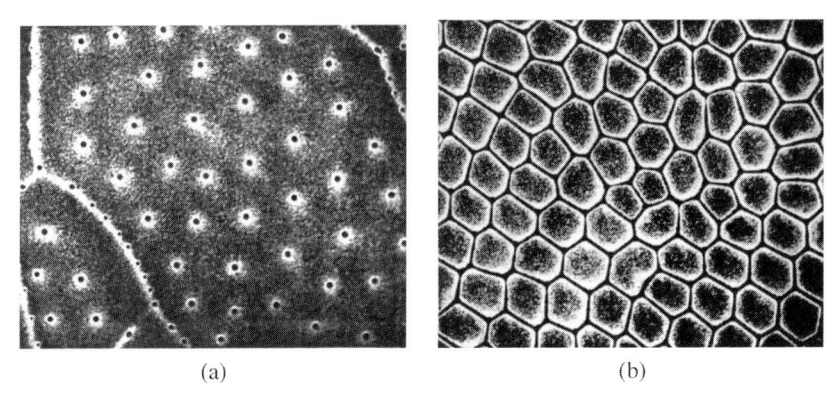

Figure 4.9. Photomicrographs of (a) nodes on a cross section taken near a quenched interface and (b) cells on a cross section taken at the quenched interface. Here the dark regions are those of highest solute concentration, the white regions those of next highest concentration, and the gray regions those of lowest concentration. From Morris and Winegard (1969).

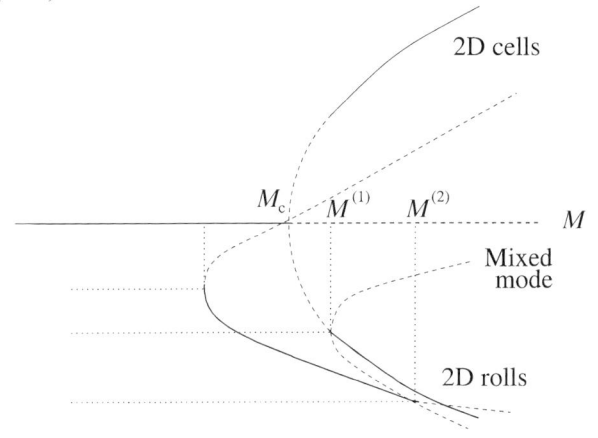

Figure 4.10. The typical bifurcation map for roll–hexagon competitions, amplitude versus M. Dashed curves represent unstable branches.

at $M = M^{(1)}$. Near $M = M_c$ and $M = M^{(2)}$ there is hysteretic behavior because two states in each case are simultaneously stable. When $b_0 < 0$, Figure 4.10 is reflected in the horizontal axis.

McFadden and Coriell (1984) and McFadden et al. (1987) used finite differences and the FTA with $k > 1$. The latter were able to obtain various patterns: hexagonal nodes and cells, two-dimensional cells, and rectangles. Figures 4.11 show two examples of the results.

Note: Bifurcation analyses yield important information when two-dimensional cells bifurcate supercritically because one can show evolution to

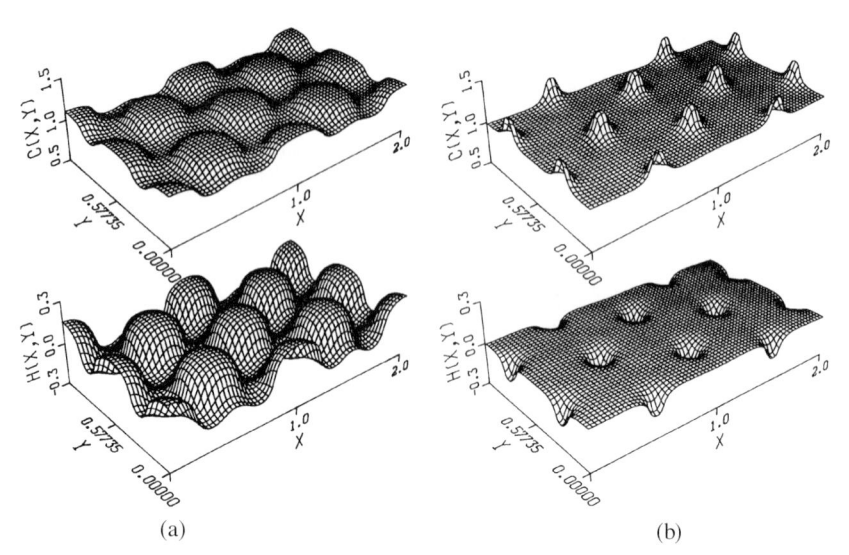

Figure 4.11. Calculated interface shapes and concentration profiles for (a) hexagonal cells and (b) hexagonal nodes. From McFadden et al. (1987).

stable secondary or tertiary states. However, the typical operation of crystal growth devices in on the lower branch, where two-dimensional instabilities bifurcate subcritically. The result, then, is that at present one must rely on careful numerical treatments for lower branch pattern and wave number selection questions even though these will involve the solution of three-dimensional, time-dependent free-boundary problems.

One means of bypassing the free-boundary nature of morphological instabilities is to examine limiting cases in which longwave instabilities are present and for which one can derive evolution equations that correctly represent the dynamics.

4.2 Long-Scale Theories

As discussed in Section 4.1, when a_c is of unit order in ϵ, then bifurcation theory and its variants give weakly nonlinear behaviors. Now, if one wishes to examine those limiting cases in which $a_c \to 0$ and a single partial differential equation can be derived that gives the dynamics without the need to solve a free-boundary problem directly. In each case the field variables are solved as functionals of h and incorporated into the boundary conditions to obtain the equation for h.

Figure 4.12 shows the results of the linear theory under the FTA for the one-sided model and generalizes Figure 3.5 into the three dimensions of the

4.2 Long-scale theories

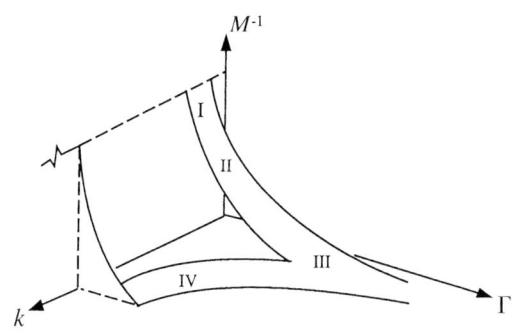

Figure 4.12. Neutral stability surface in (k, Γ, M^{-1}) space for the one-sided FTA model. The plane interface is unstable below the surface. From Riley and Davis (1990). (Reprinted by permission. Copyright 1990 by SIAM.)

control parameters M, Γ, and k. There are four distinguished limits for which the morphological instability has longwave character, as discussed by Riley and Davis (1990): I: $M \sim 1$, $\Gamma \sim 1$, $k \ll a_c^2 \ll 1$, II: $M \sim 1$, $\Gamma \gg 1$, $k \sim a_c^2 \ll 1$; III: $M \gg 1$, $a_c^2 \ll k \ll 1$, $\Gamma \gg 1$; IV: $M \gg 1$, $\Gamma \sim k^{-1}$, $a_c^2 \ll k \sim 1$.

4.2.1 Small Segregation Coefficient

Sivashinsky (1983) noticed that the characteristic equation of linearized theory under the FTA, Eq. (3.26), has $a_c \to 0$ as $k \to 0$. In particular, if $1 - M^{-1} \equiv \epsilon$, and $k \sim \epsilon^2$, $a_1 \sim \epsilon^{1/2}$, $\sigma \sim \epsilon^2$, and $\Gamma \sim 1$, then one has at leading order $\sigma = a_1^2 - \Gamma a_1^4 - k$. Thus, if one introduces into the nonlinear problem the scales

$$1 - M^{-1} \equiv \epsilon, \quad k = \epsilon^2 \bar{k}$$
$$(\tilde{x}, \tilde{y}) = \epsilon^{\frac{1}{2}}(x, y), \quad \tilde{t} = \epsilon^2 t \tag{4.16a}$$

and writes

$$C \sim C_0 + \epsilon C_1 + \epsilon^2 C_2 + \epsilon^3 C_3$$
$$h \sim \epsilon H, \tag{4.16b}$$

then system (3.19) can be solved sequentially. At $O(1)$, one finds that $C_0 = 1 - \exp(-z)$. At $O(\epsilon)$, $C_1 = 0$. At $O(\epsilon^2)$, $C_2 = \alpha \exp(-z)$ where $\alpha \equiv 1/2H^2 - H - \Gamma \nabla_1^2 H$. At $O(\epsilon^3)$, the particular solution is $C_3 = (\nabla_1^2 \alpha) z \exp(-z)$, and if this is substituted into the flux condition, one obtains the evolution equation corresponding to Region I in Figures 4.12,

$$H_{\tilde{t}} + \Gamma \nabla_1^4 H + \nabla_1^2 H + \bar{k} H = \nabla_1 \cdot (H \nabla_1 H), \tag{4.17}$$

where $\nabla_1^2 = \partial^2/\partial \tilde{x}^2 + \partial^2/\partial \tilde{y}^2$, and $\Gamma, \bar{k} = O(1)$. This *weakly nonlinear* equation can be analyzed in the linear regime by discarding the nonlinearity and writing $H = \hat{H} \exp[\sigma \tilde{t} + i(a_1 \tilde{x} + a_2 \tilde{y})]$ to obtain the characteristic equation

$$\sigma = a^2 - \Gamma a^4 - \bar{k}, \qquad (4.18)$$

where $a = \sqrt{a_1^2 + a_2^2}$. There is a finite band of wave numbers for instability.

Bifurcation in one space dimension shows that two-dimensional cells, one-dimensional solutions of Eq. (4.17), are subcritical and hence unstable. This is consistent with the work of Novick-Cohen (1980), who showed that all x-periodic, one-dimensional steady solutions of Eq. (4.17) are unstable. Bernoff and Bertozzi (1995) have shown that there is a finite-time blowup of solutions of Eq. (4.17). Attempts have been made to regularize the equation by formally replacing h_{xxxx} by $\left[h_{xx}/(1 + h_x^2)^{\frac{3}{2}}\right]_{xx}$; see Hyman, Novick-Cohen, and Rosenau (1988).

Bifurcation in two space dimensions by Humphreys, Heminger, and Young (1990) allowing the interaction of two-dimensional cells and hexagons shows that all local branches are unstable; two-dimensional cells are subcritical, and hexagons open to the left. A similar result is shown by Skelton et al. (1995), as shown in Figure 4.13.

4.2.2 Small Segregation Coefficient and Large Surface Energy

Riley and Davis (1990) noticed that there is another limit in which long-scale instability occurs; here k is small and Γ is large. They let

$$k = \epsilon^2 \bar{k}, \quad \Gamma \equiv \epsilon^{-2}, \quad M = O(1) \qquad (4.19a)$$

$$(\tilde{x}, \tilde{y}) = \epsilon(x, y), \quad \tilde{t} = \epsilon^2 t \qquad (4.19b)$$

and

$$h = H \sim 1, \quad C \sim 1, \qquad (4.19c)$$

corresponding to Region II in Figure 4.12. They derived an evolution equation of the form

$$(M^{-1} - \nabla_1^2)H_{\tilde{t}} + (1 - M^{-1} - \bar{k})\nabla_1^2 H + \nabla_1^4 H + M^{-1}\bar{k}H$$
$$= M^{-1}\nabla_1 \cdot (H\nabla_1 H) - \nabla_1 \cdot (\nabla_1^2 H \nabla_1 H), \qquad (4.20)$$

where $\bar{k} = O(1)$. Note that since $h = O(1)$, that Eq. (4.20) is a *strongly nonlinear* equation capable in principle of describing interfacial deformations of unit order in ϵ, a measure of the surface energy.

The linearized theory of Eq. (4.20) using normal modes gives

$$\sigma(M^{-1} + a^2) = a^2(1 - M - \bar{k}) - a^4 - M^{-1}\bar{k} = 0 \qquad (4.21)$$

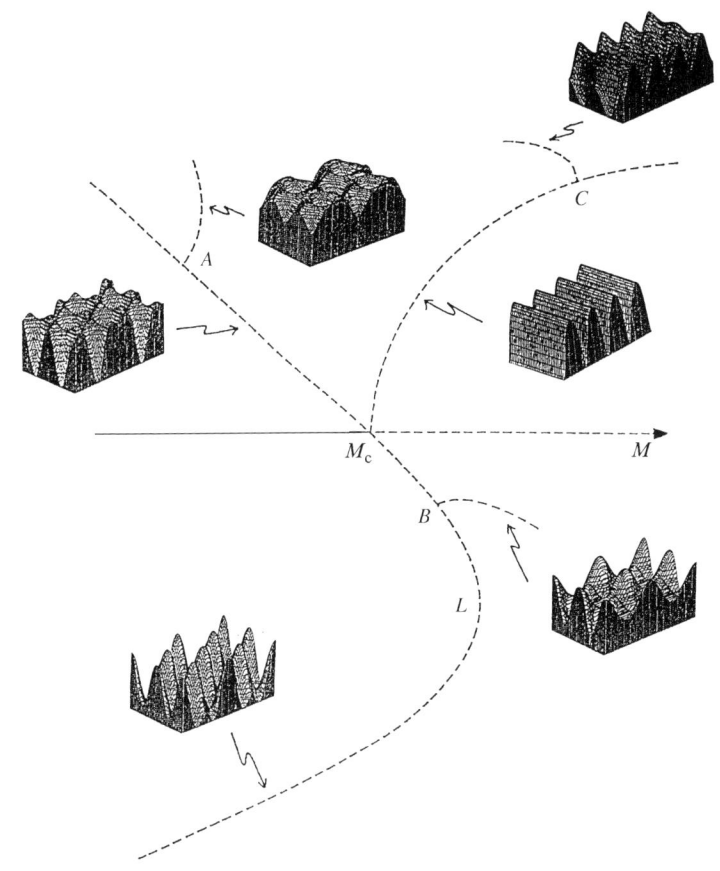

Figure 4.13. Bifurcation diagram and spatial structures. Dashed curves denote unstable branches. From Skelton et al. (1995).

so that there is for $\sigma = 0$, $\bar{k} = \bar{k}_c = (1 + M^{-1/2})^2$, thus separating conditions of growth $\bar{k} < \bar{k}_c$ from decay ($\bar{k} > \bar{k}_c$).

The bifurcating solution in one space dimension turns out to be supercritical for $M > 9/4$ and subcritical otherwise, and thus $M = 9/4$ is a transition point. Two-dimensional cells are stable for $M > 9/4$, and as M decreases through $9/4$ they begin as transcritical but turn around and restabilize at the limit point. Sarocka, Bernoff, and, Rossi (1999) showed in one space dimension that Eq. (4.20) is a natural regularization of Eq. (4.17).

In two-space dimensions (Eq. (4.20)) all local branches are unstable, and numerical results on periodic boxes valid to unit-order distortions reveal no new branches distinct from those of the local theory. Figure 4.14 shows a typical bifurcation map.

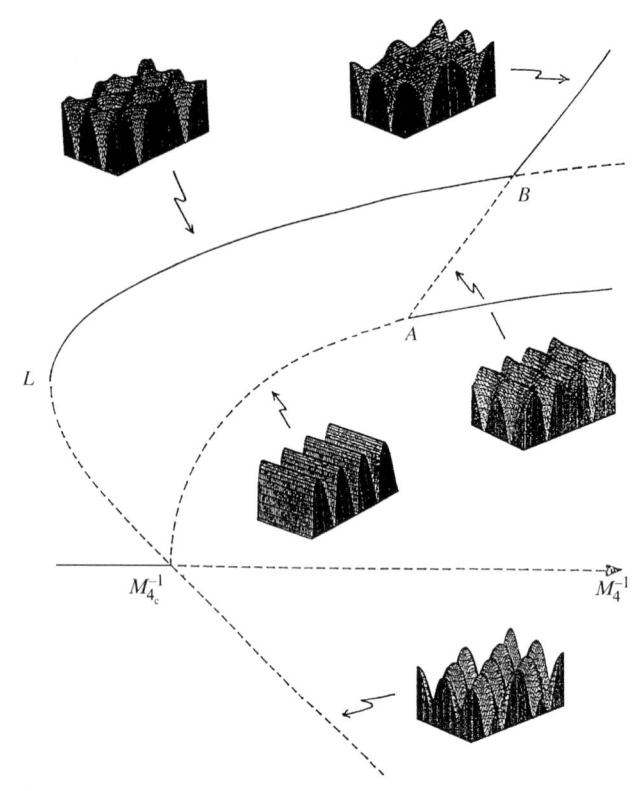

Figure 4.14. Bifurcation diagram and spatial structures. Dashed curves denote unstable branches. From Skelton et al. (1995).

Region III of Figure 4.12 gives no new information not included in Regions I and II.

4.2.3 Near Absolute Stability

The upper branch shown in Figure 3.6 asymptomatically approaches the absolute stability limit in which $a_c \to 0$. Brattkus and Davis (1988) examined $k = O(1)$ and Γ near $\Gamma_s = 1/k$ in the FTA in the one-sided model and defined for region IV in Figure 4.12. Let

$$\epsilon = \Gamma_s - \Gamma \tag{4.22a}$$

with

$$M = \epsilon^{-2}\bar{M} \tag{4.22b}$$

and

$$(\tilde{x}, \tilde{y}) = \epsilon^{\frac{1}{2}}(x, y), \quad \tilde{t} = \epsilon^2 t. \tag{4.22c}$$

Here $\bar{M} = O(1)$ and k is arbitrary. The asymptotic expansion for the interface position has the form $h \sim H$, and thus unit-order distortions are allowed. Brattkus and Davis derived the *strongly nonlinear* equation

$$H_{\bar{t}\bar{t}} - \nabla_1^2 H_{\bar{t}} + \frac{1}{4}(1-\nu^2)\nabla_1^4 H + \nabla_1^2 H + \mu^{-1} H = H_{\bar{t}} \nabla_1^2 H + |\nabla_1 H|_{\bar{t}}^2$$
$$-\frac{1}{2}(1-\nu)\nabla_1^2\left(|\nabla_1 H|^2\right) - \nu\nabla_1 \cdot \left(\nabla_1^2 H \nabla_1 H\right) - \frac{1}{2}\nabla_1 \cdot \left(|\nabla_1 H|^2 \nabla_1 H\right), \quad (4.23)$$

where

$$\bar{M} = \mu(1+2k)/k^3 \quad (4.24a)$$

and

$$\nu = (1+2k)^{-1}. \quad (4.24b)$$

In the linearized theory, the characteristic equation takes the form

$$\bar{\sigma}^2 + \bar{a}^2(2+k^{-1})\bar{\sigma} + [k\bar{M}^{-1} - k\bar{a}^2 + \bar{a}^4(1+k^{-1})] = 0. \quad (4.25)$$

Notice that the growth rate $\bar{\sigma}$ satisfies a quadratic equation, and hence both first and second derivatives are inherited by the evolution equation (4.23). Note, too, that the nonlinearities of Eq. (4.23) are both cubic and quadratic, which is possible because the equation is strongly nonlinear.

In one space dimension the bifurcations are supercritical, and two-dimensional cells equilibrate to finite-amplitude forms. In two space dimensions, all two-dimensional states are initially unstable. Brattkus and Davis found for μ fixed and $k \to 1$ that pattern selection gives a normal two-dimensional-cell–hexagonal interaction, as shown in Figure 4.10. Skelton et al. (1995) found numerically that no new branches or instabilities are obtained in a numerical domain substantially larger than that covered by the local theory and for k as small as 0.90.

That the evolution equation (4.23) contains both first and second time derivatives suggests that it might be possible to excite secondary bifurcations to time-periodic states. Brattkus (1990) attempted this in the case of a single space dimension. In this case Eq. (4.23) has the form

$$\mathcal{L}(H) = H_{\bar{t}\bar{t}} - H_{\hat{x}\hat{x}\bar{t}} + \frac{1}{4}(1-\nu^2)H_{\hat{x}\hat{x}\hat{x}\hat{x}} + H_{\hat{x}\hat{x}} + \mu^{-1}H = \mathcal{N}(H), \quad (4.26a)$$

where

$$\mathcal{N}(H) = H_{\hat{x}\hat{x}}H_{\bar{t}} + 2H_{\hat{x}}H_{\hat{x}\bar{t}} - H_{\hat{x}}H_{\hat{x}\hat{x}\hat{x}} - H_{\hat{x}}^2 - \tfrac{1}{2}\left(H_{\hat{x}}^2\right)_{\hat{x}\hat{x}}. \quad (4.26b)$$

Brattkus (1990) was not able to find time-dependent instabilities of two-dimensional cells (using Floquet theory). However, he rederived the two-dimensional evolution equation for the symmetric model. He found that

Eq. (4.26) remains intact except that the coefficient in the last term, of $\mathcal{N}(H)$ becomes 3/4 rather than 1/2. In this case he *was* able to obtain instabilities of two-dimensional cells in which the principal mode is spatially subharmonic and time-periodic. That is, pairs of neighboring cells oscillate with respect to each other, giving what is sometimes called an "optical" mode.

Kassner et al. (1991, 1994a,b) rederived the evolution equation for the symmetric model in two dimensions and found a very rich dynamics in this equation, obtaining, among other things, chaotic behavior.

Finally, the use of Eq. (4.23) and its generalizations depends on the preservation of the underlying model of local thermodynamic equilibrium at the interface. Because that absolute stability limit can involve rather large pulling speeds (e.g., 1–10 m/s in metallic alloys), some caution must be exercised in applying the results directly to experiment (see Chapter 6). However, in the case of phase transformation in liquid crystals, the absolute stability boundary occurs at rather low speeds (i.e., 1 cm/s), and thus the *symmetric model* may be applicable quantitatively.

4.3 Remarks

Both bifurcation and longwave theories of directional solidification have been discussed, culminating in the understanding of bifurcation type, pattern, and wave number selection. All of these theories involve a model in which there is local thermodynamic equilibrium at the front, and all material properties are isotropic, independent of crystallography direction. The next two chapters will relax these restrictions.

That most quantitative experiments are performed on Hele–Shaw cells to simulate two-dimensional structures leads one to ask: To what degree is two-dimensionality met?

Ajaev and Davis (1999) examined the Hele–Shaw case in the one-sided model using the FTA. They defined three regimes for the gap width d of the cell compared with the critical wavelength $\lambda_c = 2\pi/a_{*_c}$ of the Mullins and Sekerka instability for the planar front.

When $d < \frac{1}{2}\lambda_c$, no morphological mode "fits" into the gap, and only the curvature of the basic state interface, and hence the contact angle, alters the classical morphological instability. Ajaev and Davis found that the interface is more unstable than the planar case, independent of whether it is concave or convex toward the solid. This promotion of instability is consistent with the observations of de Cheveigné et al. (1986) for alloys of CB_4 and SCN in that the threshold for instability is promoted by decreasing d. When $d < 50\ \mu\text{m}$, this effect is quite pronounced, whereas for larger d no gap dependence is seen.

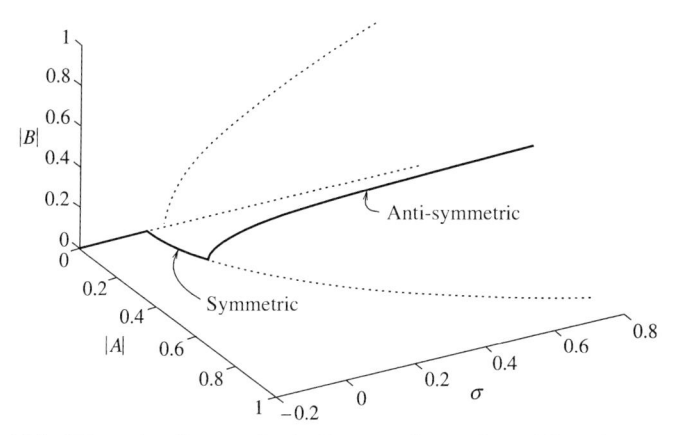

Figure 4.15. Bifurcation diagram for the "resonant" case. Dotted lines denote unstable branches. The region of stability of the center symmetric "two-dimensional" solution appears owing to the contact-angle condition. From Ajaev and Davis (1999).

Caroli, Caroli, and Roulet (1986) also showed for the *symmetric* model and contact angles close to $\pi/2$ that weak cross-gap curvatures decrease V_c for $d < 50$ μm.

When $d \approx (1/2)\lambda_c$, morphological instability is possible across the gap with an eigenfunction that is antisymmetric with respect to the centerline. There is a "resonance" between this antisymmetry and the symmetry of the curved basic-state interface, causing the concave-toward-solid interfaces to become more unstable than the planar case and a sensibly two-dimensional structure to appear first. However, there is a secondary bifurcation to a sensibly three-dimensional state, as shown in Figure 4.15. Thus, Hele–Shaw cells give only approximate two-dimensional configurations.

References

Ajaev, A. S., and Davis, S. H. (1999). Three-dimensional effects in directional solidification in Hele-Shaw cells, *Proc. Roy. Soc.* **455**, 3589–3616.

Alexander, J. I. D., Wollkind, D. J., and Sekerka, R. F. (1986). The effect of latent heat on weakly nonlinear morphological instability, *J. Crystal Growth* **79**, 849–865.

Bennett, M. J., and Brown, R. A. (1989). Cellular dynamics during directional solidification: Interaction of multiple cells, *Phys. Rev. B* **39**, 11705–11723.

Bernoff, A. J., and Bertozzi, A. L. (1995). Singularities in a modified Kuramoto–Sivashinsky equation describing interface motion for phase transition, *Physica D* **85**, 375–404.

Brattkus, K. (1990). Oscillatory instabilities in cellular solidification, *Appl. Mech. Rev.* **43**, 556–558.

Brattkus, K., and Davis, S. H. (1988). Cellular growth near absolute stability, *Phys. Rev. B.* **38**, 11452–11460.

Brattkus, K., and Misbah, C. (1990). Phase dynamics in directional solidification, *Phys. Rev. Lett.* **64**, 1935–1938.

Caroli, B., Caroli, C., and Roulet, B. (1982). On the emergence of one dimensional front instabilities in directional solidification and fusion of binary mixtures, *J. de Phys.* (Paris) **43**, 1767–1780.

———. (1986). The Mullins–Sekerka instability in thin samples, *J. Crystal Growth* **76**, 31–49.

de Cheveigné, S., Guthmann, C., and Lebrun, M.-M. (1985). Nature of the transition of the solidification front of a binary mixture from a planar to a cellular morphology, *J. Crystal Growth* **73**, 242–244.

———. (1986). Cellular instabilities in directional solidification, *J. de Phys.* (Paris) **47**, 2095–2103.

Cole, G. S., and Winegard, W. C. (1963). The transition from plane to cellular interface in solidifying tin-lead-antimony alloys, *J. Inst. Metals* **92**, 322–326.

Coriell, S. R., and McFadden, G. J. (1993). Morphological stability, D. T. J. Hurle, editor, in *Handbook of Crystal Growth* **1b**, Chap. 12, 785–857, North-Holland, Amsterdam.

Eckhaus, W. (1965), *Studies in Nonlinear Stability Theory*, Springer–Verlag, Berlin.

Humphreys, L. B., Heminger, J. A., and Young, G. W. (1990). Morphological stability in a float zone, *J. Crystal Growth* **100**, 31–50.

Hyman, J. M., Novick-Cohen, A., and Rosenau, P. (1988). Modified asymptotic approach to modeling a dilute-binary-alloy solidification front, *Phys. Rev. B* **37**, 7603–7608.

Kassner, K., Misbah, C., and Müller-Krumbhaar, H. (1991). Transition to chaos in directional solidification, *Phys. Rev. Lett.* **67**, 1551–1553.

Kassner, K., Misbah, C., Müller-Krumbhaar, H., and Valance, A. (1994a). Directional solidification at high speed. I. Secondary instabilities, *Phys. Rev. E* **49**, 5477–5494.

———. (1994b). Directional solidification at high speed. II. Transition to chaos, *Phys. Rev. E* **49**, 5495–5515.

Kessler, D. A., and Levine, H. (1989). Steady-state cellular growth during directional solidification, *Phys. Rev. A* **39**, 3041–3052.

Kopczyński, P., Rappel, W.-J., and Karma, A. (1996). Critical role of crystalline anisotropy in the stability of cellular array structures in directional solidification, *Phys. Rev. Lett.* **77**, 3387–3390.

Kurowski, P., de Cheveigné, S., Faive, G., and Guthmann, C. (1989). Cusp instability in cellular growth, *J. de Phys.* **50**, 3007–3019.

Levine, H., Rappel, W.-J., and Riecke, H. (1991). Resonant interactions and traveling solidification cells, *Phys. Rev. A* **43**, 1122–1125.

Liu, D., Williams, L. M., and Cummins, H. Z. (1994). Subcritical–supercritical bifurcation crossover in directional solidification, *Phys. Rev. E* **50**, R4286–R4289.

McFadden, G. B., and Coriell, S. R. (1984). Nonplanar interface morphologies during unidirectional solidification of a binary alloy, *Physica D* **12**, 253–261.

McFadden, G. B., Boisvert, R. F., and Coriell, S. R. (1987). Nonplanar interface morphologies during unidirectional solidification of a binary alloy II. Three-dimensional computations, *J. Crystal Growth* **84**, 371–388.

Merchant, G. J., and Davis, S. H. (1989a). Shallow cells in directional solidification, *Phys. Rev. Lett.* **63**, 573–575.

———. (1989b). Directional solidification near minimum c_∞: Two-dimensional isolas and multiple solutions, *Phys. Rev. B* **40**, 11140–11152.

Morris, L. R., and Winegard, W. C. (1969). The development of cells during the solidification of a dilute Pb–Sn alloy, *J. Crystal Growth* **5**, 361–375.

Newell, A. C., and Whitehead, J. A. (1969). Finite bandwith, finite amplitude convection, *J. Fluid Mech.* **38**, 279–303.
Novick-Cohen, A. (1980). Instability of periodic states for the Sivashinsky equation, *Q. Appl. Math.* **48**, 217–224.
Ramprasad, N., Bennett, M. J., and Brown, R. A. (1988). Wavelength dependence of cells of finite depth in directional solidification, *Phys. Rev. B* **38**, 583–592.
Rappel, W.-J., and Riecke, H. (1992). Parity-breaking in directional solidification: Numerics versus amplitude equations, *Phys. Rev. A* **45**, 846–859.
Riley, D. S., and Davis, S. H. (1989). Eckhaus instabilities in generalized Landau–Ginzburg equations, *Phys. Fluids A* **1**, 1745–1747.
——— . (1990). Long-wave morphological instabilities in the directional solidification of a dilute binary mixture, *SIAM J. Appl. Math.* **50**, 420–436.
Saito, Y., Misbah, C., and Müller-Krumbhaar, H. (1989). Directional solidification: Transition from cell to dendrites, *Phys. Rev. Lett.* **63**, 2377–2380.
Sarocka, D. S., Bernoff, A. J., and Rossi, L. F. (1999). Large-amplitude solutions to the Sivashinsky and Riley–Davis equations for directional solidification, *Physica D* **127**, 146–176.
Segel, L. A. (1969). Distant side-walls cause slow amplitude modulation of cellular convection, *J. Fluid Mech.* **38**, 203–224.
Sivashinsky, G. I. (1983). On cellular instability in the solidification of a dilute binary alloy, *Physica D* **8**, 243–248.
Skelton, A. C., McFadden, G. B., Impey, M. D., Riley, D. S., Cliffe, K. A., Wheeler, A. A., and Davis, S. H. (1995). On long-wave morphological instabilities in directional solidification, *Eur. J. Appl. Math.* **6**, 639–652.
Sriranganathan, R., Wollkind, D. J., and Oulton, D. B. (1983). A theoretical investigation of the development of interfacial cells during the solidification of a binary alloy: Comparison with the experiments of Morris and Winegard, *J. Crystal Growth* **62**, 265–283.
Ungar, L. H., and Brown, R. A. (1984). Cellular interface morphologies in directional solidification: The one-sided model, *Phys. Rev. B* **29**, 1367–1380.
——— . (1985). Cellular morphologies in directional solidification. IV. The formation of deep cells, *Phys. Rev. B* **31**, 5931–5940.
Warren, J. A., and Langer, J. S. (1993). Prediction of dendrite spacing in a directional solidification experiment, *Phys. Rev. E* **47**, 2702–2712.
Wollkind, D. J., Oulton, D. B., and Sriranganathan, R. (1984). A nonlinear stability analysis of a model equation for alloy solidification, *J. de Phys.* **45**, 505–516.
Wollkind, D. J., and Segel, L. A. (1970). A nonlinear stability analysis of the freezing of a dilute binary alloy, *Philos. Trans. Roy. Soc. London A* **268**, 351–380.
Wollkind, D. J., Sriranganathan, R., and Oulton, D. B. (1984). Interfacial patterns during plane front alloy solidification, *Physica* **12D**, 215–240.

5

Anisotropy

The modern theory of morphological instability began with the work of Mullins and Sekerka (1963, 1964), as discussed in Chapter 3, who modeled the solid as an isotropic, continuous medium whose only "memory" of its crystalline structure is the presence of latent heat. They used linear stability theory to show that adverse gradients of temperature or solute concentration at a moving front drives an instability, and the surface energy cuts off the short-scale corrugations. The nonlinear extensions of this model were discussed in Chapter 4.

The preceding developments have had broad impact on the field, which is remarkable given that no account had been taken of the details of the crystallographic properties of the solid. One would expect the crystal structure to impress geometrical symmetries on the system that compete with nonlinearities for the determination of morphological pattern and hence the microstructure of the solid. In principle, all of the bulk material properties (e.g., k_T^s, c_p^s, ρ^s, D^s) and the interfacial properties (e.g., γ, μ) possess the same orientational information, though some of the anisotropies may be more important than others in particular situations.

5.1 Surface Energy and Kinetics

Consider first the anisotropy of surface energy γ. Now, instead of γ being constant, it depends on the orientation of the interface with respect to the underlying crystal structure. This information is transmitted to the surface energy by allowing $\gamma = \gamma(\mathbf{n})$, where $\mathbf{n}$ is the unit outward normal vector to the body. The formulation of a two-dimensional drop on a substrate of Section 2.2 still

5.1 Surface energy and kinetics

holds, and Eq. (2.47) is still correct, namely,

$$w^{-1}E' = \int_{\ell_1}^{\ell_2} I\,dx + \gamma_1(\ell_2 - \ell_1) + \gamma_2[\ell - (\ell_2 - \ell_1)], \quad (5.1)$$

where now

$$I = \gamma(h_x)(1 + h_x^2)^{\frac{1}{2}} + \lambda h. \quad (5.2)$$

The geometrical quantities are shown in Figure 2.11.

To evaluate the variation of E', one needs some notation and several relations:

$$\frac{\partial}{\partial h_x}\gamma \equiv \gamma_{,x}, \quad \frac{\partial^2}{\partial h_x^2}\gamma = \gamma_{,xx}, \quad \frac{\partial}{\partial x}\gamma_{,x} = \gamma_{,xx}h_{xx}. \quad (5.3)$$

The variation of E' is then given by

$$w^{-1}\delta E' = \int_{\ell_1}^{\ell_2}\left[I_h\delta h + I_{h_x}(\delta h)_x\right] + (I + \gamma_1 - \gamma_2)\,\delta\ell|_{\ell_1}^{\ell_2} \quad (5.4)$$

and using integration by parts, one obtains

$$w^{-1}\delta E' = \int_{\ell_1}^{\ell_2}\left[I_h - \left(I_{h_x}\right)_x\right]\delta h\,dx + I_{h_x}\delta h|_{\ell_1}^{\ell_2} + (I + \gamma_1 - \gamma_2)\,\delta\ell|_{\ell_1}^{\ell_2}. \quad (5.5)$$

Again, use endpoint relations (2.51), $\delta h|_{\ell_1,\ell_2} = -h_x\delta\ell|_{\ell_1,\ell_2}$, to obtain the final form of the first variation

$$w^{-1}\delta E' = \int_{\ell_1}^{\ell_2}\left[I_h - \left(I_{h_x}\right)_x\right]\delta h\,dx + \left[I - h_x I_{h_x} + \gamma_1 - \gamma_2\right]\delta\ell|_{\ell_1}^{\ell_2}. \quad (5.6)$$

The Euler–Lagrange equation becomes

$$I_h - \left(I_{h_x}\right)_x = 0$$

or

$$\lambda = \frac{h_{xx}}{(1 + h_x^2)^{3/2}}\left\{\gamma + 2\gamma_{,x}h_x(1 + h_x^2) + \gamma_{,xx}(1 + h_x^2)^2\right\}. \quad (5.7)$$

It is convenient to define θ to be the angle between the z-axis and $\mathbf{n}$, $\mathbf{n} = (\sin\theta, \cos\theta)$. In terms of θ, $-h_x = \tan\theta$, and thus at the contact line θ is the contact angle. Further, $(1 + h_x^2)^{1/2} = \sec\theta$ and $\gamma_{,x} = \gamma_\theta \cos^2\theta$. After a bit of manipulation, form (5.7) becomes

$$\lambda = 2H(\gamma + \gamma_{\theta\theta}), \quad (5.8)$$

which is a generalization of Eq. (2.54) made by Herring (1951) to situations in which γ varies with the orientation of the front. Equation (5.8), derived in two dimensions, is valid in three dimensions if the mean curvature H is replaced by its appropriate form. The corresponding Gibbs–Thomson equation is then

$$T^I = T_m \left[1 + 2H \frac{\gamma + \gamma_{\theta\theta}}{L_v} \right]; \qquad (5.9)$$

compare this with Eq. (2.57).

If one returns to Eq. (5.6), the end conditions for the first variation are

$$I - h_x I_{h_x} + \gamma_1 - \gamma_2 = 0, \quad x = \ell_1, \ell_2$$

or

$$\frac{\gamma}{(1 + h_x^2)^{\frac{1}{2}}} = \gamma_2 - \gamma_1 + \gamma_{,x} h_x (1 + h_x^2)^{\frac{1}{2}}. \qquad (5.10)$$

From the definition of θ, Eq. (5.10) becomes

$$\gamma \cos \theta = \gamma_2 - \gamma_1 - \gamma_\theta \sin \theta, \qquad (5.11)$$

which is the generalization to orientational dependence of the Young–Laplace equation (2.59) made by Herring (1951). Again, at the contact line θ is the contact angle.

Consider a single crystal of fixed volume (no phase transformation) and seek its shape at *thermodynamic equilibrium*. From surface energy minimization, Eq. (5.8), $2H(\gamma + \gamma_{\theta\theta})$ is constant. Clearly, if γ were isotropic, as it would be for a drop of liquid or a glass, γ would be constant, and hence H would be constant. In two or three dimensions, the equilibrium figure would be a circle or a sphere, respectively.

Consider a rectangular parallelpiped shown in Figure 5.1 of dimensions 2 (ℓ_1, ℓ_2, ℓ_3) in the direction (x, y, z) in which each face parallel to a different coordinate axis has a different γ, as shown. If one minimizes $E = 2(\gamma_1 \ell_2 \ell_3 + \gamma_2 \ell_1 \ell_3 + \gamma_3 \ell_1 \ell_2)$ subject to $V = \ell_1 \ell_2 \ell_3$ fixed, one finds (Mullins 1963) that

$$\frac{\gamma_1}{\ell_1} = \frac{\gamma_2}{\ell_2} = \frac{\gamma_3}{\ell_3} = \text{constant.} \qquad (5.12)$$

Thus, the γ of any face is proportional to the distance of that face from the center of the parallelpiped. The equilibrium shape, because it minimizes surface energy, makes the face with the largest γ have the smallest area. More generally, it is found that the shape will form so as to preferentially expose crystallographic orientations of low γ.

5.1 Surface energy and kinetics

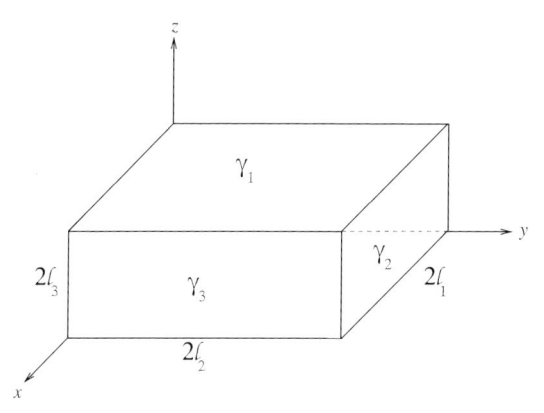

Figure 5.1. A rectangular parallelepiped of dimensions $(2\ell_1, 2\ell_2, 2\ell_3)$ in directions (x, y, z).

A typical representation for γ for a crystal with *fourfold symmetry* is

$$\gamma = \gamma_0[1 + \alpha_4 \cos 4\theta]. \tag{5.13}$$

In this case,

$$\gamma + \gamma_{\theta\theta} = \gamma_0[1 - 15\alpha_4 \cos 4\theta], \tag{5.14}$$

and thus if $|\alpha_4| < 1/15$, the coefficient of the curvature in Eq. (5.9) is positive, and the description of the interface using the Gibbs–Thomson equation is well defined mathematically. Call this the case of "small" anisotropy. When $|\alpha_4| > 1/15$, $\gamma + \gamma_{\theta\theta}$ changes sign, giving rise to orientations (i.e., angles) that are forbidden and a crystal interface with missing orientations. Call this the case of "large" anisotropy.

The extra term in the edge condition (5.11) behaves for fourfold symmetry like

$$\gamma_\theta \sin \theta = -4\gamma_0 \alpha_4 \sin 4\theta \sin \theta, \tag{5.15}$$

which is extremely small when θ is small but can become significant when the contact angle is moderate.

To determine the shape of an equilibrium state, use the Wulff (1901) construction based on property (5.12). Construct a polar plot of $\gamma(\theta)$ in which the radial vector equals $\gamma(\mathbf{n})\mathbf{n}$. Corresponding to each point on the figure, draw a ray from the origin, and where the ray intersects the figure, draw a plane perpendicular to the ray; this is called a Wulff plane. According to Mullins (1963), "the Wulff plane, considered as a potential crystal face ... has a γ proportional ... to the length of the ray." Wulff planes are drawn at every point on

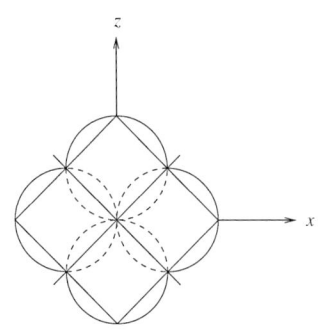

Figure 5.2. A polar plot of $\gamma = \gamma_0(1 + \alpha_4 \cos 4\theta)$ for large enough α_4. From Mullins (1961). (Reprinted with permission of Taylor & Francis.)

the figure. The locus of the minimum radii of all the intersecting planes, which contains all points accessible to the center by rays, gives the crystal shape.

Clearly, if γ is isotropic, a circle or sphere will emerge from this construction. More interestingly, when in two dimensions, $\gamma(\theta) = \gamma_0(1 + \alpha_4 \cos 4\theta)$, and $|\alpha_4|$ is large enough, the polar plot, shown in Figure 5.2, consists of a superposition of four circles. When all of the Wulff planes are drawn, the areas outside the square shown are filled, and the figure predicted is the square, a figure formed from intersecting straight lines. Whenever the polar plot has cusps, the predicted figure has straight-line or planar sides. Figure 5.3 shows the cases that can occur in two dimensions.

In the preceding derivation only the two-dimensional problem with a one-dimensional interface has been examined. If, in fact, $h = h(x, y)$, then the results above are directly extendable; see Coriell and Sekerka (1976) and Golovin and Davis (1998).

If, again in the two-dimensional case, one wishes to examine linear or weakly nonlinear theories, the area-weighted surface tension I can be expressed in

Figure 5.3. The equilibrium figures in two dimensions. From Mullins (1961). (Reprinted with permission of Taylor & Francis.)

powers of slope,

$$I(h_x) = \hat{\gamma}_0 + \hat{\gamma}_1 h_x + \hat{\gamma}_2 h_x^2 + \cdots. \tag{5.16}$$

The surface energy in a frame of reference associated with the principal axes can be written (McFadden, Coriell, and Sekerka 1988) as

$$\gamma = \gamma_0' + \gamma_4' \cos^4 \hat{\theta} + \gamma_6' \cos^6 \hat{\theta} + \cdots, \tag{5.17}$$

where $\hat{\theta}$ is the angle between **n** and one of the principal crystal orientations and the coefficients γ_i' are in principle known from experiment. Now, the coordinates must be rotated to express these coefficients in terms of the coordinates with the z-axis normal to the front. In two dimensions, Golovin and Davis (1998) found that

$$\hat{\gamma}_0 = \gamma_0' + \gamma_4' \left[\cos^4 \hat{\theta} + \sin^4 \hat{\theta}\right] + \gamma_6' \left[\cos^6 \theta + \sin^6 \hat{\theta}\right]$$

$$\hat{\gamma}_1 = 0 \tag{5.18}$$

$$\hat{\gamma}_2 = \frac{1}{2}\gamma_0' - \frac{3}{2}\gamma_4' \left[\cos^4 \hat{\theta} + \sin^4 \hat{\theta}\right] - \frac{5}{2}\gamma_6' \left[\cos^6 \hat{\theta} + \sin^6 \hat{\theta}\right].$$

The kinetic coefficient can be handled similarly, with $\hat{\mu} = \hat{\mu}_0 + \hat{\mu}_1 h_x + \hat{\mu}_2 h_x^2 + \cdots$. Consider now the kinetic coefficient for a crystal with fourfold symmetry

$$\mu^{-1} = \mu_0[1 + \beta_4 \cos 4\theta]. \tag{5.19}$$

In the principal directions,

$$\mu^{-1} = \mu_0' + \mu_4' \cos^4 \theta + \mu_6' \cos^6 \theta + \cdots, \tag{5.20}$$

where now

$$\hat{\mu}_0 = \mu_0' + \mu_4' \left[\cos^4 \hat{\theta} + \sin^4 \hat{\theta}\right] + \mu_6' \left[\cos^6 \hat{\theta} + \sin^6 \hat{\theta}\right]$$

$$\hat{\mu}_1 = 0 \tag{5.21}$$

$$\hat{\mu}_2 = -2\mu_4' \left[\cos^4 \hat{\theta} + \sin^4 \hat{\theta}\right] - 3\mu_6' \left[\cos^6 \hat{\theta} + \sin^6 \hat{\theta}\right].$$

5.2 Directional Solidification with "Small" Anisotropy

A quite general formulation for the effects of both anisotropy and departures from thermodynamic equilibrium has been given by Coriell and Sekerka (1976) in which the solidification model discussed earlier was generalized by the definition of four constitutive functions. Departures from equilibrium will be discussed in Chapter 6, and here only the effects of anisitropy will be considered. Coriell and Sekerka allowed the surface energy to be anisotropic as described above. They wrote the kinetic undercooling as

$$V_n = f(T_e - T^I, h_x, h_{xx}). \tag{5.22}$$

The function f is not known in general, but if f is smooth, and in the case of linear stability theory about a planar interface, this relation can be linearized as

$$V'_n \sim f(T_e^0 - T^{10}, 0, 0) + \mu_T(T'_e - T^{I'}) + \mu_x h_x + \mu_{xx} h_{xx}, \tag{5.23}$$

where now only constant coefficients μ_T, μ_x, and μ_{xx} need to be measured and $f(T_e^0 - T^{10}, 0, 0)$ is part of the basic state. Equation (5.23) for linear theory can then be written

$$T^{I'} = T'_e + \mu_T^{-1} V'_n + \mu_T^{-1}[\mu_x h_x + \mu_{xx} h_{xx}], \tag{5.24}$$

and T'_e contains the constitutional and capillary undercooling. In dimensional terms the linearized Gibbs–Thomson relation is now

$$Gh = mC + T_m 2H\gamma' + \mu_T^{-1} V'_n + \mu_T^{-1}[\mu_x h_x + \mu_{xx} h_{xx}]. \tag{5.25}$$

Coriell and Sekerka (1976) examined the one-sided model under the frozen temperature approximation and found, for "small" anisotropy, linear instability modes that have complex growth rates $\sigma = \sigma_R + i\sigma_I$ indicating the presence of traveling or standing waves with angular frequencies $\omega \equiv \sigma_I \neq 0$.

They explained the presence of these waves as follows: If, say, $\mu_T, \mu_x > 0$ and a normal mode $\exp[i(ax + \omega t)]$ is defined, there is a wave traveling in the negative x-direction (for $\omega > 0$). Parts of the front for which $h_x > 0$ have larger $|V'_n|$ than those for when $h_x < 0$, and a wave motion can ensue. Coriell and Sekerka argued that when the disturbance reaches finite amplitude, the cells that grow will be tilted at an angle $\tan^{-1}(c/V)$ with respect to the heat flow, where $c = \omega/a$ is the phase speed of the wave.

A nonlinear analysis in the Sivashinsky limit, $k \to 0$, $\Gamma = O(1)$ is valid for long-scale weakly nonlinear states. As shown in Section 4.3, $M^{-1} = 1 - \epsilon$, and the wave number is $O(\epsilon^{1/2})$. When the preceding model of anisotropy is inserted and

$$\hat{\mu}_T \equiv \frac{V^2}{mG_c D\mu_T}, \quad \hat{\mu}_x \equiv \frac{\mu_x}{\mu_T} \frac{V}{mG_c D}, \tag{5.26}$$

the theory can be extended (Young, Davis, and Brattkus 1987) if $\hat{\mu}_T = \epsilon^{1/2} \bar{\mu}_T$, $\hat{\mu}_x = \epsilon^{1/2} \bar{\mu}_x$, where the barred quantities are of unit order. The evolution equation, written for the two-dimensional case as a generalization of Eq. (4.17) is

$$H_{\bar{t}} + \Gamma \nabla_1^4 H + \nabla_1^2 H + \bar{\mu} \cdot \nabla \cdot (\nabla^2 H) + \bar{k} H = \nabla_1 \cdot (H \nabla_1 H) \tag{5.27}$$

where $\bar{\mu} = (\bar{\mu}_x, \bar{\mu}_y)^T$ and μ_y has the expected definition.

5.2 Directional solidification with "small" anisotropy

If Eq. (5.27) were linearized about the trivial solution, one would find that
$$\sigma = a^2 - \Gamma a^4 - \bar{k} - i(\bar{\mu}_x a_1^2 + \bar{\mu}_y a_2^2). \tag{5.28}$$
The presence of the odd derivative in Eq. (5.27) creates the traveling waves whose angular frequency is $\omega = -(\bar{\mu}_x a_1^2 + \bar{\mu}_y a_2^2)$, which is consistent with the computations of Coriell and Sekerka (1976).

When the one-dimensional version of Eq. (5.27) is integrated numerically on a periodic domain, one finds tilted cells, as shown in Figure 5.4(a), with the corresponding concentration distribution shown in Figure 5.4(b). It is also evident that the solution bifurcates subcritically and thus suffers from the same defect as does Eq. (4.17), viz. the solution shown is unstable.

Notice that V/mG_cD, is independent of V, whereas V^2/mG_cD is linear in V. Dispersion relation (5.28) thus shows that *any* kinetic anisotropy for which

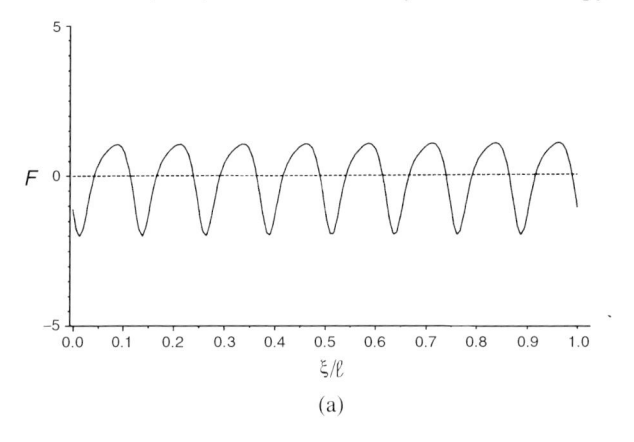

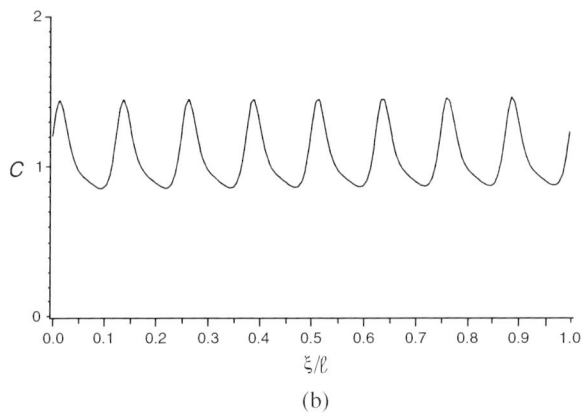

Figure 5.4. The (a) interfacial shape and (b) interfacial concentration from Eq. (5.27) in one space dimension. From Young, Davis, and Brattkus (1987). (Reprinted with permission of North-Holland.)

$(\bar{\mu}_x, \bar{\mu}_y) \neq (0,0)$ excites traveling waves. This prediction is at odds with the experimental observations of Trivedi, Seetharaman, and Eshelman (1991) and Trivedi (1990) on PVA–0.22% ethanol, who found tilted cells only when the angle between the preferred crystallographic direction and the heat flow direction exceeds a critical value. For small angles, tilting seems to occur only for large enough V.

The first attempt at pattern selection was the bifurcation analysis of McFadden, Coriell, and Sekerka (1988), who developed expressions for fourfold surface-energy anisotropy for the directions [001], [011] and [111] for the one-sided model under the FTA. Further, they calculated amplitude equations for hexagonal and roll interactions but only through second order in amplitude – a branch that is unstable. They found from linear theory for the direction [011] which contains a twofold axis, that rolls are indicated. For the direction [001], a fourfold axis, either triangles or hexagons are indicated, depending on the alignment of the array with respect to the interacting rolls.

Hoyle, McFadden, and Davis (1996) undertook a full analysis of pattern selection on the upper (high-speed) branch of the neutral curve for the one-sided model with the FTA; here, rolls bifurcate supercritically. Hoyle et al. first derived generalizations to the (strongly nonlinear) Brattkus–Davis equation valid near absolute stability, $\Gamma_s - 1/k \equiv \epsilon$, $M = \epsilon^{-2}\bar{M}$, $\partial/\partial x, \partial/\partial y \sim \epsilon^{1/2}$, $\partial/\partial t \sim \epsilon^2$, and $h \sim 1$, appropriate to surface-energy anisotropy for directions [001], [011], and [111]. They then examined bifurcations to hexagonal–rolls and rectangle–rolls. The amplitude equations that emerge are identical to Eqs. (4.14) and (4.15) except that now the symmetries among the coefficients are broken in various ways.

For growth in the direction [100] all coefficients are real, and the anisotropy enters only the cubic coefficients. For rolls versus rectangles

$$a^2 \dot{A} = (s - s_1)A - (\alpha_1 |A|^2 + \alpha_2 |B|^2)A,$$
$$a^2 \dot{B} = (s - s_2)B - (\beta_1 |B|^2 + \beta_2 |A|^2)B,$$
(5.29)

and $s_1 = s_2 = 0$. The analysis predicts for the isotropic case that whether squares or rolls are to be preferred depends on the value of the distribution coefficient k. For k near unity, rolls are preferred over squares, and there is an exchange of preference for sufficiently large departures of k from unity. When anisotropy is present, rectangles can lose existence, one roll can be stable while the other loses stability, or rolls and rectangles can all be unstable locally.

Roughly square-shaped patterns of solute segregation were observed by Flemings in the 1970's during directional solidification of Fe–Ni alloys. The observed fourfold symmetry of the patterns is likely associated with the incipient

5.2 Directional solidification with "small" anisotropy

formation of sidearms during the cell–dendrite transition, but the observed arrangement of the cells into a square array is suggestive of the preceding predictions even if the growth is probably occurring well past the onset of instability considered here.

For rolls versus hexagons,

$$a^2 \frac{dA}{dT} = (s - s_1)A + \alpha \bar{B}\bar{C} - \left[v_1 |A|^2 + v_2 |B|^2 + v_3 |C|^2\right] A,$$

$$a^2 \frac{dB}{dT} = (s - s_2)B + \alpha \bar{C}\bar{A} - \left[v_4 |B|^2 + v_5 |C|^2 + v_2 |A|^2\right] B, \qquad (5.30)$$

$$a^2 \frac{dC}{dT} = (s - s_3)C + \alpha \bar{A}\bar{B} - \left[v_6 |C|^2 + v_3 |A|^2 + v_5 |B|^2\right] C.$$

and $s_1 = s_2 = s_3 = 0$.

The hexagon–roll competition in this case leads to various situations, depending on the angle ψ that the hexagon makes with the underlying preferred directions. When aligned, $\psi = 0$, there is a loss of distinction between hexagons and other mixed modes. Hence, the loss of symmetry induces imperfections in the secondary bifurcations. Whereas a mixed mode (Class V) crosses the hexagon mode in the isotropic case, Figure 5.5(a), anisotropy breaks this bifurcation, as shown in Figures 5.5(b) and (c). When $\psi = 0$, two of the three rolls have equal amplitudes for all values of the bifurcation parameter ($s_1 = s_2$). For small misalignments, $0 < |\psi| \ll 1$, these two roll amplitudes split apart symmetrically from their $\psi = 0$ values.

For growth in the [111] direction, the anisotropy enters the amplitude equations in the quadratic coefficients, which are complex, and here $s_1 = s_2 = 0$. In the roll–rectangle competition the usual two supercritical rolls can change to both subcritical or one of each. Further, the coefficients are complex, and thus some rolls or rectangles travel.

The hexagon–roll competition in this case leads to various possibilities. There can be sixfold symmetry and threefold but not sixfold symmetry, depending on the orientation to the crystalline axes, the degree of anisotropy, and the distribution coefficient.

For growth in the [110] direction, the anisotropy enters the amplitude equations in the linear and cubic coefficients, and these are real. For the competition between rolls and rectangles, the growth rates are angle dependent, and the two rolls bifurcate at different values of s. The rectangles now bifurcate from the first roll to exist. Far from onset, the linear selection is weak, and the cubic coefficients dominate the selection process, which now resembles that of the [100] case.

96   5. Anisotropy

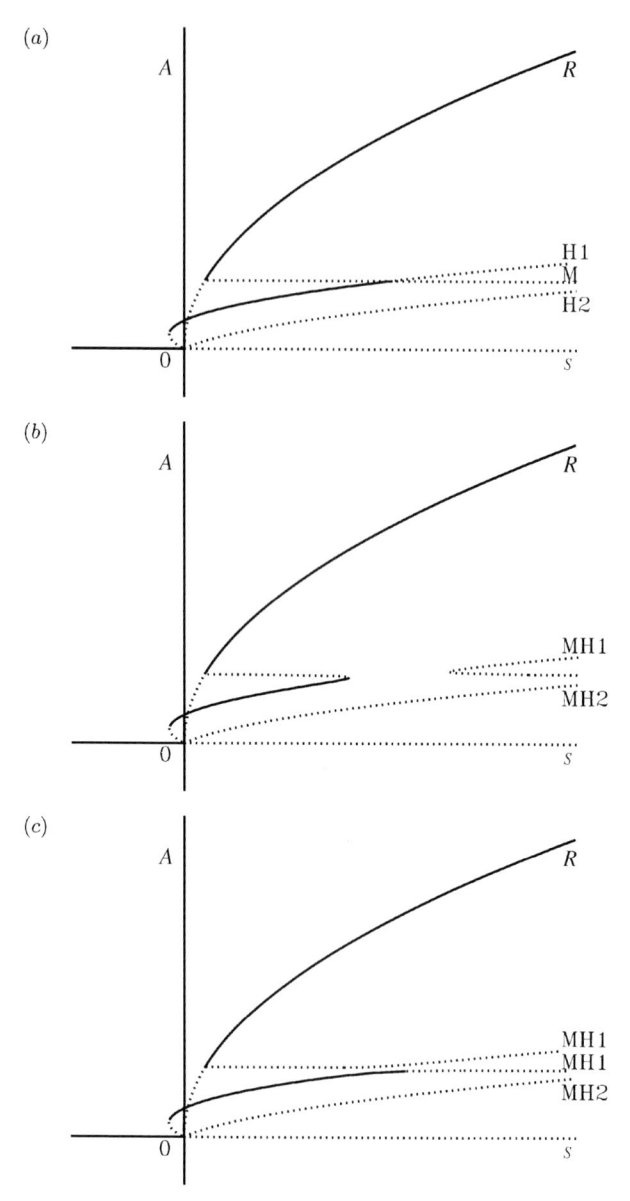

Figure 5.5. Bifurcation diagram for the roll-hexagon competition with (a) zero anisotropy, (b) some anisotropy, and (c) larger anisotropy. Solid lines indicate stable branches; R, H, and M indicate roll, hexagonal, and mixed modes, respectively. Modes HM are mixed–hexagon modes. From Hoyle et al. (1996). (Reprinted with permission of The Royal Society.)

The hexagon–roll competition contains linear selection that splits the degeneracy of the rolls so that all three bifurcate at different values of s. Hexagonal nodes bifurcate from the leftmost roll, whereas hexagonal cells bifurcate from the rightmost roll. Again, mixed modes and hexagons become indistinguishable. All hexagonal states are unstable close to the origin, but at large enough amplitudes the hexagonal node branch will turn around, becoming stable, a situation is similar to that in the [100] case.

In summary, then, the presence of kinetic anisotropy produces traveling waves in the linear stability theory and hence tilted cells in the finite amplitude regime. The presence of surface-energy anisotropy does not produce traveling waves but influences the patterns selected by the system by having crystalline properties compete with nonlinearities.

5.3 Directional Solidification with "Small" Anisotropy: Stepwise Growth

The principal directions of a crystal give faces that are atomically flat. The microscopic picture of a vicinal surface of a crystal–fluid interface consists of a series of atomically flat terraces that are interrupted by steps of roughly atomic dimensions. If the steps are well separated from each other, the mean orientation of the interface then differs only slightly from the singular orientation of the terraces, and the motion of the interface in its normal direction proceeds by layer growth that occurs at the step locations. Therefore, the limiting factors for the growth are the attachment rate at the steps and the step density. This view implies that an interface without steps, which consists entirely of the singular orientation, is stationary, resulting in the strong orientational dependence (anisotropy) of the interface–attachment coefficient for vicinal interfaces. In practice, for large enough undercooling, steps on a singular interface can arise via two-dimensional nucleation of a new layer in the interior of a terrace and via defects in the surface, such as screw dislocations that act as step sources. Once steps are present on a singular surface, the mean orientation differs from the singular orientation, and motion of the interface in its normal direction can occur at a finite rate as material is added at the steps.

For directional solidification of a binary alloy the effect of an isotropic kinetic coefficient is merely to depress the interface temperature, leaving the conditions at the onset of instability unaffected. The effect of a mild anisotropy is to generate traveling-wave instabilities at the onset (Coriell and Sekerka 1976). The conditions at the onset are again unchanged if the anisotropy is mild enough, in which case a quasi-static treatment of the diffusion field is a good approximation, and a weakly nonlinear treatment of the problem predicts the formation of tilted cells in two dimensions, as discussed in Section 5.2. If the kinetic anisotropy is strong enough, the effect of the traveling-wave instability is to produce phase differences between the perturbed interface and the solute field in which lateral solute transport can significantly stabilize the interface, as shown by Chernov, Coriell, and Murray (1993).

When a vicinal surface is present at an orientation of $\bar{p}$ with respect to the planar interface at $z = 0$, the kinetic coefficient can take a special form of Eq. (5.23), namely $\mu_x = \mu_{xx} = 0$ with $\mu_T = \beta_{ST}(\bar{p} + h_x)$ where β_{ST} is a constant. In nondimensional form

$$\mu_T = \beta(p + h_x), \qquad (5.31)$$

where

$$\beta = \frac{\beta_{ST} m G_c D}{V^2}, \qquad (5.32)$$

and it is convenient to introduce the parameter $\tilde{\alpha}$,

$$\tilde{\alpha} = (\beta p^2)^{-1}; \qquad (5.33)$$

when $\tilde{\alpha} \to 0$, attachment is instantaneous, and the kinetic effects are lost.

When this model is used, the dispersion relation from linear theory becomes

$$M^{-1} = 1 + \frac{k + \sigma}{1 - k - r} - \Gamma a^2 + \tilde{\alpha}(ia - p\sigma), \qquad (5.34)$$

where

$$r = \frac{1}{2}\left[1 + \sqrt{1 + 4a^2 + 4\sigma}\right].$$

The presence of $\tilde{\alpha} \neq 0$ gives rise to complex values of $\sigma = \sigma_R + i\omega$. In fact, for fixed a, up to three solutions, $[M^{-1}(a), \omega(a)]$ are possible, depending upon the choices of the other parameters, as shown by Grimm, Davis, and McFadden (1999).

Figures 5.6 show how the presence of anisotropy distorts the neutral curves. Figure 5.6(a) shows such curves for k fixed and $\tilde{\alpha}$ (and hence β) is varied. When $\tilde{\alpha} = 0$, the Mullins and Sekerka result holds, but as $\tilde{\alpha}$ is increased, the curve is distorted. The interface is stabilized and for large enough β^{-1}, and the neutral curve becomes multivalued and spawns an isolated piece or isola. There is still a continuous part passing through the origin valid only for long waves. For large enough $\tilde{\alpha}$ the isolas shrink and disappear. Figure 5.6(b) shows the analogous behavior for β^{-1} fixed and k varying. Figure 5.7 summarizes the behavior in the plane β^{-1} versus k; isolas exist for large enough β^{-1} and small enough k; or if β^{-1} is small, for k small enough and Γ large enough.

(a)

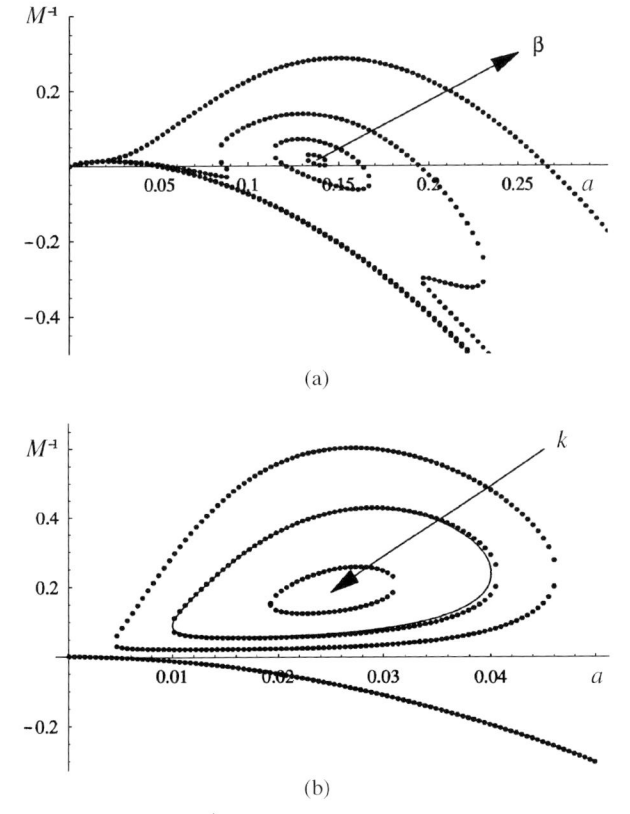

(b)

Figure 5.6. Neutral curves M^{-1} versus a for (a) $k = 0.18$, $\Gamma = 10$, $p = 0.1$, and β in the range 48–100, and (b) for $\Gamma = 100$, $p = 0.1$, $\beta = 10$, and k in the range 2×10^{-4} to 6×10^{-4}. From Grimm et al. (1999). (Reprinted with permission of the American Physical Society.)

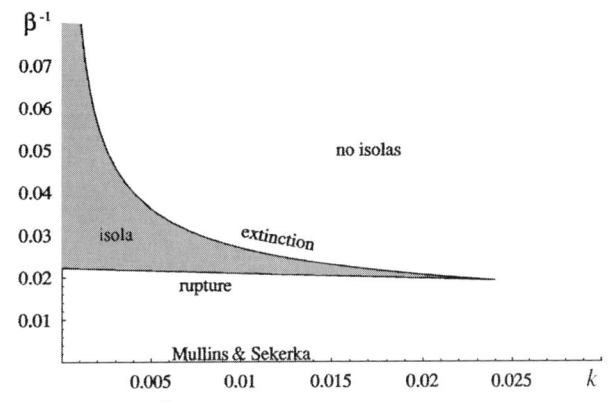

Figure 5.7. State map, β^{-1} versus k, for $\Gamma = 0.1$ and $p = 0.1$. Isolas exist in the shaded region. From Grimm et al. (1999). (Reprinted with permission of the American Physical Society.)

Necessary conditions for the prediction of isolas are small k and large Γ. As long as $\tilde{\alpha}$ is taken to be small, it is the term $-\Gamma a^2$ that dominates the neutral curve for large a. If k is even smaller (and Γ is no longer large), the stabilization at large wave numbers can be dominated by anisotropy. Or, expressed in physical terms, anisotropy plays a role analogous to that of surface energy, and its influence is strongly stabilizing. However, it is not possible to replace completely the effect of capillarity by anisotropy; if $\Gamma = 0$, large wave numbers are not stabilized.

It is worthwhile noting that in some cases linear theory predicts a "stability window," meaning that the domain of M^{-1} for which there is linear stability for all wave numbers is not necessarily the semi-infinite interval as was true in the isotropic case, but can now be composed of a semi-infinite interval *and* a finite interval (the "window"). This can be easily understood by following vertical lines in Figure 5.6(a). Note that the physically relevant part of the continuous branch of the neutral stability curve is not clearly visible in this figure.

In Section 5.2, the Sivashinsky limit, $k \to 0$, $\Gamma \sim 1$, $|M - 1| \to 0$, was extended by allowing the anisotropy to be $O(\epsilon^{1/2})$. It was found that the onset of instability is unchanged, but now traveling waves that bifurcate subcritically are predicted.

Grimm et al. (1999) have found a new limit with $\tilde{\alpha} = O(1)$. As before, they let $M^{-1} = 1 - \mu\epsilon$, $k = \kappa\epsilon^2$, $\kappa = O(1)$, $\Gamma = O(1)$, $\partial/\partial x = O(\epsilon^{1/2})$, but now because $\tilde{\alpha}$ is of unit order, they took $\partial/\partial t = \epsilon^{3/2}$ (rather than ϵ^2). They wrote $h \sim h_0 + \epsilon^{1/2} h_1$ and found that

$$h_{0_t} + \tilde{\alpha} h_{0_{xxx}} - \frac{1}{2}\left(h_0^2\right)_{xx} = 0. \tag{5.35a}$$

This is an equation for nondissipative dispersive waves that exhibit no instability. At the next order they found that

$$h_{1_t} + \tilde{\alpha} h_{1_{xxx}} - (h_0 h_1)_{xx} = -h_{0_\tau} - \kappa h_0 - \mu h_{0_{xx}} - \Gamma h_{0_{xxxx}} + \alpha h_{0_{xt}} - \tilde{\alpha}(h_{0_x}^2)_x. \tag{5.35b}$$

One can recompose Eqs. (5.35) for $h \sim h_0 + \epsilon^{1/2} h_1$ and find that

$$h_t + \tilde{\alpha} h_{xxx} - \frac{1}{2}(h^2)_{xx} + \epsilon \left\{ kh + \Gamma h_{xxxx} + \mu h_{xx} - \tilde{\alpha} h_{xt} + \alpha \left(h_x^2\right)_x \right\} = O(\epsilon^2). \tag{5.36}$$

To analyze Eq. (5.36), write $\partial/\partial t = \partial/\partial t_0 + \epsilon^{1/2}\partial/\partial t_2 + \epsilon\partial/\partial t_4 + \cdots$, $\mu = \mu_c + \epsilon^{1/2}\mu_2 + \cdots$ and $h = \epsilon^{1/4} v_1 + \epsilon^{1/2} v_2 + \cdots$. From linear theory,

$$a_c^2 = \sqrt{\frac{\kappa}{\Gamma + \tilde{\alpha}^2}}, \quad \omega_c = \tilde{\alpha} a_c^3, \quad \mu_c = 2\sqrt{\kappa\left(\Gamma + \tilde{\alpha}^2\right)}$$

with phase speed $c = \omega/a_c$.

5.3 Directional solidification with "small" anisotropy

The first orthogonality condition for $v_1 = A \exp(ia_c x + i\omega_c t) + c.c.$ is given by

$$A_{t_2} = \frac{i}{3\tilde{\alpha}} \sqrt[4]{\frac{\kappa}{\Gamma + \tilde{\alpha}^2}} A |A|^2, \qquad (5.37)$$

which determines the nonlinear frequency shift. Instability is determined at higher order and gives

$$A_{t_4} = (\mu_2 + i\mu_i)A - \frac{1}{2}\left(3 - \frac{\Gamma}{\tilde{\alpha}^2}\right)A|A|^2 + ibA|A|^4,$$

where μ_i and b are real constants that again determine a frequency shift. The term proportional to $A|A|^4$ is kept for illustration only; it is asymptotically small compared with the cubic term away from the transition points. The Landau coefficient, $-(3 - \Gamma/\tilde{\alpha}^2)/2$, changes sign, and when

$$\tilde{\alpha}^2 > \frac{1}{3}\Gamma, \qquad (5.38)$$

the bifurcating traveling waves can be supercritical and hence in principal observable, though no effort has yet been made to find such waves experimentally.

The result here for $\tilde{\alpha}^2 > \Gamma/3$ is tilted cells in the weakly nonlinear limit. One could directly seek traveling-waves solutions of Eq. (5.35a) such as $h = u(x - ct)$ and find that the equation satisfies $-cu + \tilde{\alpha}u'' = uu'$. If one writes

$$u' = Q, \qquad (5.39a)$$

then

$$Q' = \tilde{\alpha}^{-1}(Q + c)u, \qquad (5.39b)$$

and a first integral has the form (Grimm et al. 1999)

$$\frac{1}{2\tilde{\alpha}}u^2 - Q + c\ln|Q + c| = K, \qquad (5.40)$$

where K is a constant of integration.

For small amplitude, the solutions of Eq. (5.40) are nearly sinusoidal, which is consistent with bifurcation theory. For larger amplitudes, these cells become increasingly tilted, and finally, as the amplitude gets very large, they are more and more facet-like (see Figure 5.8). The front is then composed of nearly

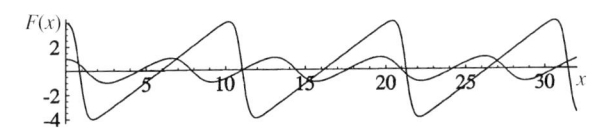

Figure 5.8. The interfacial shape for two values of anisotropy. When anisotropy is very small, the interface is nearly sinusoidal, and when it is large, the interface resembles a faceted front. From Grimm et al. (1999). (Reprinted with permission of the American Physical Society.)

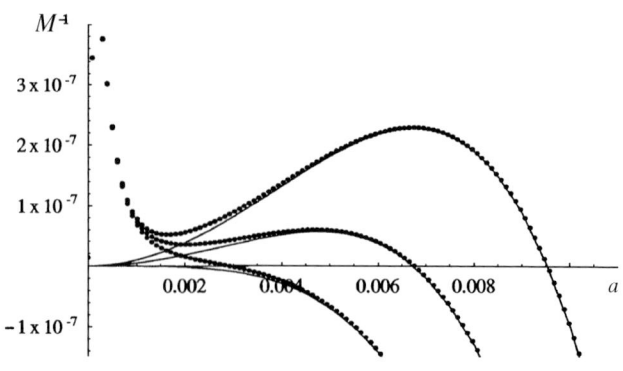

Figure 5.9. Neutral curves, M^{-1} versus a, near the absolute stability limit of the isotropic case: $K = 0.1$, $p = 0.01$, $\beta = 10^9$, and Γ is equal to 9.99, 9.995, and 10.0. The solid lines represent the isotropic case. From Grimm et al. (1999). (Reprinted with permission of the American Physical Society.)

linear pieces ($Q \approx |c|$) connected by sharp variations (where both u and Q vary "rapidly"). This reflects the competition between the interface trying to align with the crystal lattice while still remaining close to its equilibrium position. See Section 5.5 for discussions on the development of facets.

The morphology for small k has been discussed – especially the stabilization of the front due to kinetic anisotropy. When k is of unit order, the effect of kinetic anisotropy is reversed.

In the isotropic case, all modes are linearly stable above the absolute stability limit, $\Gamma > 1/k$, and near this limit the critical wave number, scaled in units of δ_c, approaches zero. Figure 5.9 shows the neutral curves in the presence of kinetic anisotropy. The neutral curve now has two maxima, one at low wave numbers that is influenced by surface energy and related to a Mullins and Sekerka mode, and another occurring at lower wave numbers that is hardly affected by Γ. It is then easily seen that the latter becomes dominant at some value of Γ near the absolute stability limit. The new instability persists even far beyond the absolute stability limit of the isotropic case.

For fixed temperature gradient G_T, the parameters M, Γ, and β depend on the dimensional parameters C_∞ and V, which are more useful for comparison with experiments. Figure 5.10 shows the domains of stability in terms of these dimensional parameters. Roughly speaking, the marginal stability curves are composed of two branches. The first one is found for low V and can be related to the isotropic case (anisotropy slightly stabilizes at very low V), and it is also related to the isolas, if they exist. The high-speed branch lies mainly above the upper part of the marginal stability curve of the isotropic case and is thus beyond the absolute stability limit. It is uniquely due to the presence of kinetic

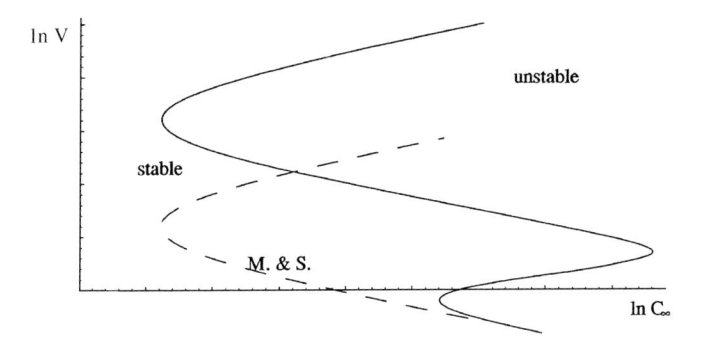

5.3 Directional solidification with "small" anisotropy

Figure 5.10. Neutral curves, V versus C_∞, G_T fixed. The lower branch is a smooth deformation of the Mullins–Sekerka theory. The dashed line and the upper branch represent the new instability directly related to kinetic anisotropy. The plot M^{-1} versus a for these parameters has an isola. From Grimm et al. (1999). (Reprinted with permission of the American Physical Society.)

anisotropy. If one considers lines of constant C_∞ in Figure 5.10, the following picture results: for C_∞ small enough, the planar front is linearly stable for any speed. For large enough C_∞, the planar front is unstable for some finite interval of speeds, that is, stable for very low and for very high speeds. In an intermediate range, there are two intervals of speeds for which the planar front is unstable: one for low speeds, reflecting the classical Mullins–Sekerka instability, and one for high speeds that owes its presence entirely to the kinetic anisotropy. The theory predicts a point of intersection of the marginal stability curves, where the two phenomena are competing. In the isotropic case, the critical wave number and the critical value of M^{-1} both tend to zero as $\Gamma \to k^{-1}$, and finally all modes are linearly stable when $\Gamma = k^{-1}$ (see Figure 5.9).

The absolute stability limit, already mentioned in the general discussion of the dispersion relation in the preceding paragraphs, was used by Brattkus and Davis (1988) for the derivation of a strongly nonlinear longwave evolution equation by setting $\epsilon = k^{-1} - \Gamma$, $M^{-1} = O(\epsilon^2)$, $k = O(1)$ and $\Gamma = O(1)$. The wave number and growth rate are then to be taken as $a = O(\epsilon^{1/2})$ and $\sigma = O(\epsilon)$. If one now tries to take into account the effect of anisotropy, two scalings seem possible for p and β. Either one takes $p = O(\epsilon^{1/2})$ and $\beta = O(\epsilon^{-5/2})$ or $p = O(1)$, and $\beta = O(\epsilon^{-3/2})$, both yielding

$$M^{-1} = a^2 - \frac{1}{k^2}[k(\sigma + a^2) + a^2](a^2 + \sigma) + i\tilde{\alpha}a. \tag{5.41}$$

A problem arises when looking at the neutral stability curve in this limit. Set $\sigma = i\omega$ and separate Eq. (5.41) into real and imaginary parts,

$$M^{-1} = a^2 - \frac{1+k}{k^2}a^4 + \frac{\omega^2}{k} \tag{5.42}$$

and
$$\omega = \frac{\tilde{\alpha}k^2}{1+2k}a^{-1}.$$

One sees that the scaling is inappropriate for $a \to 0$ because $\omega \propto a^{-1}$.

There is a new limit in the neighborhood of the second maximum of the neutral curve that can be defined and describes unstable fronts beyond the absolute stability limit, that is, for $\Gamma > 1/k$. Let $M^{-1} = O(\epsilon^2)$, $p = O(\epsilon)$, $\mu_T = O(\epsilon^{-4})$, $a = O(\epsilon)$ and $\omega = O(\epsilon)$. On the neutral curve at order ϵ^2 one obtains

$$M^{-1} = (k^{-1} - \Gamma)a^2 + \frac{\omega^2}{k}, \tag{5.43a}$$

with ω given by

$$\omega^3 + a^2\omega - \frac{\tilde{\alpha}k^2}{1+2k}a = 0. \tag{5.43b}$$

The resulting curve $\omega = \omega(a)$ then grows as $a^{1/3}$ at the origin, and, after having reached a maximum, decays like a^{-1}. The neutral stability curve in terms of M^{-1} therefore has a vertical tangent at the origin, which is rather unusual. This is corrected by terms of the following order that introduce a singular perturbation near the origin. A more important observation is that one must have $\Gamma \geq k^{-1}$ for this limit to apply (although an accurate description of the neutral stability curve near the origin when $\Gamma < k^{-1}$ may be obtained).

For $h \sim h_0 + \epsilon h_1$, and $c \sim \tilde{c} + \epsilon c_1$, and $\partial/\partial t \to \epsilon \partial/\partial t + \epsilon^2 \partial/\partial T$, Grimm et al. (1999) found that

$$h_0(t, T, x) = A(T, x)e^{i\omega_0 t} + \bar{A}(T, x)e^{-i\omega_0 t} \tag{5.44}$$

and

$$h_{1_{tt}} + kM^{-1}h_1 + \mathcal{L}h_0 + \mathcal{N}(h_0) - \frac{\tilde{\alpha}h_{1x}}{1+h_{1x}/p} = 0, \tag{5.45}$$

where

$$\omega_0 = \sqrt{kM^{-1}}$$

$$\mathcal{L}h = 2h_{Tt} + (1+2k)M^{-1}h_t - \frac{1+2k}{k}h_{txx} + kh_{xx} \tag{5.46}$$

$$\mathcal{N}(h) = h_t h_{xx} + 3kM^{-1}hh_t - 2h_x h_{tx}.$$

The amplitude A of Eq. (5.44) is determined by an orthogonality condition on Eq. (5.45). The result is

$$2\omega_0 A_T + \frac{1+2k}{k}\omega_0^3 A - \frac{1+2k}{k}\omega_0 A_{xx} + ik\tilde{\alpha}A_x R(|A_x|) = 0, \tag{5.47}$$

where R is given by the integral

$$R(x) = \frac{1}{\pi} \int_0^{2\pi} \frac{\cos^2 t}{p + 2x \cos t} dt. \tag{5.48}$$

A power series of $R(x)$ about $x = 0$ shows that there is no term linear in x, and the sign of the quadratic term is positive. This is sufficient to indicate that the underlying bifurcation problem for the critical modes gives only subcritical bifurcations.

Absent from this analysis are other nonlinearities, as appear in the Brattkus–Davis limit. They can be included by using different scalings of H and C. It turns out that the coefficients of these nonlinear terms are such that they do not significantly influence the bifurcation behavior (i.e., the corresponding contribution to the Landau coefficient remains imaginary).

5.4 Unconstrained Growth with "Small" Anisotropy

Consider first the instabilities of a growing *circle* in a pure, undercooled melt, the two-dimensional analog of a growing spherical nucleus. When anisotropy is absent and an initial condition with fourfold symmetry is posed, the circle will become unstable near $R = R_c$, and the fourfold morphology is similar to that shown in Figure 5.11. Kessler, Koplik, and Levine (1984) showed for the symmetric model that the curvature at the tips decreases with time and passes through zero, indicating that tip splitting has begun.

Consider now the influence of fourfold anisotropy of the surface energy ($\alpha_4 \neq 0$). Kessler et al. (1984) showed for $\alpha_4 \neq 0$ that tip splitting is retarded, and if α_4 is large enough, tip curvatures will always increase in time.

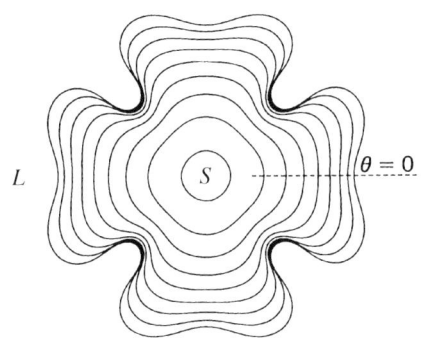

Figure 5.11. Numerical solutions for the symmetric model at various times for the shape of a two-dimensional crystal of an isotropic material. From Brush and Sekerka (1989). (Reprinted with permission of North-Holland.)

Lemieux et al. (1987) considered the two-dimensional model with both surface tension and kinetic anisotropies present, that is, in general $\alpha_4, \beta_4 \neq 0$. When surface energy $\gamma_0 = 0$ and kinetic undercooling is present, $\mu^{-1} \neq 0$, Lemieux et al. found for small ΔT, and hence small V_n, that there is a continuous "Ivantsov" family of steady-finger solutions and that, when $\mu_0^{-1} V_n$ is large, the whole family is destroyed.

When both surface energy and kinetic undercooling are present, $\mu_0^{-1}, \gamma_0 \neq 0$, there are no steady states in the isotropic case, $\alpha_4 = \beta_4 = 0$, which is obvious. However, when $\mu_0^{-1} V_n = O(1)$, which corresponds to significant values of ΔT, numerical computations can be used to obtain steady solutions.

When both surface energy and kinetic undercooling are present, $\mu^{-1}, \gamma_0 \neq 0$, and both anisotropies are present, $\alpha_4, \beta_4 \neq 0$, Lemieux et al. found selection similar to the equilibrium case ($\mu^{-1} = 0$) and $\gamma \neq 0$, and the selected V_n is shifted by the presence of kinetics.

Finally, when $\mu^{-1}, \gamma \neq 0$, and α_4, β_4 get larger (but still in the "small anisotropy" range), their numerical solutions show that the selected speed saturates to a finite value.

Brush and Sekerka (1989) introduced kinetic anisotropy instead of surface energy anisotropy and found precisely the same type of behavior. For $\beta_4 = 0.06$ and for various values of μ_0^{-1}, the tip becomes a point of a local maximum in curvature; see Figure 5.12. Thus ... "it appears that any anisotropy, whether introduced via the surface tension or the kinetics, can influence the growth of the tip ...".

Consider now the freezing of a pure melt in which a *planar* front propagates under hypercooled conditions. The front is kinetically controlled and propagates at constant speed. Let $\hat{\mu} \equiv \mu^{-1}$ be the reciprocal of the kinetic coefficient and $\hat{\gamma}$ be the surface tension. Again write

$$\hat{\mu} = \hat{\mu}_0 + \hat{\mu}_x h_x + \hat{\mu}_y h_y + \hat{\mu}_{x2} h_x^2 + \hat{\mu}_{xy} h_x h_y + \hat{\mu}_{y2} h_y^2 + \cdots,$$
$$\hat{I} = \hat{\gamma}_0 + \hat{\gamma}_x h_x + \hat{\gamma}_y h_y + \hat{\gamma}_{x2} h_x^2 + \hat{\gamma}_{xy} h_x h_y + \hat{\gamma}_{y2} h_y^2 + \cdots,$$
(5.49)

where $\hat{I} = \sqrt{1 + h_x^2 + h_y^2}\, \tilde{\gamma}(h_x, h_y)$ is the *weighted* surface tension and $\hat{\mu}_0$ and $\hat{\gamma}_0$ are the reciprocal kinetic coefficient and the surface tension of the planar crystal surface, respectively. In this notation, the *isotropic* case corresponds to $\hat{\gamma}_0 = 2\hat{\gamma}_{x2} = 2\hat{\gamma}_{y2} = \bar{\gamma}_0$, where $\bar{\gamma}_0$ is the isotropic surface tension. The surface is growing in the direction of positive z; $z = h$ is the location of the solidification front, and x and y are coordinates on the planar front that are defined by appropriate coordinate transformation of the frame aligned with the principal axes of the growing crystal; see Section 5.1.

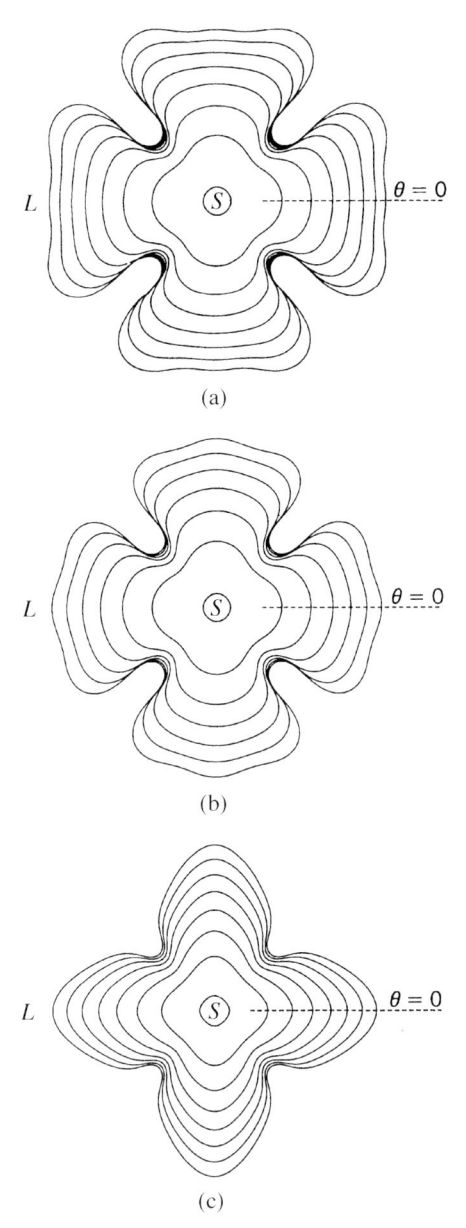

5.4 Unconstrained growth with "small" anisotropy

Figure 5.12. Numerical solutions for the symmetric model at various times for the shape of a two-dimensional crystal with (a) $\mu_0^{-1} = 13.3$, $\beta_4 = 0.06$, (b) $\mu_0^{-1} = 20$, $\beta_4 = 0.06$, and (c) $\mu_0^{-1} = 200$, $\beta_4 = 0.06$. From Brush and Sekerka (1989). (Reprinted with permission of North-Holland.)

The steady propagation of the front exists when $S^{-1} > 1$ and the planar basic state is

$$\bar{T} = Se^{-Vz}, \quad V = 1 - S. \tag{5.50}$$

Golovin and Davis (1998) considered the case when both surface energy and kinetics are fourfold anisotropic. Linear stability theory gives that

$$S\sigma - \kappa_1 \Gamma V \mathbf{a} \cdot \mathbf{A} \cdot \mathbf{a}^T + SV^2 + \Lambda(\sigma, a)[-SV + \sigma + \Gamma \mathbf{a} \cdot \mathbf{A} \cdot \mathbf{a}^T + iV\boldsymbol{\mu}' \cdot \mathbf{a}] = \mathbf{0},$$

$$\Lambda(\sigma, a) = \frac{1}{2}\left[V + \sqrt{V^2 + 4\sigma + 4a^2}\right], \quad a^2 = \mathbf{a} \cdot \mathbf{a},$$

$$\mathbf{A} = \begin{pmatrix} 2\gamma_{x2} & \gamma_{xy} \\ \gamma_{xy} & 2\gamma_{y2} \end{pmatrix}, \quad \boldsymbol{\mu}' = (\mu_x, \mu_y)^T, \tag{5.51}$$

where $\kappa_1 = L_v/c_p T_m$. Here $\mathbf{A}$ is the matrix of the coefficients describing the anisotropy of the surface energy, and $\boldsymbol{\mu}'$ is the vector of reciprocal kinetic anisotropies. Elements of $\mathbf{A}$ and $\boldsymbol{\mu}'$ are the corresponding coefficients of the expansions (5.49) divided by $\bar{\gamma}_0$ and $\hat{\mu}_0$, respectively. For a given crystal with cubic symmetry, the elements of $\mathbf{A}$ and $\boldsymbol{\mu}'$ depend on the direction of the crystal growth. In the isotropic case, $\mu_x = \mu_y = \gamma_{xy} = 0$, $2\gamma_{x2} = 2\gamma_{y2} = 1$, and thus $\boldsymbol{\mu}' = \mathbf{0}$, and $\mathbf{A}$ is the unit matrix.

According to the characteristic equation (5.51), the solidification front becomes unstable with respect to long-scale oscillatory perturbations at ($a = 0$ when $S > S_c$), propagating in different directions along the solidification front characterized by the angle ψ, $\tan \psi = a_y/a_x$, as shown in Figure 5.13. It

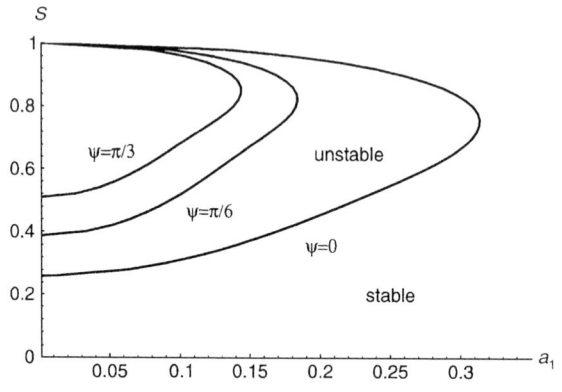

Figure 5.13. Neutral stability curves, S versus a, for a two-dimensional front with $\mu_x = 1.0$, $\mu_y = 0.5$, $\gamma_{xy} = 0.75$, $\gamma_{y2} = 1.25$, $\Gamma = 1$, and $\kappa_1 = 0.3$ for various directions ψ; S_c occurs at $a = 0$ and depends on ψ. From Golovin and Davis (1998). (Reprinted with permission of Elsevier Science B.V.)

can be seen that the critical Stefan number S_c depends on the direction of the disturbance wave vector, namely,

$$S_c = \frac{\alpha(\psi)\Gamma(1-\kappa_1)}{1+\alpha(\psi)\Gamma(1-\kappa_1)+m^2(\psi)},$$

$$\alpha(\psi) = 2\gamma_{x2}\cos^2\psi + \gamma_{xy}\sin 2\psi + 2\gamma_{y2}\sin^2\psi,$$

$$m(\psi) = \mu_x\cos\psi + \mu_y\sin\psi. \tag{5.52}$$

The (constant) front speed corresponding to the onset of the instability is $V_c = 1 - S_c$. In the general case, when both surface tension and kinetics anisotropies are present, this instability gives rise to the propagation of long waves along the solidification front with the following phase velocity, $\mathbf{v}_p$, and the group velocity $\mathbf{v}_g$:

$$\mathbf{v}_p = \omega\mathbf{a}/a^2 = V_c(\psi)(\boldsymbol{\mu}'\cdot\mathbf{a})\mathbf{a}/a^2,$$
$$\mathbf{v}_g = V_c(\psi)\boldsymbol{\mu}'. \tag{5.53}$$

These long waves manifest themselves as periodic cells traveling along the solidification front. Thus, the existence of *traveling* cells depends on the presence of kinetic anisotropy. If there is no kinetic anisotropy, $\mu' = 0$, the instability is then monotonic, leading to stationary, spatially periodic structures: *stationary* cells.

In general, with the increase of S, the traveling cells that first start to grow travel in the directions given by an angle ψ_0 that is determined by the coefficients of surface tension and kinetic anisotropies and that satisfies the equation

$$[1+m^2(\psi_0)]\alpha'(\psi_0) = 2\alpha(\psi_0)m(\psi_0)m'(\psi_0) \tag{5.54}$$

together with the following inequality:

$$[(\alpha''(1+m^2) - 2\alpha(m')^2 + mm'')][1+\alpha\Gamma(1-\kappa_1)+m^2]$$
$$-2[\alpha'(1+m^2) - 2\alpha mm'][\alpha'\Gamma(1-\kappa_1)+2mm'] > 0. \tag{5.55}$$

Here $m(\psi)$ and $\alpha(\psi)$ are defined in Eq. (5.52), and the prime denotes differentiation with respect to ψ. The angle ψ_0 gives the minimum value of $S_c = S_c(\psi_0) = S_0$, $V_0 = 1 - S_0$. In a particular case, when only a kinetic anisotropy is present, the preferred direction of the traveling-cell propagation is given by an angle ψ_0^k

$$\tan 2\psi_0^k = \frac{\mu_y}{\mu_x}. \tag{5.56}$$

In the case when only surface tension anisotropy is present, the stationary roll-like periodic structure forms in the preferred direction given by an angle ψ_0^s such that

$$\tan 2\psi_0^s = \frac{\gamma_{xy}}{\gamma_{x2} - \gamma_{y2}}. \tag{5.57}$$

In the general case there also exists a specific direction characterized by an angle ψ_0^n at which the group velocity of the waves is normal to their wave vector (in the frame moving with the planar front with the velocity $V_c(\psi_0^n)$), and the phase velocity is zero:

$$\tan \psi_0^n = -\frac{\mu_x}{\mu_y}, \tag{5.58}$$

and thus $\psi_0^n = \pi/2 + \psi_0^k$. When the surface tension and the kinetic anisotropies satisfy the relation

$$\gamma_{xy}(\mu_y^2 - \mu_x^2) + 2\mu_x\mu_y(\gamma_{x2} - \gamma_{y2}) = 0, \tag{5.59}$$

ψ_0^n can give a minimum of S_c, and the instability first occurs in the direction determined by ψ_0^n, so that the wave, whose group velocity is normal to its wave vector, is first to appear. Notice that if either anisotropy, kinetic or capillary, is zero, then relation (5.59) is an identity. If only the kinetic anisotropy is present, the angle ψ_0^n always corresponds to a *maximum* of the dependence $S_c(\psi)$.

Typically, variations of the instability threshold S_c depend on the direction of cell propagation at the front (on the angle ψ), given by Eq. (5.52).

Owing to the presence of anisotropy, the instability threshold $S_0 = S_c(\psi_0)$ given by relation (5.59), at which the traveling (or stationary) cells first appear in the preferred direction on the propagating solidification front, depends in turn on the direction of the propagation of the front compared with the principal directions. This dependence is illustrated in two cases below for crystals with cubic symmetry: when a crystal grows in a Hele–Shaw cell, so that the solidification front is effectively one-dimensional, and in a three-dimensional case with a two-dimensional front.

5.4.1 Two-Dimensional Crystal and One-Dimensional Front

It is natural to consider the dimensional surface tension and the dimensional kinetic coefficient to be dependent on the direction of a crystal growth in the following form:

$$\hat{\gamma}_0 = \bar{\gamma}_0(1 + \alpha_4 \cos 4\phi), \quad \hat{\mu}_0 = \bar{\mu}_0(1 + \beta_4 \cos 4\phi), \tag{5.60}$$

5.4 Unconstrained growth with "small" anisotropy

where now ϕ is an angle between the [10] direction and the direction of the crystal growth; $\bar{\gamma}_0$ and $\bar{\mu}_0$ are the mean surface tension and the reciprocal of mean kinetic coefficients, respectively. Consider a local perturbation of the front with the normal vector at the angle $\phi + \delta\phi$ with respect to the [10] direction, so that $\cos \delta\phi = (1 + h_x)^{-1/2}$, $\sin \delta\phi = -h_x(1 + h_x^2)^{-1/2}$. If one uses Eq. (5.59) and expands $\mathcal{E}(h)$ and $\mu(h)$ for small $|h_x|$, one gets γ_{x2} and μ_x for the dimensionless parameters (only parameters with x indices are relevant in two-dimensional case)

$$\gamma_{x2} = \frac{1}{2}(1 - 15\alpha_4 \cos 4\phi), \quad \mu_x = \frac{4\beta_4 \sin 4\phi}{1 + \beta_4 \cos 4\phi}. \tag{5.61}$$

If one sets $\psi = 0$ in Eq. (5.52) for two-dimensional case, and substitutes Eqs. (5.60) and (5.61) into Eq. (5.52), the instability threshold is obtained as a function of the growth direction ϕ and of the coefficients of surface-tension and kinetic anisotropies, α_4 and β_4.

The variation of $S_c(\phi)$ at different anisotropies of the surface tension and of the kinetic coefficient has extrema at $\phi = n\pi/4$ designating the crystallographic directions [10], [01], and [11]; in these directions $\mu_x = 0$, and morphological instability is *monotonic*, leading to formation of *stationary* cells; there are no traveling cells when a crystal grows in one of the principal crystallographic directions.

If different crystals grow in the direction [10] or [11], their kinetic anisotropy coefficients μ_x are equal to zero, and the instability is monotonic, leading to the formation of stationary cells. However, the instability threshold corresponding to different crystals still depends on the coefficients of the kinetic anisotropy β_4 because the parameter Γ depends on the anisotropic kinetic coefficient $\bar{\mu}_0$. Owing to this dependence, when crystals grow in the [10] direction, the critical Stefan number will be maximal for a crystal with minimal negative kinetic and surface-tension anisotropy coefficients β_4 and α_4, and when various crystals grow in the [11] direction, the critical Stefan number will be maximal for crystals with maximal positive values of the kinetic and surface-tension anisotropy coefficients. The critical Stefan number can be maximal for a crystal that has some intermediate value of β_4, and this value does not depend on α_4.

5.4.2 Three-Dimensional Crystal and Two-Dimensional Front

The anisotropies of the surface tension and of the kinetic coefficient are described by Eqs. (5.49), whose coefficients depend on the direction of the crystal growth, and can be expressed in terms of two angles, θ and ϕ, that are spherical

112   5. Anisotropy

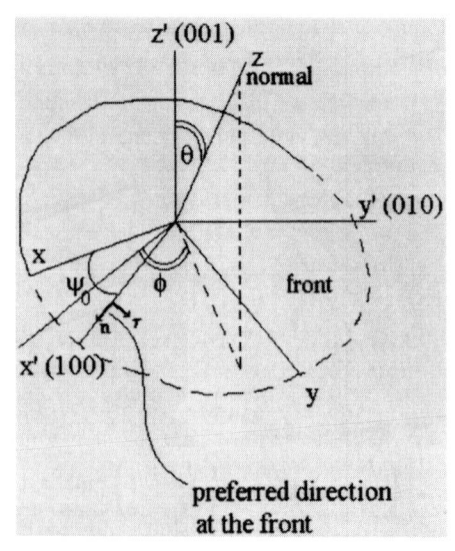

Figure 5.14. System of coordinates related to a three-dimensional crystal with a two-dimensional solidification front. The frame of reference (x', y', z') is aligned with the principal crystallographic axes [100], [010], [001], respectively. The coordinate system (x, y, z) is connected with the moving solidification front; z is the coordinate normal to the front, and (x, y) are the coordinates at the front (x-axis lies in the (x', y')-plane). The direction of the front motion (crystal growth) is characterized by two angles, ϕ and θ, with respect to the principal crystallographic directions. The angle ψ_0 defines the direction of the wave vector for the disturbance that first starts to grow as a result of the morphological instability. From Golovin and Davis (1998). (Reprinted with permission of Elsevier Science B.V.)

angles of the normal to a plane front in the frame of reference connected with the cubic crystal principal directions [100], [101], [001]; see Figure 5.14. This dependence contains "primary" anisotropy coefficients of the surface tension $\gamma'_0, \gamma'_4, \gamma'_6$ and kinetics μ'_0, μ'_4, μ'_6 in the frame of reference connected with the principal directions [100], [010], [001] that are supposed known from experiment or theory. One can obtain the dependence of the instability threshold S_0, at which traveling cells first appear, on the angles θ and ϕ characterizing the direction of the crystal growth. This dependence has extrema in the [100], [010], and [111] directions, where $\boldsymbol{\mu}' = \boldsymbol{0}$, and the instability is *monotonic*, leading to the formation of cells. Which direction will give a maximum and which a minimum to the critical Stefan number is strongly affected by the sign of the "primary" anisotropy coefficients. When all the coefficients are positive, the critical Stefan numbers are minimal in the [100], [010], and [001] directions and maximal in the [111] ($\theta = \arctan(\sqrt{2})$, $\phi = \pi/4$) and other similar

5.4 Unconstrained growth with "small" anisotropy

directions. When some of the coefficients are negative, the maximal critical Stefan number can be in the [100], [010], and [001] directions. In the following the nonlinear evolution of the waves (traveling cells) in the vicinity of the instability threshold will be considered.

To derive an equation describing *nonlinear* evolution of the traveling cells generated by the instability of the solidification front described above new coordinates are introduced connected, with the direction at the front in which the instability first appears, which is characterized by the angle ψ_0: $X = \epsilon(\mathbf{r} \cdot \mathbf{n})$, $Y = \epsilon^2(\mathbf{r} \cdot \boldsymbol{\tau})$, where $\mathbf{n} = (\cos\psi_0, \sin\psi_0)$ and $\boldsymbol{\tau} = (-\sin\psi_0, \cos\psi_0)$ are the unit vectors in the direction of the most rapidly growing perturbations ($\mathbf{n}$) and normal to the direction ($\boldsymbol{\tau}$), (see Fig. 5.14), and ϵ is defined by the degree of the supercriticality, $\epsilon^2 = (S - S_0)/S_2$; S_2 is an arbitrary number that can be fixed by choosing a convenient scaling for ϵ. Thus, the new long-scale coordinates X and Y are along the preferred direction and normal to it, respectively. For the anisotropic problem the details of the derivation are given in Appendix C of Golovin and Davis (1998). For the front deformations $h = O(\epsilon)$, in the frame of reference moving with the group velocity of the waves, one obtains the following anisotropic evolution equation;

$$h_\tau + a_1 h_{\xi\xi\xi} + a_2 h_\xi^2 + \epsilon(a_3 S_2 h_{\xi\xi} - a_4 h_{\eta\eta} + a_5 h_{\xi\xi\eta}$$
$$+ a_6 h_{\xi\xi\xi\xi} - g_1 h_\xi h_{\xi\xi} - g_2 h_\xi h_\eta) = 0, \qquad (5.62)$$

where $\tau = (\epsilon^3 + \epsilon^4)t$, $\xi = X - (\epsilon V_0 - \epsilon^2 S_2)m_0 t$, $\eta = Y - (\epsilon^2 V_0 - \epsilon^4 S_2)m'_0 t$, and the coefficients are given in Golovin and Davis (1998). It can be shown that coefficients a_4 and a_6 are always positive.

Equation (5.62) describes the generic long waves (traveling cells) brought about by the oscillatory, long-scale morphological instability of a planar solidification front that propagate in the preferred direction determined by Eqs. (5.55) and (5.56). It can be seen that the main part of the equation involves a Korteweg–de Vries (KdV) operator and is modified by smaller terms describing the effect of instability and dissipation. The specific case, when the group velocity of the waves is normal to their wave vector and the phase velocity is equal to zero, is described by the following equation with the physical $h = O(\epsilon^2)$ and $T = \epsilon^4 t$:

$$h_T + \bar{a}_5 h_{\eta XX} + \bar{a}_2 h_X^2 + \bar{a}_3 S_2 h_{XX} - \bar{a}_4 h_{\eta\eta} + \bar{a}_6 h_{XXXX} = 0, \qquad (5.63)$$

where the coefficients $\bar{a}_i$ are related to the respective coefficients a_i in Eq. (5.62), and the coordinate η is a translation in time of Y. In one dimension, Eq. (5.63) is the Kuramoto–Sivashinsky equation, and in two dimensions it describes dispersion and damping in the direction normal to the preferred direction of the instability.

When there is no kinetic anisotropy and only surface-tension anisotropy is present, the slow, long-scale spatial evolution of the instability is described, near the instability threshold, by an anisotropic Kuramoto–Sivashinsky equation with the physical $h = O(\epsilon^2)$ and $T = \epsilon^4 t$:

$$h_T + \tilde{a}_3 S_2 h_{XX} - \tilde{a}_4 h_{YY} + \tilde{a}_2 h_x^2 + \tilde{a}_6 h_{XXXX} = 0, \tag{5.64}$$

where coefficients $\tilde{a}_i$ are equal to the respective coefficients a_i in Eq. (5.62) with $\beta_4 = 0$.

Consider now the *general case* described by Eq. (5.62). Introduce a coordinate scaling

$$\xi \to a'\xi, \quad \eta \to b'\eta, \quad \tau \to c'\tau, \quad h \to s'h \tag{5.65}$$

with

$$a' = a_6 a_1^{-1}, \quad b' = a_4^{1/2} a_6^{3/2} a_1^{-2}, \quad c' = a_6^3 a_1^{-4}, \quad s' = 3a_1^2 a_6^{-1} a_2^{-1},$$

and choose the free parameter S_2 such that $S_2 = a_1^2 a_6^{-1} a_3^{-1}$. This transforms Eq. (5.62) into a simpler form

$$h_\tau + h_{\xi\xi\xi\xi} + 3h_\xi^2 + \epsilon(h_{\xi\xi} - h_{\eta\eta} + D_1 h_{\xi\xi\eta} + h_{\xi\xi\xi\xi} - G_1 h_\xi h_{\xi\xi} - G_2 h_\xi h_\eta) = 0, \tag{5.66}$$

where

$$D_1 = a_5 a_6^{-1/2} a_4^{-1/2}, \quad G_1 = 3g_1 a_1 a_6^{-1} a_2^{-1}, \quad G_2 = 3g_2 a_2^{-1} a_4^{-1/2} a_6^{-1/2}.$$

Seek solutions of Eq. (5.66) in the form of stationary waves propagating in the preferred direction of the instability with a constant velocity, that is, in the form $h(\zeta) = h(\xi - c\tau)$, (that also describe the waves at a one-dimensional solidification front of a growing two-dimensional crystal in a Hele–Shaw cell).

In order to find a solution it is more convenient to introduce a new function $u(\zeta) = h_\zeta$. Consider waves propagating in the preferred direction, and seek a solution in the form $u = u_0 + \epsilon u_1 + \cdots$, $c = c_0 + \epsilon c_1 + \cdots$. The problem for $u(\zeta)$ in this case reduces to that studied in detail in Bar and Nepomnyashchy (1995). At unit order in ϵ, one obtains periodic solutions $u_0(\zeta)$ expressed in terms of elliptic functions, with wave numbers q (normalized by 2π) and amplitudes depending on c_0 and q, which, in turn, are found from a solvability condition of the corresponding problem at order ϵ. It is easy to show that the shape h_0 of the solidification front in the main approximation is connected with

5.4 Unconstrained growth with "small" anisotropy

u_0 as follows:

$$h_0(\zeta, \tau) = H_0(\zeta) - v_0 \tau$$

$$H_0(\zeta) = \int_0^\zeta u_0 d\zeta = 2\pi^{-1} q K(p) E[am(\pi^{-1}q\zeta K(p), p), p]$$
$$- 2\pi^{-2} q^2 E(p) K(p) \zeta$$

$$v_0 = 4\pi^{-4} q^4 K^2(p)[(p-1)K^2(p) + 2(2-p)E(p)K(p)$$
$$- 3E^2(p)]. \quad (5.67)$$

Here $E(p)$ and $K(p)$ are complete elliptic integrals of the first and second kind. $E(x, p)$ is an incomplete elliptic integral, $am(x, p)$ is the Jacobi amplitude function, and $p = p(q)$ satisfies the following relation (Bar and Nepomnyashchy 1995):

$$\frac{\pi^2}{q^2 K^2(p)} = F(p, G_1) = \frac{f_0(p) + f_1(p) j(p) + f_2(p) j^2(p)}{f_3(p) j(p) + f_4(p)},$$

$$j(p) = E(p)/K(p), \quad f_0(p) = 4(p-1)[5(p^2 + 2p - 2)$$
$$+ G_1(2p^2 - 3p + 3)],$$

$$f_1(p) = 4(2-p)[5(p+1)(2p-1) + G_1(4p^2 - 5p + 5)],$$

$$f_2(p) = -28 G_1(p^2 - p + 1),$$

$$f_3(p) = 14(p^2 - p + 1), \quad f_4(p) = 7(2-p)(p-1). \quad (5.68)$$

Given a wave number q in the interval [0,1], Eqs. (5.67) and (5.68) completely define a periodic permanent wave propagating in the preferred direction. The speed of the wave, c_0 at leading order, is

$$c_0 = 4\pi^{-2} q^2 K(p)[(2-p)K(p) - 3E(p)]. \quad (5.69)$$

Note that in the frame of reference moving with the solidification front, the waves propagate in the preferred direction along the front with the group speed given by Eq. (5.53), which physically is $O(\epsilon)$, and c_0 is just a $O(\epsilon^3)$ correction to this speed. Besides, as a result of nonlinear evolution of the waves, the front itself moves in the laboratory frame with the *modified* speed, $V_0 - \epsilon^3 v_0$, with the nonlinear correction v_0 defined by Eq. (5.67).

The shape of the amplitude of the traveling cells, $A(q)$, is determined by the value of G_1. Linear stability analysis, carried out by Bar and Nepomnyashchy (1995), shows that these waves are stable in a certain interval of the wave numbers in the range $q < 1$ with respect to arbitrary bounded perturbations

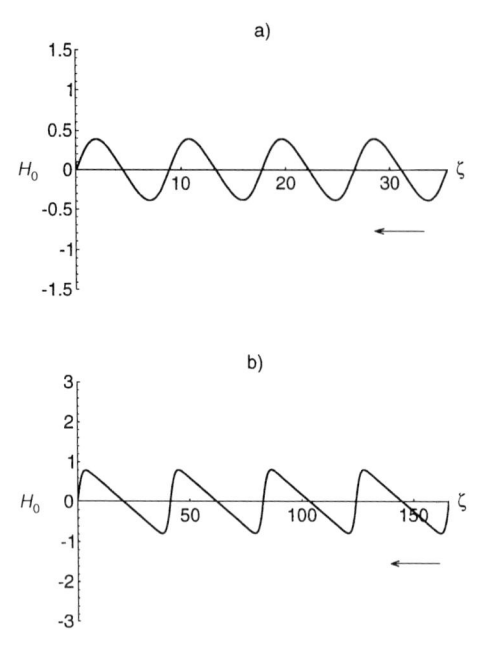

Figure 5.15. The function $H_0(\zeta)$ describing the permanent traveling wave for (a) $q = \sqrt{2}/2$, and (b) $q = 0.15$. The arrow shows the direction of propagation. From Golovin and Davis (1998). (Reprinted with permission of Elsevier Science B.V.)

in the ξ-direction only if $G_1 > -5/2$. Typical profiles of the shape of the front, $H_0(\zeta)$, corresponding to different q are shown in Figure 5.15. The larger the wavelength, the more asymmetric the cells are. These asymmetric cells can propagate on the solidification front both to the left and to the right along the preferred direction given by the angle ψ_0, depending on the sign of the coefficient a_2 in Eq. (5.62) and on the sign of the phase velocity given by Eq. (5.53); the signs of these parameters are determined by the kinetic anisotropy coefficients and by the direction of the crystal growth as well. The asymmetric cells propagating on the front during solidification lead to the formation of the *tilted cells* shown in Figure 5.16.

The wave amplitude, $A(q)$, its "correction" velocity $c_0(q)$, and the nonlinear correction for the velocity of the solidification front $v_0(q)$ are related to the wave number q. The larger the wavelength, the greater is the amplitude (see Figure 5.15). When $q \to 1$, the traveling cells are nearly sinusoidal near the bifurcation point, with $A(q) \to 0$ and $c_0 \to -1$; the case $q \to 0$ corresponds to a kink with $A(q) \to 2\sqrt{c_0}$ and $c_0 \to 7/(5 + 2G_1)$. However, the waves in both of these limiting cases are unstable (Bar and Nepomnyashchy 1995, Rednikov et al. 1995).

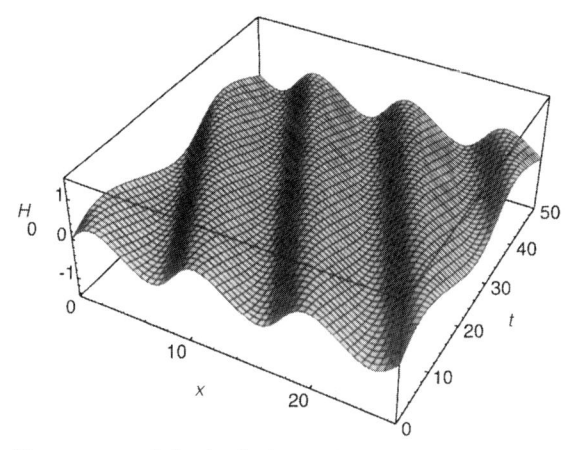

5.4 Unconstrained growth with "small" anisotropy

Figure 5.16. The structure of tilted cells formed in the nonlinear regime by the solidification front for $q = \sqrt{2}/2$. From Golovin and Davis (1998). (Reprinted with permission of Elsevier Science B.V.)

These stability analyses are related to perturbations in the preferred X-direction only. In the present case, one has to study the stability of traveling cells also with respect to transverse perturbations. This problem is addressed in Section 6 of Golovin and Davis (1998), who show that the instability is controlled by the two-dimensional case; three-dimensional instabilities do not change the stability criterion.

The analysis already presented allows one to predict in what direction of the crystal growth and for what values of the anisotropy coefficients propagating cells can be observed at the solidification front. The traveling cells can be observed in the regions of parameters where they are stable, that is, when $G_1 > -5/2$. For the *one-dimensional* front, the computation of G_1, below Eq. (5.66), requires the computation of the dimensionless anisotropy coefficients μ_{x2} and γ_{x3} that are

$$\mu_{x2} = \frac{-8\beta_4 \cos 4\phi}{1 + \beta_4 \cos 4\phi}, \quad \gamma_{x3} = -10\alpha_4 \sin 4\phi, \tag{5.70}$$

(where ϕ is the angle characterizing the direction of the crystal growth). Figure 5.11 of Golovin and Davis (1998) shows regions in the parametric plane (ϕ, β_4) at different values of the surface tension anisotropy coefficient α_4, where the traveling cells can be observed. The regions where the traveling cells are unstable are bounded by the loci of $a_2 = 0$ and are given by

$$\cos 4\phi = \beta_4^{-1}\left[3/5 - \sqrt{32/15}(1/5 + \beta_4^2)^{1/2}\right]. \tag{5.71}$$

If $0 < |\beta_4| < 1/17$, stable traveling cells exist at any direction of a crystal growth. If $1/17 < |\beta_4| < 1$, there exist certain intervals of the directions of a crystal growth at which traveling cells are unstable and where the instability of a one-dimensional solidification front results in a more complex behavior. When $\beta_4 \to 0$, the traveling cells are still stable, but their velocity tends to zero, and thus at $\beta_4 = 0$ traveling cells are replaced by stationary cells, which are described by a Kuramoto–Sivashinsky equation.

For a two-dimensional solidification front, the parameter G_1 is computed using the expressions for anisotropy coefficients given in Golovin and Davis (1998). One can exhibit regions in the parametric plane of spherical angles (ϕ, θ) defining the crystal growth direction at which the traveling cells are stable (at fixed particular values of the primed anisotropy coefficients).

Characteristics of traveling cells resulting from the morphological instability of a rapid solidification front depend on the degree of supercriticality (the distance from the instability threshold), the direction of the crystal growth, and the amplitude of the wave, which is the function of its wave number. If one supposes that the system selects a most rapidly growing mode of linear theory with the *dimensionless* wave number $q = \sqrt{2}/2$, then, in the simplest case of one-dimensional solidification fronts, *dimensional* characteristics of the waves – their speed v (the phase and the group speeds are the same in one-dimensional case), the wavelength λ, and the amplitude A can, in the main approximation, be expressed as follows:

$$v = \frac{\Delta T}{\bar{\mu}_0} W(\phi), \quad \lambda = \frac{\chi \bar{\mu}_0}{\Delta T} \tilde{\epsilon}^{-1} \Lambda(\phi), \quad A = \frac{\chi \bar{\mu}_0}{\Delta T} \tilde{\epsilon}^{-1} Q(\phi), \tag{5.72}$$

and

$$\tilde{\epsilon}^2 = \frac{S - S_0}{S_0} = \epsilon^2 \frac{S_2}{S_0}, \quad W(\phi) = \frac{4 V_0(\phi) \beta_4 \sin 4\phi}{(1 + \beta_4 \cos 4\phi)^2},$$

$$\Lambda(\phi) = \frac{2\pi \sqrt{2}}{\sqrt{S_0(\phi)}} (1 + \beta_4 \cos 4\phi) \sqrt{\frac{a_6}{a_3}},$$

$$Q(\phi) = 3\sqrt{S_0(\phi)}(1 + \beta_4 \cos 4\phi) \frac{a_1}{a_2} \sqrt{\frac{a_3}{a_6}} A\left(1/\sqrt{2}\right). \tag{5.73}$$

A plot of $Q(\phi)$ (Golovin and Davis 1998, Figure 13) for $q = \sqrt{2}/2$ shows that the cells are unstable in certain intervals of ϕ, depending on α_4 and β_4; this is where $G_1 < -5/2$.

When there is no kinetic anisotropy and only anisotropy of surface tension is present, the morphological instability of a rapidly solidifying front is monotonic and yields the formation of a pattern of stationary cells (rolls) with the wave

5.4 Unconstrained growth with "small" anisotropy

vector oriented at the solidification front in the direction given by Eq. (5.57). These cells are described by a stationary periodic solution of the anisotropic Kuramoto–Sivashinsky Eq. (5.64) that can be written in a rescaled form as

$$h_T + h_{XX} - h_{YY} + h_X^2 + h_{XXXX} = 0 \tag{5.74}$$

with the physical $h = O(\epsilon^2)$, $T = O(\epsilon^4)$, and the coordinates X and Y defined along and normal to the preferred direction of the instability, respectively. Equation (5.74) is a particular case of the following more general anisotropic Kuramoto–Sivashinsky equation of

$$h_T + h_{XX} + \alpha h_{YY} + \nabla_1^4 h + h_X^2 + \beta h_Y^2 = 0 \tag{5.75}$$

studied by Chang, Demekhin, and Kopelevich (1993) and Rost and Krug (1995) as a model for nonlinear evolution of three-dimensional perturbations of a fluid film flowing down an inclined plane, nonlinear evolution of sputter-eroded surfaces in the course of an electron beam treatment, and epitaxial growth of a vicinal surface. Equation (5.75) transforms into Eq. (5.74) as $\alpha \to -\infty$ with the rescaling $Y \to Y/\sqrt{-\alpha}$.

It was shown in Chang et al. (1993) and Rost and Krug (1995) that, in spite of the apparently stabilizing diffusional term in the Y-direction, the one-dimensional solutions of Eq. (5.76) periodic in the X-direction can, owing to the nonlinearity, become unstable with respect to perturbations normal to the preferred direction. Numerical simulations (Kawahara and Toh 1988) show that this instability leads to formation of stripes whose widths are modulated in the Y-direction that slowly vary in space and time – often in a chaotic manner. Such modulated stripes can manifest themselves in experiment as a structure of strongly elongated cells ("sausages") with the *larger* side aligned *normal* to the preferred direction of morphological instability.

In summary, then, results have been obtained on the effects of anisotropy of surface tension and kinetic coefficients on the morphological instability of a solidification front rapidly propagating in a hypercooled melt in which the solidification process is controlled by kinetics. Because of the anisotropy, the threshold of the instability (the critical undercooling or the critical Stefan number) depends on the direction of the crystal growth and on the anisotropy coefficient. In the general case, when both anisotropies are present, the instability generates traveling cells (waves) that travel on the two-dimensional front in a preferred direction determined by the coefficients of anisotropy. In a particular case, when there is no kinetic anisotropy and only the surface tension anisotropy is present, the instability is monotonic and leads to the formation of stationary cells (that are also oriented on the front in the preferred direction. It should be noted that the anisotropy coefficients in Eqs. (5.49) *depend on the direction of the crystal growth.* For instance, when a crystal with cubic symmetry grows in one of the principal cyrstallographic directions, [100], [110], or [111] (or in the directions [10] and [11] in the two-dimensional case), $\mu_x = \mu_y = 0$, and the morphological instability is monotonic and will lead to formation of *stationary* cells in this case, not to traveling ones.

A weakly nonlinear analysis of the instability near the threshold shows that, in the general case, the evolution of the traveling cells is described by a dissipation-modified anisotropic KdV equation, Eq. (5.62). When there is no kinetic anisotropy, the nonlinear evolution of the instability is governed by anisotropic Kuramoto–Sivashinsky Eq. (5.63). In an "exotic" case, when the anisotropy coefficients satisfy relation (5.59), the morphological instability creates cells traveling along the solidification front with the group velocity normal to the wave vector. These cells are described by Eq. (5.64), which is a Kuramoto–Sivashinsky equation "in the preferred direction," but one that encompasses dispersion and diffusional damping in the direction normal to the preferred one.

In the general case when both anisotropies are present, the dissipation-modified anisotropic KdV equation has the solutions in the form of cells that travel at constant speed. Propagation of the cells along the front changes the speed of the front itself.

Stability analysis shows that these waves are stable only if the nonlinear coefficient in Eq. (5.67) G_1 is greater than $-5/2$ and if the wave number is in a certain finite interval depending on G_1. This interval does not include the zero wave number; hence, the *solitary wave,* corresponding to a single "step" propagating along the solidification front, is *unstable*. The stability condition imposes restrictions on the values of the anisotropy coefficients as well as on the directions of the crystal growth at which these cells can be observed in experiments. Owing to anisotropy, the characteristics of stable traveling cells depend on the direction of the crystal growth. If the traveling cells are unstable, one would expect more complex wavy patterns to appear such as wave trains or chaotic wavy patterns.

In the particular case when only surface tension anisotropy is present, the nonlinear evolution of stationary cells governed by the anisotropic Kuramoto–Sivashinsky equation can lead to formation of a spatially irregular structure of strongly elongated cells ("sausages") whose long sides are aligned normal to the preferred direction of the morphological instability.

It is important to say that the analysis presented is valid only if the anisotropy of surface tension (the coefficients γ_0', γ_2', γ_4' for a three-dimensional cubic crystal or the coefficient α_4 for a two-dimensional cubic crystal growing in a Hele–Shaw cell) is smaller than a certain threshold; in the two-dimensional case with fourfold symmetry, $|\alpha_4| < 1/15$. Otherwise, the weighted surface energy becomes negative, and the solidification front becomes unstable at all values of undercooling. It can be shown using the dispersion relation that in this case the linear growth rate of the perturbations with the wave vector a is $\sigma = |\Gamma| a^2$, and there is no shortwave cutoff in the linear terms. This situation is believed to cause the formation of facets and corners that cannot be described by this weakly nonlinear theory but will be discussed in the next section.

5.5 Unconstrained Growth with "Large" Anisotropy – One-Dimensional Interfaces

In the previous discussion of anisotropic surface tension, the anisotropy was "small" in the sense that $\mathcal{F} \equiv \gamma + \gamma_{\theta\theta} > 0$ everywhere, and hence the capillary-undercooling effect is well defined. In this section it will be assumed that $\mathcal{F} < 0$ for certain angles, which will lead in the case of static equilibrium to solid particles with missing orientations. The figures will have sharp corners, and their shapes will be describable by the Wulff construction (Wulff 1901, Herring 1951), as discussed in Section 5.1.

When the interface is not in equilibrium but is evolving with time, the condition $\mathcal{F} < 0$ somewhere makes the evolution ill-conditioned; the coefficient of the curvature in the Gibbs–Thomson equation changes sign. One must "regularize" the model in some way, the most natural means being the introduction of new physics by endowing the edges with "line tension," which can be ascribed to the dynamics of steps (Nozieres 1992).

The energy of steps is usually considered to be a function of the step density, the latter depending on the orientation of the crystal surface with respect to the principal crystal direction (x-axis); see Figure 5.17. If λ is the crystal lattice spacing, and d is the step length, then for the density of steps $n(x)$ and their energy $E(n)$ for a unit length in the x-direction one has (Nozieres 1992)

$$n\lambda = \frac{\lambda}{d} = \tan\theta = h_x, \quad E(n) = \frac{\gamma}{\cos\theta} = \gamma\sqrt{1 + h_x^2}. \qquad (5.76)$$

Equations (5.76) are valid for vicinal surfaces when $|\theta|$ is small and the steps are well-defined (Nozieres 1992). In the corner region, where the mean curvature H is high, the orientation of the surface varies very rapidly, leading to a large variation of $n(x)$; in fact

$$\frac{dn}{dx} = \frac{1}{\lambda} h_{xx} = \frac{1}{\lambda} \frac{2H}{\cos^3\theta} \qquad (5.77)$$

(the curvature is positive for a convex crystal surface). In this case, when $|n_x|$

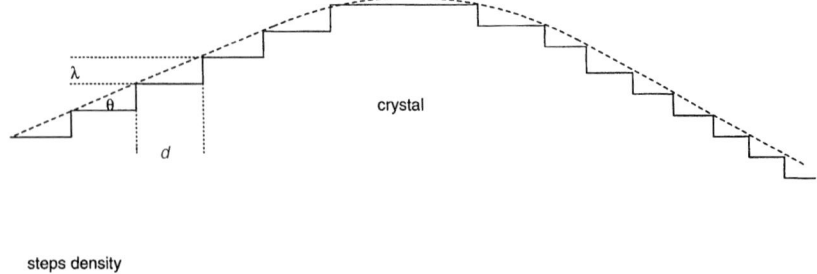

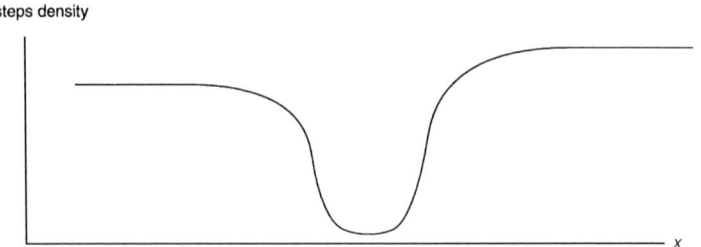

Figure 5.17. Steps on the crystal surface in the corner region where the gradient of the steps density is large. From Golovin et al. (1998). (Reprinted with permission of Elsevier Science B.V.)

is very large, it is natural to suggest that the step energy is a function of not only n but also of n_x. For a repulsive interaction of steps one can write

$$E(n, n_x) = E(n, 0) + \frac{1}{2}\frac{\partial^2 E(n, 0)}{\partial n_x^2} n_x^2 + \cdots = E_0(n) + \frac{1}{2}v(n)n_x^2 + \cdots, \quad (5.78)$$

where $v(n) > 0$, and $E_0(n)$ is the step energy corresponding to a plane (low curvature) surface. For small angles θ one can neglect the dependence $v(n)$ in (5.78) and, neglecting higher-order terms, set $v(n) \approx v(0) \equiv v_0$, and write

$$E(n, n_x) = E_0(n) + \frac{1}{2}v_0 n_x^2. \quad (5.79)$$

Equation (5.79) means that an additional energy is required to form a corner between two plane surfaces with different orientations (Stewart and Goldenfeld 1992). This additional energy is the "surface tension" between two different "phases," the two crystal facets with different orientations. Thus, step interaction in the corner region yields the dependence of the crystal surface tension on curvature, which, for small angles θ, can be written as

$$\gamma \approx \gamma_0(\theta) + \frac{v_0}{2\lambda^2}(2H)^2, \quad (5.80)$$

where $\gamma_0(\theta)$ is the anisotropic surface tension corresponding to a plane (or low curvature) crystal surface. It should be noted that this increase of the surface tension with curvature in the corner region is in contrast to the decrease of the

5.5 Unconstrained growth with "large" anisotropy

surface tension of a very small liquid droplet when the surface-layer thickness becomes comparable with the droplet radius (Rowlinson and Widom 1989).

The equilibrium at the crystal surface can be viewed as an equilibrium of a system of steps (Nozieres 1992). The "chemical potential" μ_s of such nonuniform system of steps can be computed as $\mu_s = \delta F/\delta n$, where $F = \int E(n, n_x)dx$ is the free energy of a surface with a nonuniform distribution of steps (Cahn and Hilliard 1958). Using Eq. (5.79), one gets for the chemical potential per unit length in the x-direction

$$\bar{\mu}_s = \frac{dE_0}{dn} - v_0 n_{xx}. \tag{5.81}$$

In equilibrium, the thermodynamic force caused by the repulsion of steps, $-d\bar{\mu}_s/dx$, should be balanced by a "supercooling" force, $-\lambda L_v \delta T/T_m$, that makes the steps move in such a way that the surface would grow; here $\delta T = T_e - T_m$ is the departure of the equilibrium temperature T_e corresponding to a curved surface from the equilibrium melting temperature of the plane surface T_m. Using Eqs. (5.76) and (5.77), as well as $h_x = \tan\theta$, one obtains, as an equilibrium condition, the following modified Gibbs–Thomson relation:

$$\frac{(T_m - T_e)}{T_m} L_v = -\frac{1}{\lambda}\frac{d^2 E_0}{dn^2} n_x + \frac{v_0}{\lambda} n_{xxx} = -\frac{\partial^2 \left[\gamma_0(h_x)\sqrt{1+h_x^2}\right]}{\partial h_x^2} h_{xx} + \frac{v_0}{\lambda^2} h_{xxxx}. \tag{5.82}$$

The modified Gibbs–Thomson relation (5.82) can be also derived, if, as it is found in Mullins (1963), one considers a variation δh of the crystal surface in equilibrium that conserves the volume, so that $\int_0^l \delta h\, dx = 0$. In this case, the variation of the surface free energy,

$$F = \int_0^l \left[E_0(n) + \frac{1}{2} v_0 n_x^2\right] dx = \int_0^l \gamma_0(h_x)\sqrt{1+h_x^2}\, dx + \frac{v_0}{2\lambda^2}\int_0^l h_{xx}^2\, dx.$$

is

$$\begin{aligned}\delta F &= \int_0^l \left[\frac{\partial\left(\gamma_0(h_x)\sqrt{1+h_x^2}\right)}{\partial h_x}\delta h_x + \frac{v_0}{\lambda^2} h_{xx}\delta h_{xx}\right] dx \\ &= \int_0^l \left[-\left(\frac{\partial^2\left[\gamma_0\sqrt{1+h_x^2}\right]}{\partial h_x^2}\right) h_{xx} + \frac{v_0}{\lambda^2} h_{xxxx}\right] \delta h\, dx \\ &\quad + \left[\left(\frac{\partial\left(\gamma_0\sqrt{1+h_x^2}\right)}{\partial h_x}\right) - \frac{v_0}{\lambda^2} h_{xxx}\right]\delta h\bigg|_0^l + \frac{v_0}{\lambda^2} h_{xx}\delta h_x\bigg|_0^l. \end{aligned} \tag{5.83}$$

In equilibrium, the sum of the variations of the surface free energy (Eq. (5.83)) and the bulk free energy must be zero. Because the volume is conserved and $\int_0^l \delta h\, dx = 0$, the variation of the bulk free energy is equal to the total change of the bulk chemical potential, which is $\int_0^l L_v(\delta T/T_m)\delta h\, dx$. (The bulk chemical potential plays the role of the Lagrange multiplier; Coriell and Sekerka 1976). Taking $\delta h(0) = \delta h(l) = 0$ and neglecting the term with h_{xx} at the boundaries of the corner region, one obtains the modified Gibbs–Thomson relation (5.82) valid for $|h_x| \ll 1$.

Consider now a planar interface moving at constant speed in a hypercooled melt. The kinetic undercooling is taken to be isotropic, and the expression for the fourfold symmetric surface energy $\mathcal{E}(h_x)$ is written for small $|h_x|$ as

$$\mathcal{E} = f_0(\theta) + f_1(\theta)h_x + f_2(\theta)h_x^2 + \cdots \qquad (5.84)$$

with

$$f_0 = 1 - 15\alpha_4 \cos 4\theta, \quad f_1 = 60\alpha_4 \sin 4\theta, \quad f_2 = \frac{3}{2}(95\alpha_4 \cos 4\theta - 1).$$

In scaled coordinates the governing system has the form

$$\Theta_t - V\Theta_z = \nabla^2 \Theta \quad \text{for} \quad z > h(x, t).$$

On $z = h(x, t)$

$$S(V + h_t) = -\Theta_z + h_x \Theta_x \qquad (5.85)$$

$$\Theta = 1 + \Gamma \mathcal{E}(h_x) h_{xx} - K h_{xxxx} - \frac{V + h_t}{(1 + h_x^2)^{1/2}}.$$

As $z \to \infty$,

$$C \to 0.$$

Notice that the Gibbs–Thomson relation now contains a fourth-order spatial derivative that regularizes the "large" anisotropy case where $\mathcal{E}$ has zeros. In the preceding

$$\mathcal{E}(h) = \frac{\partial^2}{\partial h_x^2}\left(\sigma_0 \sqrt{1 + h_x^2}\right) \qquad (5.86)$$

and

$$K = \nu_0 T_m / \lambda^2 \kappa \mu_0 L_v.$$

A weakly nonlinear evolution equation can be derived by introducing the long-scale $X = \epsilon x$, the hierarchy of slow times $T_2, T_3, T_4, \ldots$ such that $T_k = \epsilon^k t, k = 2, 3, \ldots$, and consider $h(X, T_2, T_3, T_4, \ldots) = O(1)$, $\Theta(Z, X, T_2, T_3, T_4, \ldots) = \Theta^{(0)}(Z, X, T_2, T_3, T_4, \ldots) + \epsilon^2 \Theta^{(2)}(Z, X, T_2, T_3, T_4, \ldots) +$

5.5 Unconstrained growth with "large" anisotropy

$\epsilon^3 \Theta^{(3)}(Z, X, T_2, T_3, T_4, \ldots) + \epsilon^4 \Theta^{(4)}(Z, X, T_2, T_3, T_4, \ldots) + \cdots$; functions $\Theta^{(i)}$ must decay at infinity; the expansion of the effective surface tension $\mathcal{E}(h_x)$ is given by Eqs. (5.84).

After a good deal of analysis, Golovin, Davis, and Nepomnyashchy (1998) found that, in reconstituted form,

$$h_t - \frac{1}{2}V h_x^2 + \frac{1}{8}V h_x^4 = (-\tilde{\mu} + \alpha h_x + \beta h_x^2)h_{xx} - (|s_4| + K)h_{xxxx}, \quad (5.87)$$

where

$$\tilde{\mu} = \frac{S}{V} - \Gamma f_0 = \frac{S}{V} - \Gamma(1 - 15\alpha_4 \cos 4\theta),$$

$$\alpha = \Gamma f_1 = 60\Gamma \alpha_4 \sin 4\theta, \quad (5.88)$$

$$\beta = \frac{S}{V} - \Gamma\left(\frac{1}{2}f_0 + f_2\right) = \frac{S}{V} + \Gamma(135\alpha_4 \cos 4\theta - 1),$$

and s_4 is defined by $\sigma \sim s_2 a^2 + s_4 a^4$ as $a \to 0$.

Notice that the regularization imposed by the interaction of steps in the corner region leads to the curvature-dependent surface tension, which augments the coefficient of the fourth-order term in Eq. (5.87). It does not alter the structure of the equation. The *form* of Eq. (5.87) has been derived by Angenent and Gurtin (1989) and Gurtin (1993).

Note that the mean value of h over one wavelength is nonzero because there is continuous crystal growth. If the effect of step interaction in the corner region is large enough, then Eq. (5.87) governs the dynamics of facet formation in hypercooled solidification. If the effect is small, as might be expected on physical grounds, the band of unstable wave numbers is not necessarily small, and we should consider this as a *model equation* for describing the process of the formation of corners and facets caused by anisotropic surface tension.

Equation (5.87) can be written in a canonical form by introducing new variables

$$\xi = \frac{3V}{\beta\sqrt{2}}x, \quad \tau = \frac{3V^2}{\beta}t, \quad q = h_x/\sqrt{2},$$

and finding the following equation for the *slope q* of the interface:

$$q_\tau - qq_\xi + q^3 q_\xi = (-mq + bq^2 + q^3 - \chi q_{\xi\xi})_{\xi\xi}, \quad (5.89)$$

where

$$m = \frac{3\tilde{\mu}}{2\beta}, \quad b = \frac{3\alpha}{2\sqrt{2\beta}}, \quad \chi = \frac{27V^2(|s_4| + K)}{4\beta^3}. \quad (5.90)$$

Equation (5.89) is essentially a *convective* Cahn–Hilliard equation 1958. The Cahn–Hilliard equation, absent the convective terms, describes the process of spinodal decomposition and has the following form:

$$u_\tau = (-u + bu^2 + u^3 - u_{\xi\xi})_{\xi\xi} = 0, \tag{5.91}$$

where u is the order parameter (e.g., the concentration of one of the components of an alloy). This equation has been intensively studied (see Novick-Cohen 1998 for the review). It has the solution (Novick-Cohen and Segel 1984)

$$u = -b/3 \pm (b^2/3 + 1)^{1/2} \tanh\left[(b^2/3 + 1)^{1/2}\xi/\sqrt{2}\right] \tag{5.92}$$

that describes the final stage of the spinodal decomposition, the formation of the interface between two thermodynamically stable states of an alloy with different concentrations. The form of Eq. (5.91) originates from the alloy double-well free energy $F(u)$ given by $F = -u^2/2 + bu^3/3 + u^4/4$ and also takes into account the stabilizing effect of the excess energy of the interface between the two separated phases (Novick-Cohen and Segel 1984). Solution (5.92) describes the interface between the two stable compositions of the alloy corresponding to the *binodal*, that is, the points of the bitangent to the graph $F(u)$, that give the values of the concentrations at which the chemical potentials of each component are equal in both phases.

With convective terms, $-qq_\xi + q^3 q_\xi$, Eq. (5.89) is similar to the so-called driven Cahn–Hilliard equation (Emmott and Bray 1996)

$$u_t - E(u^2)_x = (-u + u^3 - u_{xx})_{xx}, \tag{5.93}$$

that describes a spinodal decomposition in the presence of an external (e.g., gravitational or electric) field when the dependence of the mobility factor on the order parameter is important (Leung 1990 and Yeung et al. 1992). Equation (5.93) has an exact solution $u = u_\infty \tanh(u_\infty x/\sqrt{2})$ with $u_\infty^2 = 1 + \sqrt{2}E$ (Leung 1990, Emmott and Bray 1996) that shows the broken symmetry due to the convective term, leading to different amplitudes of kinks and antikinks (a kink has a positive derivative, and an antikink has a negative one).*

In the present case, the "order parameter" corresponds to the slope of the interface, and convective terms $-qq_\xi + q^3 q_\xi$ stem from the effect of kinetics (the finite rate of atoms or molecules attaching to the crystal surface) that

---

* By using another scaling, $\xi = [(|s_4| + K)/\mu]^{1/2} x$, $\tau = \left[(|s_4| + K)/\mu^2\right] t$, $q = [3\mu/\beta]^{1/2} h_x$, Eq. (5.87) can be written in the form $q_\tau - ((1/2)v_1 q^2 - (1/4)v_2 q^4)_\xi = (-q + bq^2 + q^3 - q_{\xi\xi})_{\xi\xi}$, where $v_1 = (V/\mu)[3|s_4| + K)/\beta]^{1/2}$, $v_2 = 3v_1\mu/\beta$, $b = (\alpha/2)[3/(\beta\mu)]^{1/2}$, that is analogous to Eq. (5.93). Thus, the convective terms are seen to be proportional to the speed V of the unperturbed front. The form of Eq. (5.89) is physically clearer and more convenient for numerical computations.

provides an independent flux of the order parameter similar to the effect of an external field in spinodal decomposition of a driven system. Equation (5.89) can be written in the conservation form

$$q_\tau + \left[-\frac{1}{2}q^2 + \frac{1}{4}q^4\right]_\xi + \left[-\frac{\partial G}{\partial q} + \chi q_{\xi\xi}\right]_{\xi\xi} = 0, \quad (5.94)$$

where the nonlinear term with the first derivative represents the flux of the "order parameter" caused by kinetics, and $G(q)$ is the *effective* double-well "free energy"

$$G = -\frac{1}{2}mq^2 + \frac{1}{3}bq^3 + \frac{1}{4}q^4. \quad (5.95)$$

The coefficients m and b, defining the shape of G, depend not only on the functions $f_0(\theta)$, $f_1(\theta)$, and $f_2(\theta)$ from the expansion of the anisotropic surface tension but also on the Stefan number S characterizing the undercooling, and hence on the front velocity $V = 1 - S$. This is because the local thermodynamic equilibrium is absent on the surface of a crystal whose rapid growth from the hypercooled melt is controlled by the attachment kinetics. The effective double-well "free energy" is symmetric when the direction of the crystallization front motion coincides with one of the principal crystal directions ($b = 0$), and it is asymmetric otherwise.

The similarity between the process of spinodal decomposition and the formation of facets from energetically unstable crystal surfaces was noticed long ago (Mullins 1961, Cabrera 1963, Cahn and Hoffman 1974) as well as in some recent works (Gurtin 1993; Carlo, Gurtin, and Podio-Guidugli 1992; Stewart and Goldenfeld 1992; Liu and Metiu 1993). It was shown that the slope of the equilibrium facets grown from the unstable surface corresponds to bitangent (binodal) points of the surface free energy in a complete analogy with the spinodal decomposition. However, when convective effects are present, this bitangent construction is destroyed (see also Leung 1990).

In Eq. (5.89), the parameter m measures the magnitude of the instability, and the coefficient b characterizes the anisotropic effect of the crystal growth direction. The relative effects of the attachment kinetics and of the surface tension dependence on curvature (the "line tension" between two facets) is characterized by the parameter χ.

Golovin et al. (1998) found a solution of Eq. (5.89) in the form

$$q = A + B \tanh\left[\frac{\phi B(\xi - c\tau)}{\sqrt{\chi}}\right] \quad (5.96)$$

corresponding to a kink moving with a constant speed c. Such a kink in the *slope* of a surface would correspond to two almost flat faces (facets) separated by a corner. By substituting Eq. (5.96), one finds the following equation for ϕ:

$$\phi^3 - \frac{1}{2}\phi - \frac{\sqrt{\chi}}{24} = 0, \qquad (5.97)$$

and the following relations for A and B:

$$A = -\frac{2b\phi}{6\phi + \sqrt{\chi}}, \quad B^2 = \frac{2m\phi + \sqrt{\chi}}{2(\phi + \sqrt{\chi}/3)} + \frac{2b^2\phi^2(6\phi - \sqrt{\chi})}{(6\phi + \sqrt{\chi})^2(\phi + \sqrt{\chi}/3)} \qquad (5.98)$$

(it is sufficient to take $B > 0$), and the speed of the kink motion c:

$$c = A(A^2 + B^2 - 1). \qquad (5.99)$$

The case when the coefficient χ is small, $\chi \ll 1$, corresponds to either small speed of the crystallization front or to weak interaction of steps in the corner region (i.e., weak surface-tension dependence on curvature). For $\chi \ll 1$, Eq. (5.97) has the following approximate solutions:

$$\phi_\pm = \pm\frac{1}{\sqrt{2}} + \frac{\sqrt{\chi}}{24} \mp \frac{\chi}{3 \cdot 2^{13/2}} + O(\chi^{3/2}), \qquad (5.100)$$

$$\phi_0 = -\frac{\sqrt{\chi}}{12} - \frac{\chi^{3/2}}{864} + O(\chi^{5/2}). \qquad (5.101)$$

Solutions $\phi_\pm$ for $\chi \ll 1$ defined by Eq. (5.100) correspond to kinks (ϕ_+) and antikinks (ϕ_-) described by Eqs. (5.98) and (5.99), and the slowly varying solution, Eq. (5.101) corresponding to ϕ_0, divides the regions of their attraction. Consider solutions $\phi_\pm$.

For $\chi \ll 1$,

$$A = -\frac{b}{3} \pm \frac{b}{9\sqrt{2}}\sqrt{\chi} - \frac{5b}{216}\chi + O(\chi^{3/2}), \qquad (5.102)$$

$$B = \sqrt{m + b^2/3} \pm \frac{9 - 5b^2 - 6m}{6\sqrt{18m + 6b^2}}\sqrt{\chi}$$

$$+ \frac{126m^2 + 189mb^2 - 135m + 40b^4 + 9b^2 - 81}{12\sqrt{2}(18m + 6b^2)^{3/2}}\chi + O(\chi^{3/2}), \qquad (5.103)$$

and the speed of the kinks c is

$$c = \frac{b}{3}\left(1 - m - \frac{4}{9}b^2\right) \mp \frac{b}{27\sqrt{2}}(12 - 7b^2 - 9m)\sqrt{\chi}$$

$$+ \frac{b}{216}(44 - 33b^2 - 31m)\chi + O(\chi^{3/2}). \qquad (5.104)$$

In Eqs. (5.102)–(5.104), the upper signs correspond to kinks, and the lower ones correspond to antikinks.

The limiting values of the surface slopes at $\pm\infty$, $q_{\pm}^{(k,a)}$ are *different* for kinks (k) and antikinks (a), namely,

$$q_{\pm}^{(k)} = q_{\pm}^{(0)} + \sqrt{\chi}q_{\pm}^{(1)} + \chi q_{\pm}^{(2)} + O(\chi^{3/2}),$$
$$q_{\pm}^{(a)} = q_{\mp}^{(0)} - \sqrt{\chi}q_{\mp}^{(1)} + \chi q_{\mp}^{(2)} + O(\chi^{3/2}),$$
(5.105)

where

$$q_{\pm}^{(0)} = \pm\sqrt{m + b^2/3} - b/3, \quad q_{\pm}^{(1)} = \pm\frac{9 - 5b^2 - 6m}{6\sqrt{18m + 6b^2}} + \frac{b}{9\sqrt{2}},$$
$$q_{\pm}^{(2)} = \pm\frac{126m^2 + 189mb^2 - 135m + 40b^4 + 9b^2 - 81}{12\sqrt{2}(18m + 6b^2)^{3/2}} - \frac{5b}{216}.$$
(5.106)

By integrating Eq. (5.89) from $-\infty$ to ∞, one can easily show that the speed of the kinks and antikinks motion is related to $q_\pm$ by

$$c = \frac{1}{4}(q_+ + q_-)\left(q_+^2 + q_-^2 - 2\right).$$
(5.107)

The kink solution (5.96) for the slope of the crystal surface corresponds to the surface in the form of two plane facets connected by a rounded corner. The kink with $q_- < 0$, $q_+ > 0$ corresponds to a valley, and the antikink with $q_- > 0$, $q_+ < 0$ corresponds to a hill. The shape of the solid–liquid interface can easily be obtained by integrating Eq. (5.89) and solution (5.96) as

$$h = A(\xi - c\tau) + \frac{\sqrt{\chi}}{\phi}\ln\left[\cosh\left(\frac{\phi B(\xi - c\tau)}{\sqrt{\chi}}\right)\right] + C\tau,$$
(5.108)

where $C = (B^2 - A^2)(2 - 3A^2 - B^2)/4$ is the correction to the mean speed of the solidification front.

Figure 5.18 shows the formation of two facets connected by a rounded corner. The angle between the facets is $\varphi = \pi - \arctan\sqrt{2q_-} + \arctan\sqrt{2q_+}$, and the corner radius of curvature ρ is of the order of the kink transitional layer, $\rho \sim 2\pi\sqrt{2\chi/m}$, which gives the length of the most rapidly growing perturbation. The kinks (antikinks) are symmetric, and they correspond to symmetric valleys (hills) in the crystallization front shape, if $b = 0$, that is, if the direction of the crystal growth (orientation of the crystallization front) coincides with one of the principal crystal directions. Otherwise, the kinks and antikinks are asymmetric and correspond to valleys and hills whose sides have different slopes. Such asymmetric hills and valleys move along the solidification front

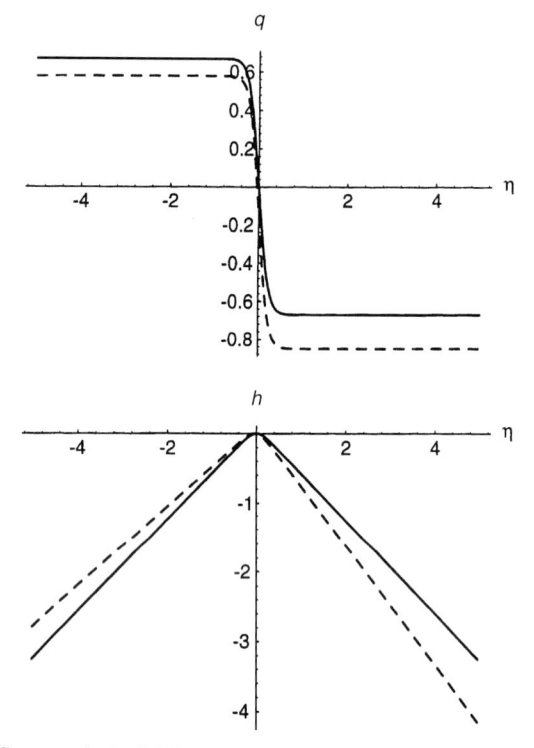

Figure 5.18. Symmetric (solid lines) and asymmetric (dashed lines) antikinks, $q(\eta)$, for $m = 0.5$, $b = 0.4$, $\chi = 0.01$ and the corresponding symmetric (solid lines) and asymmetric (dashed lines) hills that are formed by two facets divided by a rounded corner, (a) $q(\eta)$ and (b) $h(\eta)$. From Golovin et al. (1998). (Reprinted with permission of Elsevier Science B.V.)

with the speeds given by Eqs. (5.99) and (5.104). Note that in the dimensional form this speed is proportional to the speed of the solidification front V.

The kink solutions $q(\eta')$, $\eta' = \xi - c\tau$, correspond to the "phase separated" state of a "two-component system" in which the "order parameter" is the slope of the crystal surface q and in which "effective free energy" is given by Eq. (5.95). When $\chi \to 0$ (infinitely sharp corner), the slopes of the facets in the main approximation, $q_{\pm}^{(0)}$ given by Eq. (5.106), correspond, as in spinodal decomposition, to the points of the binodal. The points where $\partial^2 G/\partial q^2 = -m + 2bq + 3q^2 = 0$ correspond to the slopes $q_{s\pm} = \pm\sqrt{m/3 + b^2/9} - b/3$ that belong to the *spinodal*. Any planar crystallization front with the slope (with respect to [10] direction) in the interval $[q_{s-}, q_{s+}]$, where $\partial^2 G/\partial q^2 < 0$, is unstable with respect to infinitesimal fluctuations of the crystal shape, which can easily be seen from the dispersion relation that can be obtained by linearizing

Eq. (5.94). This is the instability that, in the case of spinodal decomposition, yields the decay of a binary alloy into two phases, and different concentrations of the alloy components correspond to the stable binodal region. In the case of a growing crystal, this instability yields the decay of the surface with an unstable orientation, corresponding to the "spinodal" region, into the facets with two stable orientations that belong to the "binodal."

The convective terms in Eq. (5.89) break the mirror symmetry of the Cahn–Hilliard equation. This leads to the motion of an asymmetric "phase-separated" kink corresponding to two facets with different stable orientations in the directions prescribed by the kinetics, that is, a hill moves in the direction *from* the smaller to larger slope, and the valley moves in the direction *from* the larger *to* the smaller slope. It should be noted that Eq. (5.104) is valid only for $m < 1 - 4b^2/9$. Otherwise, the amplitude of the kink (the slopes of the facets) is so large the two-term longwave expansion of the projection of the normal velocity that has been used leads to an artifact. Figure 5.19 shows the result of the numerical solution of Eq. (5.89) on the ξ-interval $[-l, l]$, $l \gg \rho$, by means of a finite difference Crank–Nickolson scheme with the boundary conditions $q_\xi = q_{\xi\xi\xi} = 0$ at the ends of the interval appropriate to the simulating of the Cahn–Hilliard equation (Bates and Fife 1993). Both the numerical solution $q(\xi, t)$ for the surface slope and the corresponding surface shape $h(\xi, t) = \int_{-l}^{\xi} q\, d\xi$ are shown. The dotted horizontal lines show the values of $q_\pm^{(0)}$ given by Eq. (5.106). The velocity of the kink motion computed from the numerical solution of Eq. (5.89) is in a good agreement with Eq. (5.104).

Besides, owing to the mirror symmetry broken by the convective terms in Eq. (5.99), the kinks and antikinks cannot be obtained one from the other by changing $\xi \to -\xi$. As can be seen from Eqs. (5.105) and (5.106), the convective effect of kinetics makes kinks steeper than antikinks. The amplitude of the kink is larger than the binodal value, and the amplitude of the antikink is smaller than the binodal value. Therefore, the corner of a valley formed as a result of the instability of the crystal surface is sharper than the corner of a hill. This difference between the amplitudes of the kinks and antikinks can be also in Figure 5.19 where the dotted horizontal lines show the binodal values of the surface slope, $q_\pm^{(0)}$. Thus, owing to the convective effect of kinetics, the slopes of the stable surfaces resulting from the "spinodal decomposition" of the unstable surface depart from the binodal. Using the definition of the small parameter χ, it is easy to see from Eqs. (5.105) and (5.106) that these deviations grow with the speed of the front V.

Such an effect of kinetics is natural because the evolution of a curve whose points move with the same constant normal velocity tends to produce caustics

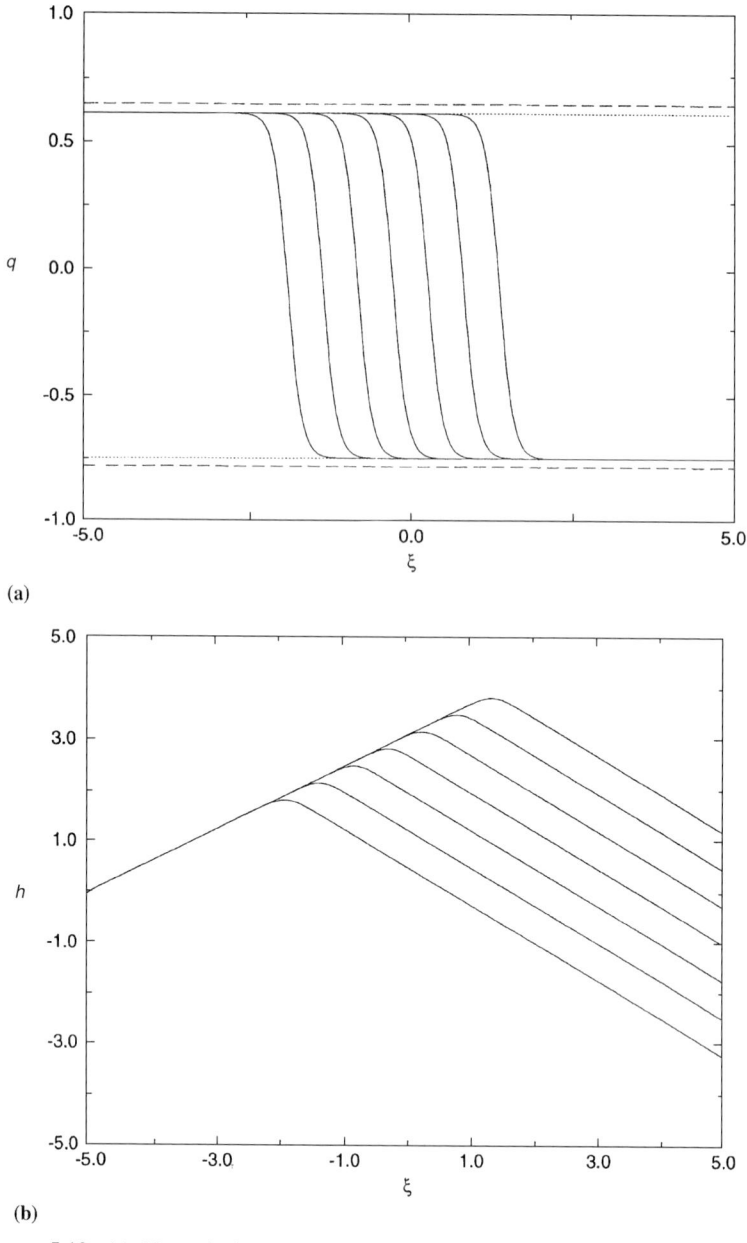

Figure 5.19. (a) Numerical solutions for slope $q(\xi, t)$ with the no-flux boundary conditions in the form of asymmetric antikinks moving from left to right, $m = 0.5$, $\chi = 0.01$, $b = 0.2$. Dashed horizontal lines correspond to the binodal slopes; dotted horizontal lines correspond to analytical solution. (b) Asymmetric hill, $h(\xi, t) = \int_{-1}^{\xi} q \, d\xi$, moving from left to right corresponding to the numerical solution shown in (a). From Golovin et al. (1998). (Reprinted with permission of Elsevier Science B.V.)

5.5 Unconstrained growth with "large" anisotropy

in the case of a concave curve and tends to reduce the curvature when the curve is convex.

Equation (5.89) is nonlinear, and its solution in a finite domain is very close to the analytical solution of Eq. (5.96) if the width of the interfacial region is small in comparison with the width of the computational domain, though this solution is not necessarily unique. As shown by Emmott and Bray (1996) for Eq. (5.93), if the symmetry conditions are used, $q_{xx}(0) = 0$, $q(l) = -q(-l)$, the antikink is unique, but there is a family of kinks parameterized by the value $q(l)$. In the computations shown above, the boundary conditions used were $q_x = q_{xxx} = 0$ at $x = \pm l$, and these ensure the uniqueness of the solutions.

In order to study the dynamics of the formation of facets on the surface of a crystal growing in a hypercooled melt, a numerical solution of Eq. (5.89) was obtained by means of a pseudospectral code with periodic boundary conditions. Figure 5.20 shows different stages of the instability of a plane crystal surface

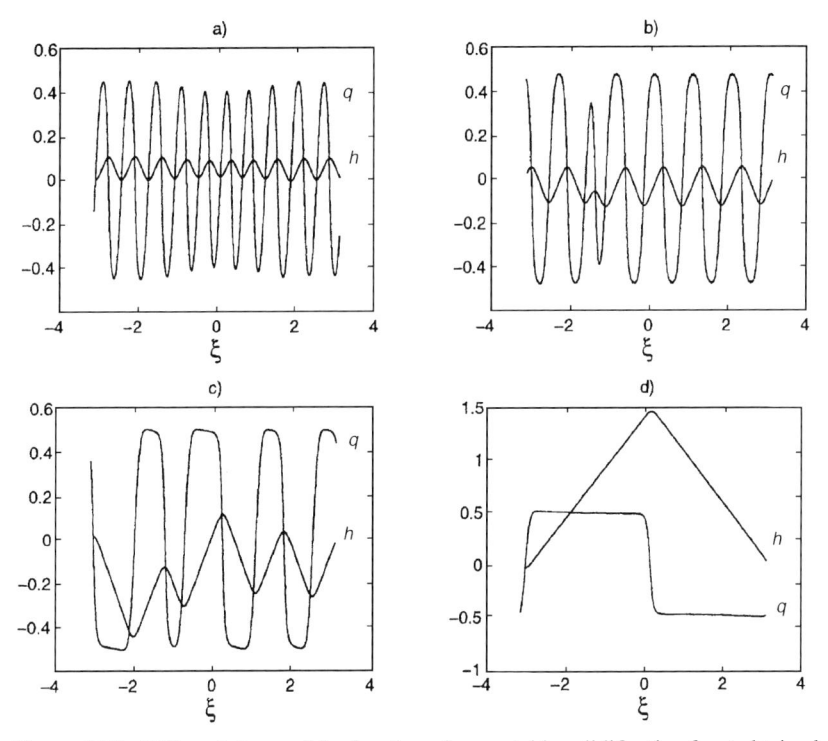

Figure 5.20. Different stages of the faceting of an unstable solidification front obtained by numerical solution for q and h with periodic boundary conditions, $m = 0.25$, $\chi = 0.001$, and $b = 0$. (a) Formation of periodic structure of hills and valleys, (b) and (c) the coarsening process, (d) final state; two facets separated by a rounded corner. From Golovin et al. (1998). (Reprinted with permission of Elsevier Science B.V.)

when the direction of the growth coincides with one of the principal crystal directions, and thus in Eq. (5.90) $b = 0$. Both the numerical solution $q(\xi, t)$ for the surface slope and the corresponding surface shape $h(\xi, t) = \int q \, d\xi$, are shown. Starting from small random initial perturbations, the system first develops into a spatially periodic hill-and-valley structure ("teeth") (Figure 5.20(a)) with the spatial period corresponding to the wavelength of the most rapidly growing mode. Each tooth consists of two facets corresponding to the wavelength of the most rapidly growing mode. Each tooth of this structure consists of two facets corresponding to the stable orientation close to the "binodal" one. In the course of further evolution the system exhibits coarsening: small teeth merge (Figure 5.20(b)) and form larger ones with larger spatial period (Figure 5.20(c)). This corresponds to the formation of several larger kinks of the surface slope. In a system with periodic boundary conditions, the final state is a single kink–antikink pair corresponding to a periodic structure of hills and valleys composed of facets with stable orientations with the period equal to the period of the computation domain (Figure 5.20(d)).

If the crystal growth direction differs from one of the crystal principal directions ($b \neq 0$), the "teeth" resulting from the instability will be tilted, and the whole hill-and-valley system will move along the crystal surface. In the course of this drift of the whole structure, the coarsening proceeds, and thus finally there will be one moving hill and valley.

It is important to know how the convective effect of kinetics influences the *rate of coarsening*. Usually, when the coarsening process is governed by diffusion and capillarity, as described by the Cahn–Hilliard equation, the rates of coarsening in two-dimensional and three-dimensional problems obey the well-known Lifshitz–Slyozov–Wagner law (Lifshitz and Slyozov 1961, Wagner 1961), namely, at the late stage of coarsening, the mean scale of the new phase growing from the other, larger phase (e.g., liquid droplets growing from their vapor) increases as $t^{1/3}$, where t is time (see Shinozaki and Oono 1993, Küpper and Masbaum 1994 and references therein). In one dimension, the solutions of the Cahn–Hilliard equation exhibit logarithmically slow rates of coarsening (see Kawakatsu and Munakata 1985; Alikakos, Bates, and Fusco 1991; Bates and Xun 1994, 1995).

However, the coarsening can go much faster if the flux of the order parameter is governed by some independent velocity scale present in the system (e.g., resulting from hydrodynamic effects); in this case the mean scale of the new phase can increase $\propto t^n$, $1/3 < n \leq 1$ (Siggia 1979; Tanaka 1997; Ratke and Thieringer 1985; Wan and Sahm 1990; Akaiwa, Hardy, and Voorhees 1991). In the present case the kinetics controlling the crystal growth from the hypercooled melt sets an independent velocity scale for the order parameter, the surface slope, and this leads to much faster coarsening dynamics. In a one-dimensional case the initial stage the coarsening goes like $\sim t$, and at the last stage the mean scale of the system increases $\sim t^{1/2}$. This result coincides with the results of the numerical simulation of a model dynamical system introduced in Emmott and Bray (1996) for governing the interaction of kinks in the driven Cahn–Hilliard equation (Eq. (5.93)). This coarsening dynamics implies that the interaction of kinks in Eq. (5.89) at the late stages, when the distance between the kinks is large, leads to their attractive motion with the speed inversely proportional to the distance between the kinks. At the initial stage, when the distance between the kinks is small, the speed of their attractive motion probably tends to a constant value as is suggested by the coarsening $\sim t$ at this stage. However, a separate investigation of the interaction of the kinks is required in the spirit of the theory developed in Kawasaki and Ohta (1982).

5.6 Unconstrained Growth with "Large" Anisotropy – Two-Dimensional Interfaces

Consider now the case of three-dimensional crystal growth in which $h = h(x, y, t)$. Golovin et al. (1999) obtained the evolution equation

$$h_t \left(1 + |\nabla h|^2\right)^{-1/2} = \mu \left[\Delta(h) + \mathcal{E}(h)\right] \tag{5.109}$$

where $\mathcal{E}$ given by

$$\mathcal{E}(h) = \frac{\partial^2 I}{\partial (h_x)^2} h_{xx} + 2 \frac{\partial^2 I}{\partial h_x \partial h_y} h_{xy} + \frac{\partial^2 I}{\partial (h_y)^2} h_{yy} - \frac{\partial^2 I}{\partial (h_{xx})^2} h_{xxxx}$$

$$- 2 \frac{\partial^2 I}{\partial h_{xy} \partial h_{xx}} h_{xxxy} - \left(\frac{\partial^2 I}{\partial (h_{xy})^2} + 2 \frac{\partial^2 I}{\partial h_{xx} \partial h_{yy}}\right) h_{xxyy}$$

$$- 2 \frac{\partial^2 I}{\partial h_{xy} \partial h_{yy}} h_{xyyy} - \frac{\partial^2 I}{\partial (h_{yy})^2} h_{yyyy}, \tag{5.110}$$

where

$$I = \gamma \sqrt{1 + |\nabla h|^2}, \tag{5.111}$$

and Δ represents the *driving force* of the crystal growth. Golovin et al. (1999) considered two cases, (1) Δ is a constant that holds when the driving force is independent of the local temperature field, and (2) Δ is *local* in the sense that the boundary-layer thickness is small compared with the curvature of the crystal. When the driving force is local, then $\Delta(h)$ would be represented by a nonlinear differential operator. When it is nonlocal, it is represented by a nonlinear integro–differential operator.

For constant driving force, $\Delta(h) = \Delta_0$, and for a cubic crystal

$$\gamma = \gamma_0 \left[1 + \alpha_4 \left(n_x^4 + n_y^4 + n_z^4\right) + \alpha_6 \left(n_x^6 + n_y^6 + n_z^6\right) + \cdots\right] + \frac{1}{2}\delta(2H)^2, \tag{5.112}$$

where n_x, n_y, n_z are the coordinates of the local unit normal to the crystal surface in the frame of reference of the principal directions [100], [101], and [001] that can be expressed in terms of the angular spherical coordinates θ and ϕ as $n_x = \sin\theta\cos\phi$, $n_y = \sin\theta\sin\phi$, $n_z = \cos\theta$, γ_0 is a constant characterizing mean surface tension of the crystal, $\alpha_4, \alpha_6, \ldots$ are the coefficients of the surface tension anisotropy, and $\delta = const > 0$. Consider a planar crystal surface growing with the orientation characterized by a normal vector $\mathbf{e}_z = (\sin\theta\cos\phi, \sin\theta\sin\phi, \cos\theta)$ and choose the other two basis vectors in the surface plane to be $\mathbf{e}_x = \mathbf{e}_z \times \mathbf{e}_z^0 / \left|\mathbf{e}_z \times \mathbf{e}_z^0\right|$, $\mathbf{e}_y = \mathbf{e}_z \times \mathbf{e}_x / |\mathbf{e}_z \times \mathbf{e}_x|$, where $\mathbf{e}_z^0$ is aligned along the [001] direction. In the coordinate system $\{\mathbf{e}_x, \mathbf{e}_y, \mathbf{e}_z\}$ the position of the crystal surface is given by $z = h(x, y, t)$. The weighed surface tension I defined by Eq. (5.2) with $\lambda = 0$ can be expanded for $|h_x| \ll 1$, $|h_y| \ll 1$ as

$$I(h_x, h_y, h_{xx}, h_{xy}, h_{yy}) = \gamma_0 \sum_{m,n=0}^{\infty} e_{mn}(\alpha_4, \alpha_6, \theta, \phi) h_x^m h_y^n$$

$$+ \frac{1}{2}\delta \left(h_{xx}^2 + 2h_{xx}h_{yy} + h_{yy}^2\right) + \cdots. \tag{5.113}$$

Here the coefficients e_{ij} are the functions of the direction of the crystal growth and of the anisotropy coefficients which were computed in Golovin and Davis (1998).

After the transformation $t \to t/(\kappa\Delta_0)$, Eq. (5.109) reads

$$h_t = \sqrt{1 + |\nabla h|^2}[1 + \Gamma E(h)], \tag{5.114}$$

where $\Gamma = \gamma_0/\Delta_0$.

Equation (5.114) has a solution corresponding to a planar, uniformly growing surface (uniformly propagating crystallization front) $h = t$ (the dimensional speed of the front is $\kappa\Delta_0$). Transform to the frame moving with the planar front, change the variable $h \to h + t$, and expand Eq. (5.114) for $|h_t| \ll 1$, $|\nabla h| \ll 1$.

5.6 Unconstrained growth with "large" anisotropy

Keeping the terms up to the second order, one obtains

$$h_t - \frac{1}{2}|\nabla h|^2 - \Gamma \mathcal{E}_2(h) = 0, \tag{5.115}$$

where $\mathcal{E}_2$ is the linear differential operator

$$\mathcal{E}_2 = \nabla A \nabla^T, \quad A = \begin{pmatrix} 2e_{20} & e_{11} \\ e_{11} & 2e_{02} \end{pmatrix}. \tag{5.116}$$

Equation (5.115) is an anisotropic Burgers equation. It can also be called the anisotropic, deterministic Kardar–Parisi–Zhang (AKPZ) equation (Wolf 1991; Barabasi, Araujo, and Stanley 1992). With noise added, this equation describes kinetic roughening of vicinal surfaces (Wolf 1991) and epitaxial growth (see Krug 1997 for review). If the eigenvalues of the anisotropy matrix A are positive, Eq. (5.116) describes a hill-and-valley structure propagating along the interface with the amplitude determined by initial conditions (Fogedby 1998). In this case the anisotropic surface tension prevents the formation of sharp corners in the course of kinetically controlled crystal growth (Uwaha 1987, Yokohama and Sekerka 1992).

When at least one of the eigenvalues of A is negative, the crystal surface with the orientation characterized by the anisotropy matrix A is *thermodynamically unstable*. This instability results in the formation of facets and corners. However, in this case Eq. (5.115) is ill-posed because it does not have a shortwave cutoff. In order to describe the dynamics of the formation of facets and corners, one has to keep higher-order terms in the evolution equation for the crystal surface.

Keeping the terms up to the fourth order, one obtains for the shape of the crystal surface in the moving frame the following evolution equation:

$$h_t = \frac{1}{2}|\nabla h|^2 - \frac{1}{8}|\nabla h|^4 + \Gamma\left(1 + \frac{1}{2}|\nabla h|^2\right)\mathcal{E}_2(h) + \Gamma\mathcal{E}_3(h)$$
$$+ \Gamma\mathcal{E}_4(h) + O(|\nabla h|^6). \tag{5.117}$$

The nonlinear terms $|\nabla h|^2/2 - |\nabla h|^4/8$ on the right-hand side of Eq. (5.118) stem from the projection of the local normal velocity of the crystallization front on the z-axis. Usually, only a $|\nabla h|^2$ term is present in this type of equation because it already describes the kinematic effect of the interfacial slope (Krug 1997). Moreover, although the two-term expansion that includes also the $|\nabla h|^4$ term is more accurate for small surface slopes, it can lead to artifacts if the slope of the interface is too large. Therefore, the fourth-order kinematic term in the evolution equation is omitted. Thus, Eq. (5.117) is written in the following form:

$$h_t = \frac{1}{2}|\nabla h|^2 + h_{xx}[\mu_{11} + \kappa_{11}h_x + \lambda_{11}h_y + a_{11}h_x^2 + b_{11}h_y^2 + c_{11}h_xh_y]$$
$$+ h_{xy}[\mu_{12} + \kappa_{12}h_x + \lambda_{12}h_y + a_{12}h_x^2 + b_{12}h_y^2 + c_{12}h_xh_y]$$
$$+ h_{yy}[\mu_{22} + \kappa_{22}h_x + \lambda_{22}h_y + a_{22}h_x^2 + b_{22}h_y^2 + c_{22}h_xh_y] - \nu\nabla^4 h. \tag{5.118}$$

Equation (5.118) is related to those derived in Stewart and Goldenfeld (1992) and Liu and Metiu (1993) for the faceting of crystal surfaces with unstable orientations when there is no surface growth. The coefficients μ_{ij} characterize the linear faceting instability of the thermodynamically unstable surface, and the coefficients of the nonlinear terms determine the stable orientations of the appearing facets and the symmetry of the faceted structure. The linear damping coefficient ν characterizes the stabilizing effect of the additional energy of edges and determines their widths. The new feature is that Eq. (5.118) describes the faceting instability in the course of kinetically controlled crystal *growth*, which accounts for the presence of the "convective" term $\frac{1}{2}|\nabla h|^2$. On the other hand, this feature can be considered as a generalization of isotropic growth models reviewed by Krug for the case of an evaporation–condensation growth mechanism. Equation (5.118) is considered in the present case to be a phenomenological model for the formation of facets and corners in the kinetically controlled growth of a *thermodynamically unstable* surface in the simplest case of a constant driving force.

A related evolution equation is derived for a more complicated case, when the difference between the crystal and liquid chemical potentials is not constant, but it is *locally* coupled to the temperature field in a thermal diffusion boundary layer near the crystal surface.

As an example, consider the solidification of a hypercooled melt. For a local $\Delta(h)$, the growth is controlled by attachment kinetics

$$V_n \propto T^e - T^l$$

valid for small deviations from equilibrium. (Bates et al. 1997 have discussed models valid at substantial undercoolings.) Golovin et al. (1999) obtained the following equation:

$$h_t - \frac{1}{2}V|\nabla h|^2 = h_{xx}[\mu_{11}^s + \kappa_{11}h_x + \lambda_{11}h_y + a_{11}^s h_x^2 + b_{11}h_y^2 + c_{11}h_xh_y]$$
$$+ h_{xy}[\mu_{12} + \kappa_{12}h_x + \lambda_{12}h_y + a_{12}h_x^2 + b_{12}h_y^2 + c_{12}^s h_xh_y]$$
$$+ h_{yy}[\mu_{22}^s + \kappa_{22}h_x + \lambda_{22}h_y + a_{22}h_x^2 + b_{22}^s h_y^2 + c_{22}h_xh_y]$$
$$+ g\frac{\partial(h_x, h_y)}{\partial(x, y)} - \chi_{ijkl}h_{x_ix_jx_kx_l}, \tag{5.119}$$

where all coefficients are exhibited there and with terms $O(|\nabla h|^4)$ in V_n neglected.

Equation (5.119) is asymptotically correct if $\mu_{ij} = O(\epsilon^2)$ and $\kappa_{ij}, \lambda_{ij} = O(\epsilon)$. Otherwise, it should be considered as a model equation for describing formation of corners and facets caused by anisotropic surface tension in the course of kinetically controlled crystal growth when the effect of the thermal boundary layer near the crystal surface is important. The coefficients are determined not only by the anisotropy of the crystal surface tension but also by the temperature field near the crystal surface, that is, they depend also on the undercooling S^{-1}.

Equation (5.119) also contains a new term $g\partial(h_x, h_y)/\partial(x, y)$, which is the main part of the Gaussian curvature $\mathcal{G} = (h_{xx}h_{yy} - h_{xy}^2)/(1 + |\nabla h|^2)^2$ in the long-wave approximation. This term was first seen in Sarocka and Bernoff (1995) as part of the intrinsic equation of interfacial motion for solidification in a hypercooled melt in the case of isotropic surface tension; in the isotropic case the coefficient g coincides with the corresponding coefficient derived there. Note that the evolution equation derived by Sarocka and Bernoff (1995) also contains a nonlinear term proportional to the square H^2 of the mean curvature, which arises if one takes into account thermal diffusion in the crystal (the two-sided model). In the one-dimensional case the term with the Gaussian curvature disappears, and Eq. (5.119) is reduced to the evolution equation derived earlier.

On the one hand, physical effects that would lead to the appearance of the term with the Gaussian curvature in the equation describing the evolution of the crystal surface are not clear. On the other hand, the Gaussian curvature is the surface invariant, and one might expect it to appear naturally in the expansion with respect to small surface slopes. In the next section it is shown that the presence of this term leads, within the framework of the present model, to the formation of some of the patterns observed in crystal-growth experiments.

5.7 Faceting with Constant Driving Force

Consider now the faceting of kinetically growing crystal surfaces with different orientations when the effects of the temperature variations in the diffusion boundary layer can be neglected (*constant driving force*).

For growth in direction [001], Eq. (5.118) becomes

$$h_t = -m\nabla^2 h - \nu \nabla^4 h + \frac{1}{2}|\nabla h|^2 + h_{xx}\left[ah_x^2 + bh_y^2\right]$$
$$+ h_{yy}\left[bh_x^2 + ah_y^2\right] + ch_{xy}h_x h_y,$$

where

$$m = \Gamma(3\alpha_4 + 5\alpha_6 - 1), \quad a = \Gamma(33\alpha_4 + 50\alpha_6 - 1),$$
$$b = \Gamma(6\alpha_4 + 15\alpha_6), \quad c = \Gamma(30\alpha_4 + 70\alpha_6 - 2), \tag{5.120}$$

and $\nu = \delta/\Delta_0 > 0$. Consider the anisotropy coefficients α_4, α_6 to be such that $m > 0$, and thus the [001] surface is thermodynamically unstable and faceting occurs. The coefficients a, b, and c characterizing the stable orientation of facets are taken to be positive, and the coefficient ν is taken to be small in order to get sharp corners. Equations (5.120) do not have a rotational symmetry but are symmetric with respect to the transformations $x \to -x$, $y \to -y$, $x \to y$. This corresponds to the fourfold symmetry of the [001] crystal surface.

Figure 5.21 shows the results of the numerical solution of Eqs. (5.120) for $b/a \ll 1$, that is, for large anisotropy, and exhibits the shapes of the crystal surface at different times. First one observes the formation of a hill-and-valley structure in the form of square pyramids. Note that the square shape of the pyramids can be distorted by their interaction. Such pyramids are often seen on the [001] surfaces growing by chemical-vapor deposition (Bloem et al. 1983), liquid phase epitaxy (Bauser and Löchner 1981), molecular beam epitaxy (van Nostrand et al. 1995, and van Nostrand, Chey and Cahill 1995), and so forth. The pyramids with a characteristic horizontal scale λ_* corresponding to the most rapidly growing mode given by the linear stability analysis, $\lambda_* = 2\pi\sqrt{2\nu/m}$, are formed in a characteristic time, $\tau_* = m^2/(4\nu)$, and they have a characteristic *slope*. This slope is determined by the anisotropy of the surface energy as well as by the surface growth speed; the curvatures of the pyramid edges and vertex are determined by their additional energy. After the pyramids have been formed, their slopes no longer change, but the structure coarsens in time, forming square pyramids with larger horizontal scales, which is typical of faceting instability of thermodynamically unstable surfaces whose slopes are determined by anisotropic free energy.

In the case of a *kinetically controlled growth*, when the evolution equation for the surface shape contains the nonlinear growth term $|\nabla h|^2$, the double tangent construction is destroyed, but the slopes of facets remain fixed. (Golovin et al. 1999). In the present case far from the vertices, the slopes can be obtained analytically.

Consider a square pyramid oriented in such a way that the projections of its edges on the basis plane coincide with x and y axes. In this case, the pyramidal

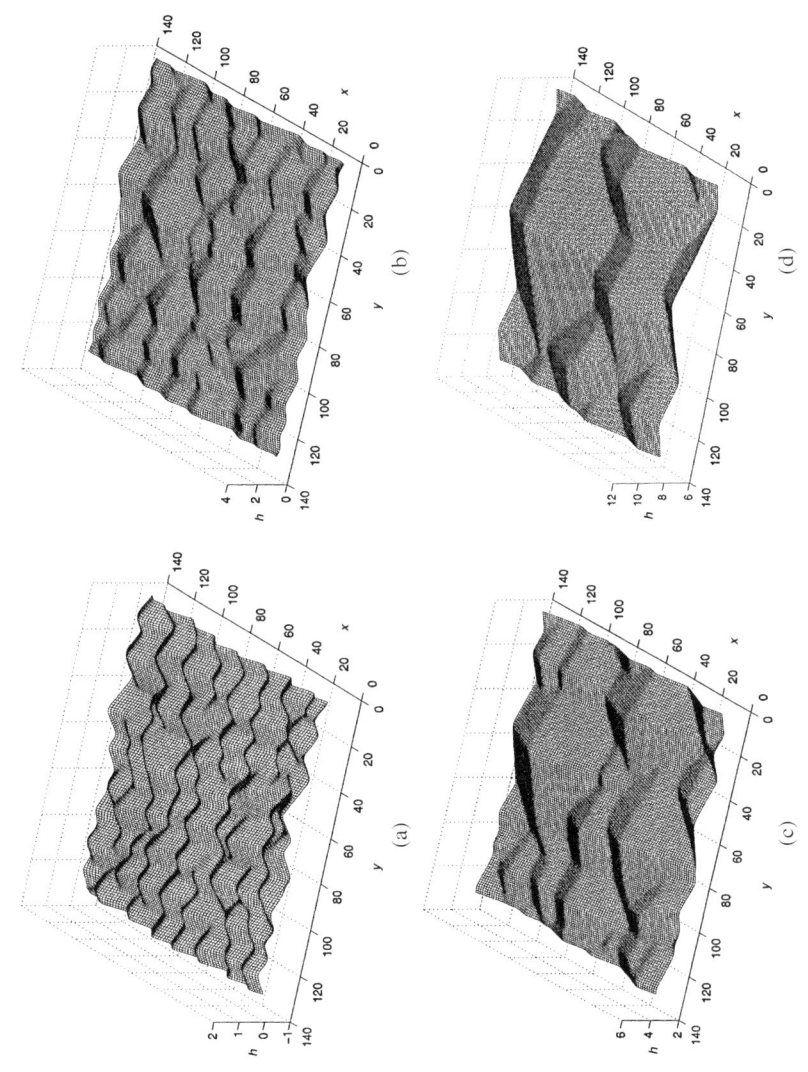

Figure 5.21. Hill-and-valley structure on a growing crystal surface in the form of square pyramids: numerical solution of Eq. (5.120) at different moments of time: a,b, $t = 10^4$; c,d, $t = 2 \times 10^4$; e,f, $t = 5 \times 10^4$; g,h, $t = 10^5$, $m = 0.5$, $\nu = 0.001$, $a = 1.0$, $b = 0.1$, $c = 0.3$. The spatial scale is arbitrary. From Golovin et al. (1999). (Reprinted with permission of Elsevier Science B.V.)

shape $h(x, y, t)$ has the following asymptotics:

$$h \sim Ay + f(x) + vt \quad \text{as} \quad y \to -\infty, \tag{5.121}$$

where v is the speed of the surface growth in the z direction (i.e., the nonlinear correction of the unit speed of planar surface growth in the laboratory frame), A is the slope of the pyramidal edges, and $f(x)$ is a function to be determined. For $x \to -\infty$ one has $h \sim Ax + f(y) + vt$. The function f must satisfy the compatibility condition

$$f'(\pm\infty) = \mp A \tag{5.122}$$

(the prime denotes differentiation).

By substituting Eq. (5.121) into Eq. (5.120), one obtains the following equation for $f(x)$:

$$\left(\frac{1}{2}A^2 - v\right) + \frac{1}{2}(f')^2 - (m - bA^2)f'' + a(f')^2 f'' - vf'''' = 0. \tag{5.123}$$

Taking an *ansatz* from Leung (1990)

$$f' = Q \tanh kx, \tag{5.124}$$

one obtains from Eq. (5.123) that

$$Q^2 = \frac{3(m - bA^2)}{a} + \frac{3}{2a}\sqrt{\frac{6v}{a}} \operatorname{sgn} Q, \quad k = \sqrt{\frac{a}{6v}} |Q|, \tag{5.125}$$

where negative Q corresponds to a pyramid (hill) and positive Q corresponds to an antipyramid (hole), respectively; k is positive by definition. Equations (5.125) show that the radius of curvature of the pyramidal edges $\sim \sqrt{\delta/\gamma_0}$, is determined by their additional energies. From the compatibility condition $Q^2 = A^2$, one finds the slope of the square pyramid edge far from the vertex,

$$A^\pm = \pm\sqrt{\frac{m \mp \sqrt{3v/(2a)}}{b + a/3}}, \tag{5.126}$$

where A^+ corresponds to hills and A^- corresponds to holes. Thus, the shape of a square pyramid (h^+) or antipyramid (h^-) far from the vertex, for $y \to -\infty$, is

$$h^\pm(x, y, t) \sim A^\pm y \mp \sqrt{\frac{6v}{a}} \ln\left[\cosh\left(\sqrt{\frac{a}{6v}}|A^\pm|x\right)\right] + (A^\pm)^2 t. \tag{5.127}$$

5.7 Faceting with constant driving force

One can see that the slopes of the pyramids depend not only on the anisotropic surface free energy characterized by the coefficients m, a, b but also on the surface growth rate. To do this one rewrites Eq. (5.126) for the asymptotic slope of the pyramid (antipyramid) edge in the dimensional variables

$$(A^\pm)^2 = \frac{\bar{m} \mp \Delta_0 \sqrt{3\delta/(2\bar{a}\gamma_0^3)}}{\bar{b} + \bar{a}/3}, \qquad (5.128)$$

where $\bar{m}, \bar{a}, \bar{b}$ are the respective coefficients defined in Eq. (5.120) divided by Γ, (i.e., the functions of the anisotropy coefficients α_4 and α_6 *only*), Δ_0 is the surface-growth driving force, and δ is the coefficient characterizing the additional energy of the pyramid edges and vertex. It is important that these *dynamical slopes* of pyramids and antipyramids be *different*; the slope of a hill is *smaller* than the slope of a hole. This is due to the *convective* effect of the kinetically controlled surface growth discussed for a one-dimensional case in Golovin et al. (1999); in the case of a concave surface (hole), the kinetically controlled growth tends to produce caustics and steepens the slope, whereas it tends to smooth the sharp corners and to reduce the slope in the case of a convex surface (hills). At the same time, the correction to the dimensionless unit growth speed of a planar surface in the laboratory frame, $v^\pm = (A^\pm)^2$, is smaller for pyramids and larger for antipyramids. Thus, faces of antipyramids propagate in the z direction faster than those of pyramids. This leads to the asymmetry between the pyramids and antipyramids that can be seen in Figure 5.21. At the beginning of the structure formation, both pyramids and antipyramids are present, but in the course of the surface growth antipyramids "grow out" and gradually disappear. A convective nature of the pyramid selection mechanism can be understood better if one considers infinitesimal perturbations of an almost planar pyramid face far from the vertex, where the surface is locally described as $h = h_0 \sim A(x+y)$. Indeed, taking $h = h_0 + \tilde{h}$ (there the infinitesimal perturbation $\tilde{h} \sim \exp[\sigma t + i(a_x x + a_y y)]$, where σ is the perturbation growth rate and (α_x, α_y) is the perturbation wave vector), one obtains the following dispersion relation:

$$\sigma = a^2 \left[m - A^2(a+b) \right] - c a_x a_y A^2 - v a^4 + i A(a_x + a_y), \qquad (5.129)$$

where $a^2 = a_x^2 + a_y^2$. The last (imaginary) term on the right-hand side of the dispersion relation shows that the disturbances of a pyramid (hill) travel from the vertex to the periphery, whereas the disturbances of an antipyramid move from the periphery to the vertex. Thus, pyramids are sources of the perturbation, and antipyramids are sinks. This accumulation of the perturbations descending from the pyramids into antipyramids eventually leads to the disappearance of the latter.

Note that, because the slopes of the pyramids do not change, their heights increase proportionally to their horizontal spatial scale in the course of the surface growth. The similar behavior caused by the existence of a "magic" slope that remains constant during the coarsening of square pyramidal mounds was observed experimentally in molecular-beam epitaxial growth (Ernst et al. 1994) and reproduced in numerical simulations of the model proposed in Siegert and Plishka (1994) for the growth and coarsening of pyramidal structures caused by an anisotropic, slope-dependent step current.

It should also be mentioned that if $a \approx b$ in Eq. (5.120), the main nonlinear terms in Eq. (5.120) become almost rotationally invariant, and its solution exhibits a labyrinthian hill-and-valley pattern similar to that produced by a two-dimensional isotropic Cahn–Hilliard equation. Moreover, as one can see from Eqs. (5.127) and (5.128), solutions in the form of pyramids do not exist if $m < \sqrt{3\nu/(2a)}$, that is, if

$$\Delta_0 > \bar{m}\sqrt{\frac{2\bar{a}\gamma_0^3}{3\delta}}. \tag{5.130}$$

This corresponds to the case when the energy of the edges or the growth driving force is very large. The preliminary numerical simulations in this case show irregular patterns.

Consider growth in the [111] direction. Now Eq. (5.118) has the form

$$h_t = \frac{1}{2}|\nabla h|^2 - m'\nabla^2 h - \nu\nabla^4 h + h_{xx}\left[d'h_y + a'h_x^2 + b'h_y^2\right] \\ + h_{yy}\left[-d'h_y + b'h_x^2 + a'h_y^2\right] + 2h_{xy}\left[d'h_x + (a' - b')h_x h_y\right], \tag{5.131}$$

where

$$m' = -\Gamma\left(1 + 3\alpha_4 + \frac{25}{9}\alpha_6\right), \quad a' = -\Gamma\left(1 + 21\alpha_4 + \frac{115}{9}\alpha_6\right),$$

$$b' = -\Gamma\left(6\alpha_4 + \frac{10}{3}\alpha_6\right), \quad d' = -4\sqrt{2}\Gamma(\alpha_4 + 5\alpha_6/3).$$

It can easily be checked that this equation is invariant with respect to rotations at the angles $2\pi n/3$. This corresponds to the threefold symmetry of [111] surface of a cubic crystal. Consider the surface to be thermodynamically unstable, and thus one chooses α_4 and α_6 in such a way that $m' > 0$; a and b are taken to be positive.

Figure 5.22 shows the numerical solution in a rectangular box with the y/x aspect ratio equal to $\sqrt{3}/2$. This aspect ratio was chosen to diminish the effect of the rectangular boundaries, breaking the threefold rotational symmetry, on the formation of the faceted structure. One can see the formation of the faceted structure consisting of triangular pyramids. At the initial stages the pyramids are slightly elongated in the y direction owing to the effect of the boundaries, and their shape is distorted by mutual interaction. At the late stages the bases of the pyramids become equilateral. Such structure of triangular pyramids is observed in experiments on the faceting of unstable [111] surfaces caused by thermal annealing (Knoppik and Lösch 1976, Madey et al. 1996, Nien and Madey 1997) during the growth of Si(111) by gas phase epitaxy at high temperatures and by chemical vapor deposition (Nishizawa, Terasaki, and Shimbo 1972, Mazumder et al. 1995), as well as during liquid phase epitaxial growth of LiNbO$_3$ thick single-crystal films (Hibiya et al. 1994). Evolution of the triangular, pyramidal, faceted structure on a growing [111] surface is the same as those for the faceting of the growing [001] surface; after the structure with the horizontal length scale corresponding to the most unstable mode has been established, it coarsens with time; the pyramid slopes remain unchanged. Note that Eq. (5.131) is invariant with respect to transformation $d \to -d$, $y \to -y$, and the pyramid orientation depends on the sign of the coefficient d.

As in the case of square pyramids considered earlier, the shape of the triangular pyramids with equilateral bases and their final slopes far from the vertices can be found analytically in a similar way. Indeed, consider a triangular pyramid with an equilateral basis oriented in such a way that the projection of one of its edges on the basis plane coincides with the negative part of the y-axis, and thus the shape of the pyramid far from the vertex for $y \to -\infty$ can be described by Eq. (5.121). A simple geometric consideration of the triangular pyramid with equilateral basis gives the following compatibility condition for the function $f(x)$ in this case:

$$f'(\pm\infty) = \mp\sqrt{3}A.$$

Substitute Eq. (5.121) into Eq. (5.131) and obtain the following equation for $f(x)$:

$$\left(\frac{1}{2}A^2 - v\right) + \frac{1}{2}(f')^2 - (m - d'A - b'A^2)f'' + a'(f')^2 f'' - vf'''' = 0. \tag{5.132}$$

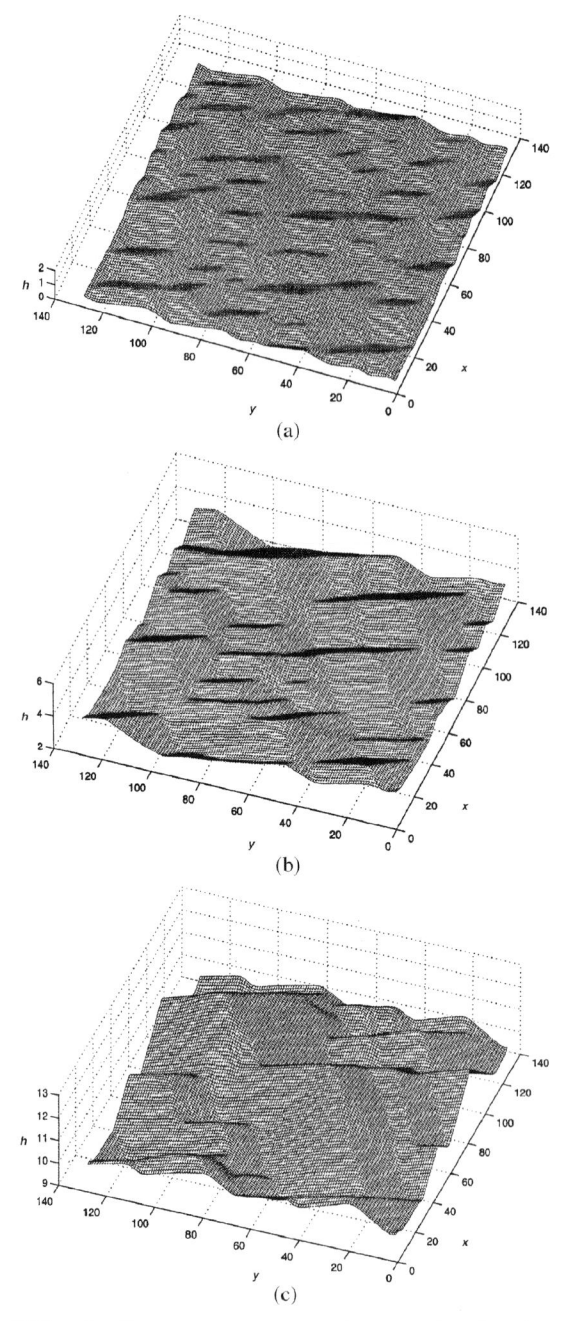

Figure 5.22. Hill-and-valley structure on a growing crystal surface in the form of triangular pyramids: numerical solution of Eq. (5.131) at different moments of time: a,b, $t = 10^4$; c,d, $t = 4 \times 10^4$; e,f, $t = 10^5$; $m' = 0.5$, $v = 0.001$, $a' = 1.0$, $b' = 0.1$, $c' = 1.8$, $d' = 0.5$. The spatial scale is arbitrary. From Golovin et al. (1999). (Reprinted with permission of Elsevier Science B.V.)

5.7 Faceting with constant driving force

Using the same *ansatz* (5.124) as before for f', one obtains from Eq. (5.132)

$$Q^2 = \frac{3(m - d'A - b'A^2)}{a'} + \frac{3}{2a'}\sqrt{\frac{6v}{a'}} \operatorname{sgn} Q,$$

$$k = \sqrt{\frac{a'}{6v}} |Q|, \qquad (5.133)$$

where, as in the case considered in the preceding section, negative Q corresponds to a pyramid, and positive Q corresponds to an antipyramid, respectively. From the compatibility condition, $Q^2 = 3A^2$, one obtains a quadratic equation for the slope of the triangular pyramid edge far from the vertex. A stable solution of this equation is determined by the sign of the coefficient d', namely,

$$A^{\pm} = \frac{-d' - \sqrt{(d')^2 + 4(a' + b')\left[m' \mp \sqrt{3v/(2a')}\right]}}{2(a' + b')}$$

for $d' > 0$,

$$A^{\pm} = \frac{-d' + \sqrt{(d')^2 + 4(a' + b')\left[m' \mp \sqrt{3v/(2a')}\right]}}{2(a' + b')} \qquad (5.134)$$

for $d' < 0$,

where $A^{\pm}$ corresponds to the hills (A^+) and holes (A^-), respectively. Thus, the shape of the triangular pyramids (antipyramids) for $y \to -\infty$ is

$$h^{\pm}(x, y, t) \sim A^{\pm}y \mp \sqrt{\frac{6v}{a}} \ln\left[\cosh\left(\sqrt{\frac{a}{2v}}|A^{\pm}|x\right)\right] + 2(A^{\pm})^2 t. \quad (5.135)$$

One can see that the parameters and evolution of triangular pyramids and antipyramids are distinguished, as discussed in the preceding section for the case of square pyramids. The solution in the form of triangular pyramids exists if $(d')^2 + 4(a' + b')[m' - \sqrt{3v/(2a')}] > 0$, that is, as in the case of the square pyramids, the triangular pyramids cannot form for large v corresponding to either very fast surface growth or to a large edge energy; these cases are not considered here.

Consider now growth in the [110] direction. Equation (5.118) reads

$$h_t = \frac{1}{2}|\nabla h|^2 + h_{xx}\left[-m_x + a_x h_x^2 + b_x h_y^2\right] + h_{yy}\left[-m_y + a_y h_x^2 + b_y h_y^2\right]$$
$$+ \tilde{c}h_{xy}h_x h_y - v\nabla^4 h,$$

where

$$m_x = -\Gamma\left(\frac{9}{2}\alpha_4 + \frac{25}{4}\alpha_6 + 1\right), \quad m_y = \Gamma\left(\frac{3}{2}\alpha_4 + \frac{5}{4}\alpha_6 - 1\right),$$

$$a_x = -\Gamma\left(\frac{69}{2}\alpha_4 + \frac{205}{4}\alpha_6 + 1\right), \quad b_x = -\Gamma\left(3\alpha_4 + \frac{45}{4}\alpha_6\right),$$

$$a_y = -\Gamma(6\alpha_4 + 15\alpha_6), \quad b_y = \Gamma\left(\frac{45}{2}\alpha_4 + \frac{25}{2}\alpha_6 - 1\right),$$

$$\tilde{c} = -\Gamma\left(21\alpha_4 + \frac{115}{2}\alpha_6 + 2\right). \tag{5.136}$$

Here consider the anisotropy coefficients α_4 and α_6 to be such that the [110] surface is thermodynamically unstable in both x and y directions, that is, $m_y > 0$. The other coefficients are also taken to be positive in the simulations.

Equation (5.136) is invariant with respect to transformations $x \to -x$, $y \to -y$, but it is not invariant with respect to the transformation $x \to y$. This reflects the twofold symmetry of the [110] surface. Thus, one can expect the faceting instability of this surface to result in the structure of either rhombic pyramids or grooves. One observes that the resulting structure depends on the degree of anisotropy. Figure 5.23 shows the evolution of the faceted surface in the case when the asymmetry between the x and y directions is not very large but the anisotropy is still large so that b_x/a_x and a_y/b_y are small. First, one observes the formation of a system of grooves (Figure 5.23(a)). Later, the grooves decay into islands in the form of rhombic pyramids (Figure 5.23(b)). Similar rhombic pyramids are observed such as during the epitaxial growth of an In–Ga–As alloy on the InP(001) surface (Jacob et al. 1997). The rhombic pyramids coarsen in time (Figure 5.23(c)), but although the shapes of the square and triangular pyramids remain self-similar during the coarsening, the coarsening rates of the rhombic pyramids are different in the x and y directions. Namely, one observes that the coarsening in the y direction goes a little faster, and thus the rhombic pyramids gradually thicken in the y direction and tend to a limiting rhombic shape that can be found analytically for regions far from the vertex in a way similar to that described in the preceding sections for the square and triangular pyramids. Indeed, consider a rhombic pyramid oriented in such a way that the projections of its edges on the basis plane coincide with the x- and y-axes. For the asymptotics of the pyramid shape far from the vertex, one has

$$\begin{aligned} h &\sim Ax + f(y) + vt, \quad f'(\pm\infty) = \mp B \text{ as } x \to -\infty,\\ h &\sim By + g(x) + vt, \quad g'(\pm\infty) = \mp A \text{ as } y \to -\infty. \end{aligned} \tag{5.137}$$

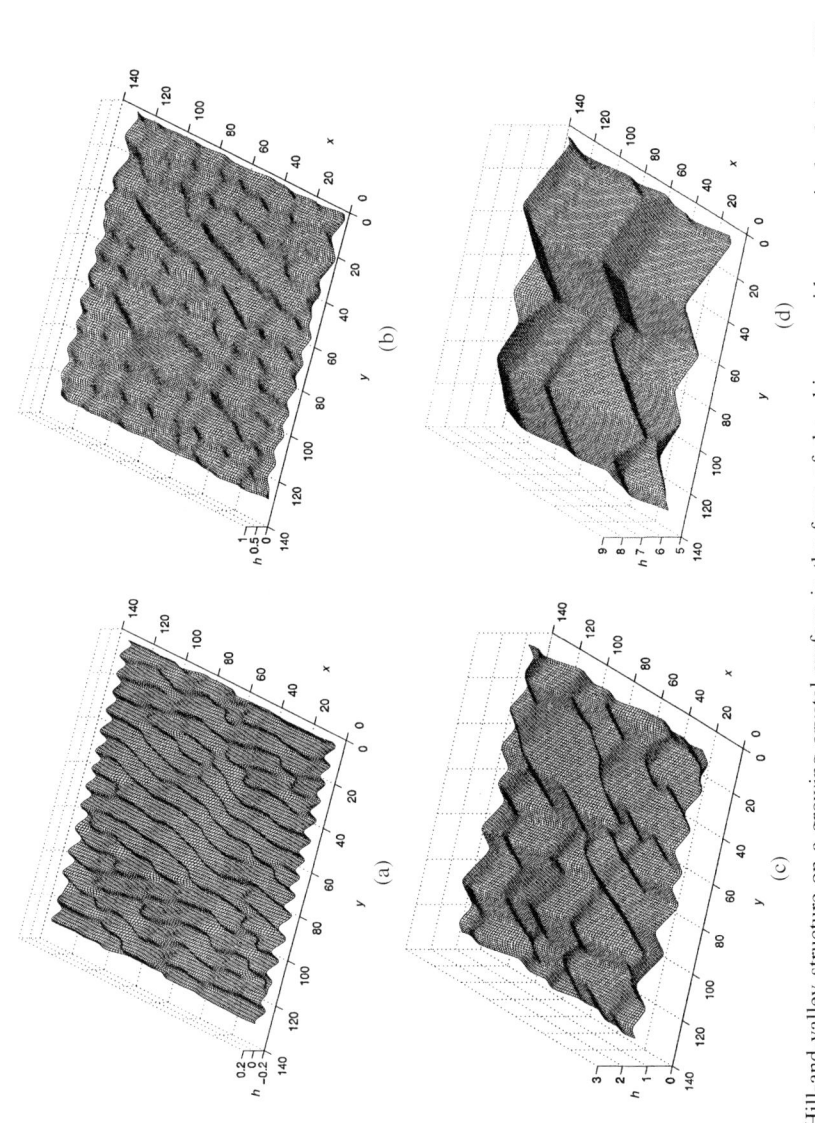

Figure 5.23. Hill-and-valley structure on a growing crystal surface in the form of rhombic pyramids; numerical solution of Eq. (5.136) at different moments of time: (a) $t = 2 \times 10^3$, (b) $t = 10^4$, (c) $t = 3 \times 10^4$, (d) 10^5 for $m_x = 0.3$, $m_y = 0.5$, $a_x = 1.0$, $b_x = 0.1$, $a_y = 0.2$, $b_y = 0.8$, $\tilde{c} = 0.3$, $\nu = 0.001$. The spatial scale is arbitrary. From Golovin et al. (1999). (Reprinted with permission of Elsevier Science B.V.)

Taking $f'(y) = Q_y \tanh k_y y$, $g'(x) = Q_x \tanh k_x x$, one obtains from Eqs. (5.137)

$$k_y = \sqrt{\frac{b_y}{6v}}|Q_y|, \quad Q_y^2 = \frac{3(m_y - a_y A^2)}{b_y} + \frac{3}{2b_y}\sqrt{\frac{6v}{b_y}}\operatorname{sgn} Q_y,$$

$$k_x = \sqrt{\frac{a_x}{6v}}|Q_x|, \quad Q_x^2 = \frac{3(m_x - b_x B^2)}{a_x} + \frac{3}{2a_x}\sqrt{\frac{6v}{a_x}}\operatorname{sgn} Q_x,$$
(5.138)

where, as before, negative (positive) $Q_{x,y}$ corresponds to pyramids (antipyramids). The compatibility conditions $Q_x^2 = A^2$, $Q_y^2 = B^2$ yield a system of linear equations for squares of the limiting slopes of the pyramid edges in the x and y directions, A^2 and B^2, respectively, whose solution gives

$$(A^{\pm})^2 = \frac{9b_x\left[m_y \mp \sqrt{3v/(2b_y)}\right] - 3b_y\left[m_x \mp \sqrt{3v/(2a_x)}\right]}{9a_y b_x - a_x b_y},$$

$$(B^{\pm})^2 = \frac{9a_y\left[m_x \mp \sqrt{3v/(2a_x)}\right] - 3a_x\left[m_y \mp \sqrt{3v/(2b_y)}\right]}{9a_y b_x - a_x b_y}.$$
(5.139)

The superscripts $+$ and $-$ correspond to pyramids and antipyramids, respectively. Thus, the functions $f(y)$ and $g(x)$ describing the asymptotic shape of the rhombic pyramids in Eqs. (5.137) are

$$f^{\pm}(y) = \mp\sqrt{\frac{6v}{b_y}} \ln\left[\cosh\left(\sqrt{\frac{b_y}{6v}}|B^{\pm}|x\right)\right],$$

$$g^{\pm}(x) = \mp\sqrt{\frac{6v}{a_x}} \ln\left[\cosh\left(\sqrt{\frac{a_x}{6v}}|A^{\pm}|x\right)\right],$$
(5.140)

and the nonlinear correction for the growth speed $v^{\pm} = 1/2[(A^{\pm})^2 + (B^{\pm})^2]$. Obviously, the solution for the square pyramids described above is the particular case of the rhombic pyramids for $m_x = m_y = m$, $a_x = b_y = a$, $a_y = b_x = b$, $A = B$.

Note that Eqs. (5.139) have real solutions provided that the right-hand sides of the two equations are positive. Thus, for $v \ll 1$, the rhombic pyramids can be formed if

$$\frac{9b_x m_y - 3b_y m_x}{9a_y b_x - a_x b_y} > 0, \quad \frac{9a_y m_x - 3a_x m_y}{9a_y b_x - a_x b_y} > 0. \tag{5.141}$$

Otherwise, one obtains the solution in the form of grooves rather than rhombic pyramids. Numerical solution of Eq. (5.136) in this case is shown in Figure 5.24, which illustrates the crystal surface shape at different times and shows the

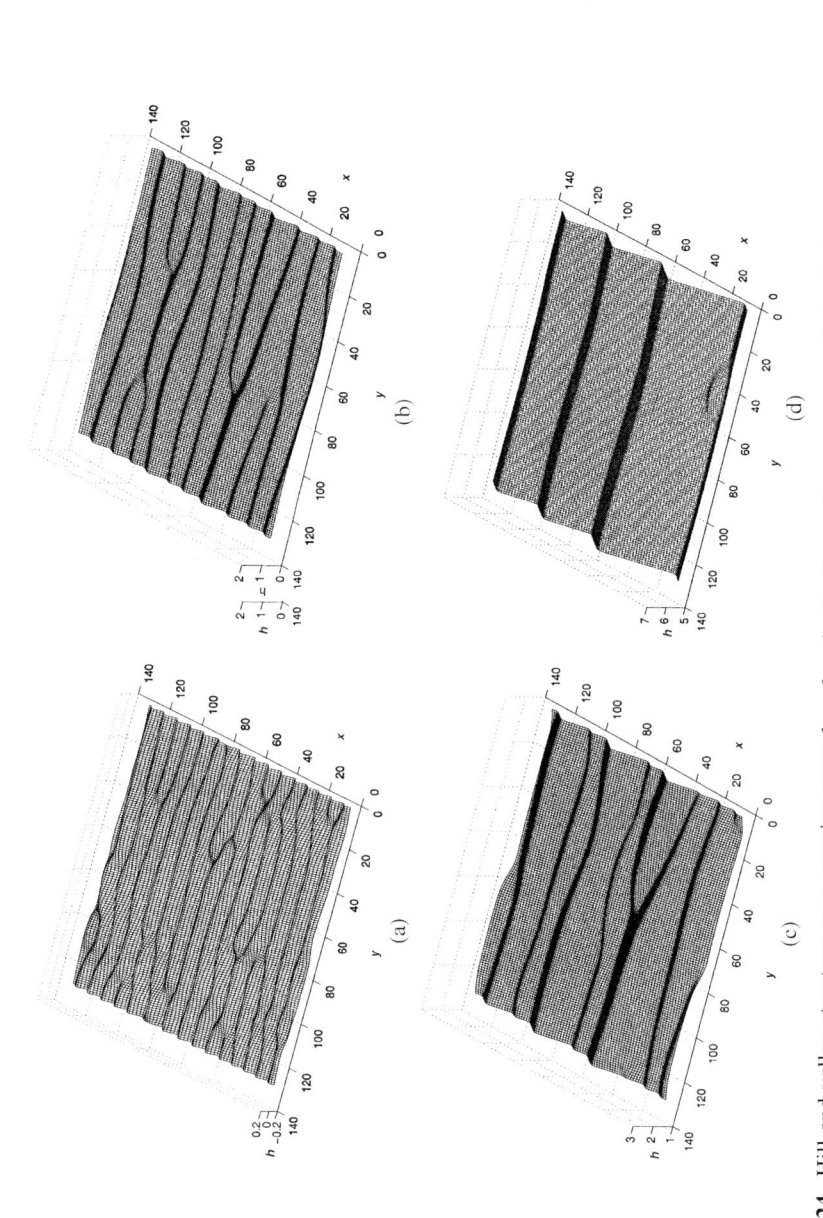

Figure 5.24. Hill-and-valley structure on a growing crystal surface in the form of grooves: numerical solution of Eq. (5.138) at different moments of time: (a) $t = 2 \times 10^3$; (b) $t = 10^4$ (c) $t = 4 \times 10^4$; (d) $t = 10^5$ for $m_x = 0.5$, $m_y = 0.1$, $a_x = 1.0$, $b_x = 0.1$, $a_y = 0.2$, $b_y = 0.8$, $\bar{c} = 0.3$, $\nu = 0.001$. The spatial scale is arbitrary. From Golovin et al. (1999). (Reprinted with permission of Elsevier Science B.V.)

corresponding surface contour plots. The grooves do not decay into pyramidal islands, and in a finite periodic box the structure ultimately become quasi-one-dimensional. The groove slope does not change during the coarsening. Formation and coarsening of the system of grooves similar to those shown in Figure 5.24 are observed, for example during the faceting of thermodynamically unstable {0001} and {1010} surfaces of Al_2O_3 caused by thermal annealing (Heffelfinger and Carter 1997) as well as in the course of unstable homoepitaxial growth of GaAs(001) (Orme et al. 1995). The coarsening rates in the x and y directions differ from each other more than in the case of rhombic pyramids.

In the preceding, the formation of square and rhombic pyramids was studied in the cases when either b/a or b_x/a_x and a_y/b_y were numerically small (large anisotropy). It turns out that, in Eq. (5.120), if $b = 0$, or if in Eq. (5.136) $b_x = a_y = 0$, it is possible to construct the *exact* solutions of these equations. First, consider the latter case, which is more general. In this case one can seek the solution of Eq. (5.136) in the form

$$h(x, y, t) = f(x) + g(y) + vt. \tag{5.142}$$

Substituting Eq. (5.142) in Eq. (5.136), one again obtains two decoupled equations for f and g. Thus, one obtains the following exact solution of Eq. (5.136) for $b_x = a_y = 0$ in the form of a rhombic pyramid or antipyramid:

$$h^\pm(x, y, t) = \mp\sqrt{\frac{6v}{a_x}} \ln(\cosh k_x^\pm x) \mp \sqrt{\frac{6v}{b_y}} \ln(\cosh k_y^\pm y) + v^\pm t,$$

$$k_x^\pm = \sqrt{\frac{m_x}{2v} \mp \sqrt{\frac{3}{8va_x}}}, \quad k_y^\pm = \sqrt{\frac{m_y}{2v} \mp \sqrt{\frac{3}{8vb_y}}}, \tag{5.144}$$

$$v^\pm = \frac{3}{2}\left(\frac{m_x}{a_x} + \frac{m_y}{b_y}\right) \mp \frac{3\sqrt{6v}}{4}\left(a_x^{-3/2} + b_y^{-3/2}\right).$$

The exact solution (5.144) is in the form of a square pyramid.

5.8 Coarsening

In many systems the coarsening of the structures resulting from instabilities is known at late stages to obey a power law in time (i.e., $L(t) \sim t^\alpha$, where L is the characteristic length scale of the structure and α is the coarsening exponent). These have been estimated for the growth of thermodynamically unstable [001],

5.8 Coarsening

[111], and [110] surfaces. Square and triangular pyramids corresponding to the faceting of growing [001] and [111] surfaces, respectively, remain self-similar during the coarsening, and thus the coarsening rates in the x and y directions are the same. For this case the characteristic scale of the structure can be computed in two ways: $L_1(t) + N^{-1} \sum_{i=1}^{N} Z_i^{-1}(t)$, where N is the number of collocation points ($N = 128$ in our case) and $Z_i(t)$ is the number of zeros of the function $h(x, y, t)$ on the ith y layer, and $L_2(t) = N_+(t)/N_0(t)$, where N_+ is the number of spatial points when $h(x, y, t) - \bar{h}(t) > 0$, and N_0 is the number of points when $h - \bar{h} = 0$ ($\bar{h}$ is the spatially mean value of h, which is equal to the zeroth Fourier model). It is found that $L_1(t)$ and $L_2(t)$ are proportional to each other, and thus both measures of the characteristic spatial scale of the structure are equivalent.

Figure 5.25 shows the growth of the characteristic spatial scale of square and triangular pyramids (squares and triangles, respectively). The results with random initial data are averaged over 10 realizations. One can see the formation of

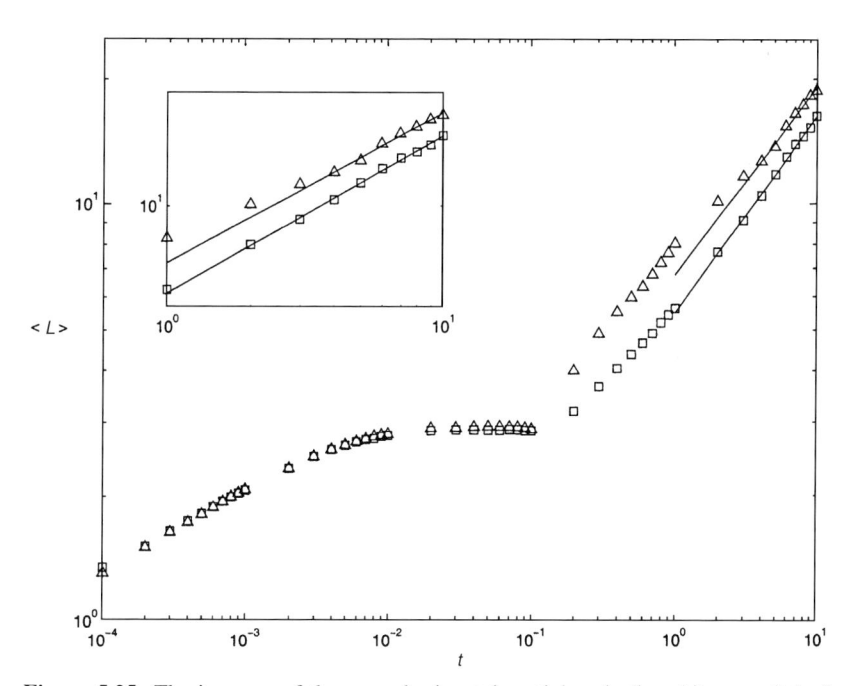

Figure 5.25. The increase of the mean horizontal spatial scale (in arbitrary units) of the faceted surface structures in the form of square (squares) and triangular (triangles) pyramids corresponding to the numerical solutions of Eqs. (5.120) and (5.131), respectively. The inset shows the power-law regime at the late stage of the coarsening. From Golovin et al. (1999). (Reprinted with permission of Elsevier Science B.V.)

the initial periodic structure after the characteristic time of linear instability and the transition to the power-law coarsening at the late stage shown in detail in the inset. The coarsening exponent α was measured to be 0.47 for square pyramids and 0.45 for triangular pyramids, which are practically the same. Similar exponents, 0.4 and 0.42 ± 0.4, were observed in experiments on coarsening of square pyramids in homolayer and multilayer epitaxial growth of Ge(001) (Van Nostrand et al. 1995, van Nostrand, Chey, and Cahill 1995). The exponents obtained in the computations are considerably larger than those corresponding to the coarsening of faceted thermodynamically unstable surfaces when there is *no* kinetically controlled surface growth. In that case the coarsening rate depends on the mechanism of the surface reconstruction, and theoretical predictions give 1/4 for the evaporation–condensation mechanism and 1/6 for the surface-diffusion mechanism (Krug 1997), which is confirmed by numerical computations (Liu and Metiu 1993). Slow coarsening with the exponents 1/4 and 1/6 was also obtained, theoretically and experimentally, in different problems of epitaxial growth and molecular-beam epitaxy. The present exponents are closer to those predicted by Mullins for the evaporation–condensation mechanism, that is, 1/2 (Mullins 1961). One can attribute the fast coarsening here to the convective effect of kinetics, which governs the growth of the crystal surface. Convective effects are known to increase the rate of coarsening in various problems of spinodal decomposition in phase-separating systems.

As has already been mentioned, it is found that the rate of coarsening of the faceted structure resulting from the faceting instability of the growing [110] surface is anisotropic, that is, the growth rates of the structure spatial scales are different in x and y directions, respectively, and are separately measured as $\langle L_{x,y} \rangle = N^{-1} \sum_{i=1}^{N} 2(n_x, n_y)_i^+ / Z_i$, where $(n_x)_i^+ [(n_x)_i^+]$ are the number of points in the ith $x[y]$ layer where $h - \bar{h}$ is positive, and Z_i is the number of zeros of $h - \bar{h}$ in the ith x or y layer. Figure 5.26 shows the coarsening of anisotropic faceted structures for rhombic pyramids (diamonds) and for grooves (circles). The two curves for each structure correspond to the x and y directions in the surface plane; the results are averaged over 10 realizations with different random initial data. The power-law coarsening regime is shown in detail in the inset. The coarsening exponents were found to be 0.61 and 0.41 for rhombic pyramids and 0.57 and 0.23 for grooves. Note that the coarsening of anisotropic facets coarsening was studied theoretically (Song et al. 1997) and experimentally (Yoon et al. 1998) for the case of the faceting transition of a thermally quenched crystal surface when the rate-limiting mechanism of facet growth is collisions between step bunches. The coarsening exponents were computed to be 1/6 and 1/2 for the grooves' characteristic width and length, respectively. This was

5.8 Coarsening

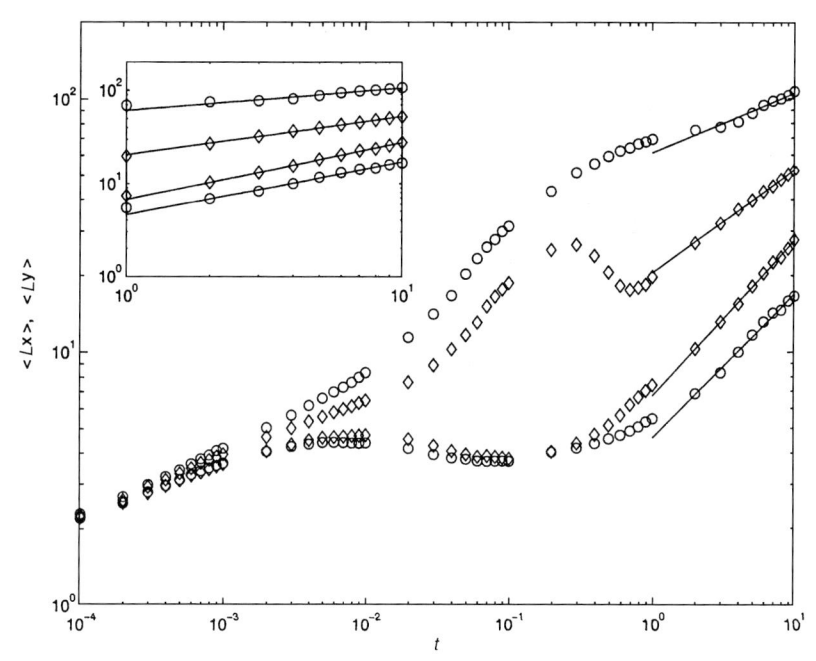

Figure 5.26. The increase of the mean horizontal spatial scales (in arbitrary units) in the x and y directions of the faceted surface structures in the form of rhombic pyramids (rhombs) and grooves (circles) corresponding to the numerical solutions of Eq. (5.136). The inset shows the power-law regime at the late stage of the coarsening. From Golovin et al. (1999). (Reprinted with permission of Elsevier Science B.V.)

confirmed in recent experiments with the faceting of the Si(113) surface for which the exponents for the groove coarsening were found to be 0.164 for the groove width and 0.44 for the groove length. In the present case the groove length also grows much faster in the beginning of the groove formation, but in the power-law regime it grows slower than the groove widths, approaching the limit determined by the computational domain. In the power-law regime the coarsening exponents are found to be $\alpha = 0.23$ for groove length and $\alpha = 0.57$ for groove width. After the groove length has reached the computational domain limit, the structure becomes effectively one-dimensional, and its coarsening is governed by the exponent characterizing the growth of the groove width. This exponent in this highly anisotropic case is close to the value of 1/2 found for the one-dimensional case (Golovin et al. 1998).

Consider now the effect of the driving force Δ being affected by the thermal boundary layer, as described by Eq. (5.119). For the case of growth in the [001] direction, Eq. (5.119) differs from Eq. (5.118) by only the single term containing the Gaussian curvature, and in this case its coefficient is negative for hypercooled solidification in a thermodynamically unstable direction. However, for other systems the coefficient may be positive.

For $g > 0$ the pyramids are more rounded and look more like "square" cones. Such rounded square pyramids as well as almost rounded cones are observed in some epitaxial growth systems (van Erk et al. 1980). It is thus conceivable that the rounding of the pyramid edges in the course of epitaxial growth can be caused by the effect of interaction between the thermal or concentration field and the shape of the crystal surface (if the growth is controlled by the evaporation–condensation mechanism).

For $g < 0$, instead of cones there are now square rounded holes forming on a growing thermodynamically unstable surface. Square, triangular, and spiral holes on the crystal surface are observed in experiments when the crystal surface is evaporating (Bethge and Keller 1974, Lampert and Reichelt 1981, Surek, Hirth, and Pound 1973), and they are believed to be caused by the dislocations meeting the crystal surface. It seems that there are no experimental observations of the formation of holes on a growing crystal surface. However, because the considered model of surface growth is proposed for the evaporation–condensation mechanism, such effects of mass or heat transfer in diffusion boundary layers near the crystal surface are conceivable.

5.9 Remarks

The influence of morphological instability on the anisotropy of surface energy γ and attachment kinetics μ is induced by the presence of crystallographic structure and sets the stage for a competition with the imposed heat-flow direction.

In the case of equilibrium figures, the case of surface-energy anisotropy being "small," $\mathcal{F} = \gamma + \gamma_{\theta\theta} > 0$ always, gives rise to distorted but smooth shapes. When the anisotropy is "large," $\mathcal{F} < 0$ somewhere, the shapes have missing orientations and hence corners where the slope is discontinuous.

The remaining discussion concerned instability and dynamic evolution of interfaces. For "small" anisotropies the bifurcations to squares or hexagons involve imperfections, traveling cells, and new branches that depend on the anisotropy for the existence and stability. Anisotropy of either γ and μ can delay or remove the tendency in two dimensions for tip splitting present in the isotropic case.

For a plane front in one dimension in a hypercooled melt, evolution equations valid in the longwave limit show traveling, asymmetric cells that display anomolous coarsening due to the convective nature of the evolution driven by phase transformation. Here the coarsening rate is $t^{1/2}$ for very long times and t for earlier times, compared with the classical result $l_n t$, for nonconvective systems.

For a planar front in a hypercooled melt of a material with "large" surface tension anisotropy, convective Cahn–Hilliard equations govern the evolution of pyramids and grooves and again lead to anomalously rapid coarsening.

In the process of analyzing such evolutions, Golovin et al. (1998) obtained a general expression for the Gibbs–Thomson equation for pure materials that display surface tension anisotropy. They found that the equilibrium temperature T^e on the interface has the form

$$T^e = T_m \left[1 - \frac{K}{L_v} (\mathcal{F} - \mathcal{F}_1) \right], \quad (5.145a)$$

where

$$\mathcal{F} = \gamma + \gamma_{\theta\theta} \quad (5.145b)$$

$$\mathcal{F}_1 = K \left(1 + \frac{\partial^2}{\partial \theta^2} \right) \frac{\partial \gamma}{\partial K} + \frac{\partial^2 \gamma}{\partial K^2} \left(\frac{\partial K}{\partial \theta^2} \right)^2 \quad (5.145c)$$

and

$$K = -2H \quad (5.145d)$$

This form can be used to write down the chemical potential and hence can be used in problems other than hypercooled solidification such as surface diffusion on thin films.

References

Akaiwa, N., Hardy, S. C., and Voorhees, P. W. (1991). The effect of convection on Ostwald ripening in solid–liquid mixtures, *Acta Metall. Mater.* **39**, 2931–2942.

Alikakos, N., Bates, P. W., and Fusco, G. (1991). Slow motion for the Cahn–Hilliard equation in one space dimension, *J. Diff. Eqs.* **90**, 81–135.

Angenent, S., and Gurtin, M. E. (1989). Multiphase thermomechanics with interfacial structure. 2. Evolution of an isothermal interface, *Arch. Rational Mech. Anal.* **108**, 323–391.

Bar, D. E., and Nepomnyashchy, A. A. (1995). Stability of periodic waves governed by the modified Kawahara equation, *Physica D* **86**, 586–602.

Barabasi, A.-L., Araujo, M., and Stanley, H. E. (1992). Three-dimensional Toom model: Convection to the anisotropic Kardar–Parisi–Zhang equation, *Phys. Rev. Lett.* **68**, 3729–3732.

Bates, P. W., and Fife, P. C. (1993). The dynamics of nucleation for the Cahn–Hilliard equation, *SIAM J. Appl. Math.* **53**, 990–1008.

Bates, P. W., and Xun, J. P. (1994). Metastable patterns for the Cahn–Hilliard equation, Part I, *J. Diff. Eqs.* **111**, 421–457.

———. (1995). Metastable patterns for the Cahn–Hilliard equation, Part 2: Layer dynamics and slow invariant manifold, *J. Diff. Eqs.* **117**, 165–216.

Bates, P. W., Fife, P. C., Gardner, R. A., and Jones, C. K. R. T. (1997). Phase field models for hypercooled solidification, *Physica D* **104**, 1–31.

Bauser, E., and Löchner, K. S. (1981). Steps on facets of solution grown GaAs epitaxial layers, *J. Crystal Growth* **55**, 457–464.

Bethge, H., and Keller, K. W. (1974). Evaporation rate of NaCℓ–H in a vacuum, *J. Crystal Growth* **23**, 105–112.

Bloem, J., Oei, Y. S., de Moor, H. H. C., Hanssen, J. H. L., and Giling, L. J. (1983). Near equilibrium growth of silicon by CVD. 1. The Si-Cℓ–H system, *J. Crystal Growth* **65**, 399–405.

Brattkus, K., and Davis, S. H. (1988). Cellular growth near absolute stability *Phys. Rev. B* **38**, 11452–11460.

Brush, L. N., and Sekerka, R. F. (1989). A numerical study of two-dimensional crystal growth forms in the presence of anisotropic growth kinetics, *J. Crystal Growth* **96**, 419–441.

Cabrera, N. (1963). On stability of structure of crystal surfaces, in *Symposium on Properties of Surfaces*, 24-31, American Society for Testing and Materials, Philadelphia.

Cahn, J. W., and Hilliard, J. E. (1958). Free energy of a nonuniform systems. III. Nucleation in a two-component incompressible fluid, *J. Chem. Phys.* **31**, 688–699.

Cahn, J. W., and Hoffman, D. W. (1974). A vector thermodynamics for anisotropic surfaces–II. Curved and facetted surfaces, *Acta Metall. Mater.* **22**, 1205–1214.

Carlo, A. D., Gurtin, M. E., and Podio-Guidugli, P. (1992). A regularized equation for anisotropic motion-by-curvature, *SIAM J. Appl. Math.* **52**, 1111–1119.

Chang, H.-C., Demekhin, E. A., and Kopelevich, D. I. (1993). Laminarizing effect of dispersion in an active–dissipative nonlinear medium, *Physica D* **63**, 299–320.

Chernov, A. A., Coriell, S. R., and Murray, B. T. (1993). Morphological stability of a vicinal face induced by step flow, *J. Crystal Growth* **132**, 405–413.

Coriell, S. R., and Sekerka, R. F. (1976). The effect of the anisotropy of surface tension and interface kinetics on morphological instability, *J. Crystal Growth* **34**, 157–163.

Emmott, C. L., and Bray, A. J. (1996). Coarsening dynamics of a one-dimensional driven Cahn–Hilliard system. *Phys. Rev. E* **54**, 4568–4575.

Ernst, H.-J., Fabre, F., Folkerts, R., and Lapujoulade, J. (1994). Observation of a growth instability during low-temperature molecular-beam epitaxy, *Phys. Rev. Lett.* **72**, 112–115.

Fogedby, H. C. (1998). Solitons and diffusive modes in the noiseless Burgers equation: Stability analysis, *Phys. Rev. E* **57**, 2331–2337.

Golovin, A. A., and Davis, S. H. (1998). Effect of anisotropy on morphological instability in the freezing of a hypercooled melt, *Physica D* **116**, 363–391.

Golovin, A. A., Davis, S. H., and Nepomnyashchy, A. A. (1998). A convective Cahn–Hilliard formation of facets and corners in crystal growth, *Physica D* **122**, 202–230.

Golovin, A. A., Davis, S. H., and Nepomnyashchy, A. A. (1999). Model for faceting in a kinetically controlled crystal growth, *Phys. Rev. E* **59**, 803–825.

Grimm, H. P., Davis, S. H., and McFadden, G. B. (1999). Steps, kinetic anisotropy, and long-wave instabilities in directional solidification, *Phys. Rev. E* **59**, 5629–5640.

Gurtin, M. E. (1993). *Thermomechanics of Evolving Phase Boundaries in the Plane*, Clarendon Press, Oxford.

Heffelfinger, J. R., and Carter, C. B. (1997). Mechanisms of surface faceting and coarsening, *Surf. Sci* **389**, 188–200.

Herring, C. (1951). Surface tension as a motivation for sintering, in W. E. Kingston, editor, *Physics of Powder Metall.*, Chap. 8, McGraw-Hill, New York.

Hibiya, T., Suzuki, H., Yonenaga, I., Kimura, S., Kawaguchi, T., Shishido, T., and Fukuda, T. (1994). Liquid phase epitaxial growth and characterization of $LiNO_3$ single crystal films, *J. Crystal Growth* **144**, 213–217.

Hoyle, R. B., McFadden, G. B., and Davis, S. H. (1996). Pattern selection with anisotropy during directional solidification, *Phil. Trans. Roy. Soc. Lond. A* **354**, 2915–2949.

Jacob, D., Androussi, Y., Benabbas, T., Francois, P., Ferre, D., Lefebvre, A., Gendry, M., and Robach, Y. (1997). Elastic misfit stress relaxation in $In_{\frac{1}{4}}Ga_{\frac{3}{4}}$ As layers grown under tension on InP(001), *J. Crystal Growth* **179**, 331–338.

Kawahara, T., and Toh, S. (1988). Pulse interactions in an unstable dissipative–dispersive nonlinear system, *Phys. Fluids* **31**, 2103–2111.

Kawakatsu, T., and Munakata, T. (1985). Kink dynamics in a one-dimensional conserved TDGL system, *Prog. Theor. Phys.* **74**, 11–19.

Kawasaki, K., and Ohta, T. (1982). Kink dynamics in one-dimensional nonlinear systems, *Physica A* **116**, 573–593.

Kessler, D., Koplik, J., and Levine, H. (1984). Geometrical models of interface evolution. II. Numerical simulation, *Phys. Rev. A* **30**, 3161–3174.

Knoppik, D., and Lösch, A. (1976). Surface structure and degree of coarsening of {111} NaCl surfaces near the thermodynamic equilibrium between crystal and vapor, *J. Crystal Growth* **34**, 332–336.

Krug, J. (1997). Origin of scale invariance in growth processes, *Adv. Phys.* **46**, 139–282.

Küpper, T., and Masbaum, N. (1994). Simulation of particle growth and Ostwald ripening via the Cahn–Hilliard equation, *Acta Metall. Mater.* **42**, 1847–1858.

Lampert, B., and Reichelt, K. (1981). The investigation of cleavage faces of Sb_2Te_3 by surface decoration, *J. Crystal Growth* **51**, 203–212.

Lemieux, M.-A., Lin, J., and Kotliar, G. (1987). Effects of nonequilibrium kinetics on velocity selection in dendritic growth, *Phys. Rev. A* **36**, 1849–1854.

Leung, K. (1990). Theory of morphological instability in driven systems, *J. Stat. Phys.* **61**, 345–364.

Lifshitz, I. M., and Slyozov, V. V. (1961). The kinetics of precipitation from supersaturated solid solutions, *J. Phys. Chem. Solids* **19**, 35–50.

Liu, F., and Metiu, H. (1993). Dynamics of phase separation of crystal surfaces, *Phys. Rev. B* **48**, 5808–5817.

Madey, T. E., Guan, J., Nien, C.-H., Dong, C.-Z., Tao, H.-S., and Campbell, R. A. (1996). Faceting induced by ultra-thin metal films on W(111) and Mo(111): structures, reactivity, and properties, *Surf. Rev. Lett.* **3**, 1315–1328.

Mazumder, M. K., Mashiko, Y., Koyama, M. H., Takakuwa, Y., and Miyamoto, N. (1995). Generation kinetics of pyramidal hillock and crystallographic defect on Si(111) vicinal surfaces growth with $SiH_2C\ell_2$, *J. Crystal Growth* **155**, 183–192.

McFadden, G. B., Coriell, S. R., and Sekerka, R. F. (1988). Effect of surface tension anisotropy on cellular morphologies, *J. Crystal Growth* **91**, 180–198.

Mullins, W. W. (1961). Theory of linear facet growth during thermal etching, *Philos. Mag.* **6**, 1313–1341.

———. (1963). Solid surface morphologies governed by capillarity, in *Metal Surfaces*, pages 17–66, American Society for Metals, Metals Park, OH.

Mullins, W. W., and Sekerka, R. F. (1963). Morphological stability of a particle growing by diffusion or heat flow, *J. Appl. Phys.* **34**, 323–329.

———. (1964). Stability of a planar interface during directional solidification of a dilute binary alloy *J. Appl. Phys.* **35**, 444–451.

Nien, C.-H., and Madey, T. E. (1997). Atomic structures on faceted W(111) surfaces induced by ultra thin films of Pd, *Surf. Sci.* **380**, L527–L532.

Nishizawa, J.-I., Terasaki, T., and Shimbo, M. (1972). Silicon epitaxial growth, *J. Crystal Growth* **17**, 241–248.

Novick-Cohen, A. (1998). The Cahn–Hilliard equation: Mathematical and modeling perspectives, *Adv. Math. Sci. Appl.* **8**, 965–985.

Novick-Cohen, A., and Segel, L. A. (1984). Nonlinear aspects of the Cahn–Hilliard equation, *Physica D* **10**, 277–298.

Nozieres, P. (1992). Shape and growth of crystals, in C. Godreche, editor, *Solids Far from Equilibrium*, pages 1–154, Cambridge University Press, Cambridge.

Orme, C., Johnson, M. D., Leung, K.-T., Orr, B. G., Smilauer, P., and Vvedensky, D. (1995). Studies of large-scale unstable growth formed during GaAs (001) homoepitaxy, *J. Crystal Growth* **150**, 128–135.

Ratke, L., and Thieringer, W. K. (1985). The influence of particle motion on Ostwald ripening in liquids, *Acta Metall. Mater.* **33**, 1793–1802.

Rednikov, A. Ye., Velarde, M. G., Ryazantsev, Yu. S., Nepomnyashchy, A. A., and Kurdjumov, V. N. (1995). Cnoidal wave-trains and solitary waves in a dissipative-modified Kortweg–de Vries equation, *Acta Appl. Math.* **39**, 457–475.

Rost, M., and Krug, J. (1995). Anisotropic Kuramoto–Sivashinsky equation for surface growth and erosion, *Phys. Rev. Lett.* **75**, 3894–3897.

Rowlinson, J. S., and Widom, B. (1989). *Molecular Theory of Capillarity*, Clarendon Press, Oxford.

Sarocka, D. C., and Bernoff, A. J. (1995). An intrinsic equation of interfacial motion for the solidification of a pure hypercooled melt, *Physica D* **85**, 348–374.

Shinozaki, A., and Oono, Y. (1993). Dispersion relation around the kink solution of the Cahn–Hilliard equation, *Phys. Rev. E* **47**, 804–811.

Siegert, M., and Plishke, M. (1994). Slope selection and coarsening in molecular bean epitaxy, *Phys. Rev. Lett.* **73**, 1517–1520.

Siggia, E. D. (1979). Late stages of spinodal decomposition in binary mixtures, *Phys. Rev. A* **20**, 595–605.

Song, S., Yoon, M., Mochrie, S. G. J., Stephenson, G. B., and Milner, S. T. (1997). Faceting kinetics of stopped Si(113) surfaces: Dynamic scaling and nano-scale grooves, *Surf. Sci.* **372**, 37–63.

Stewart, J., and Goldenfeld, N. (1992). Spinodal decomposition of a crystal surface, *Phys. Rev. A* **46**, 6505–6512.

Surek, T., Hirth, J. P., and Pound, G. M. (1973). The back-force effect at repeated nucleation sources in crystal evaporation and dissolution processes, *J. Crystal Growth* **18**, 20–28.

Tanaka, H. (1997). New mechanism of droplet coarsening in phase-separating fluid mixtures, *J. Chem. Phys.* **107**, 3734–3737.

Trivedi, R. (1990). Effects of anisotropy properties on interface pattern formation, *Appl. Mech. Rev.* **43**, 579–584.

Trivedi, R., Seetharaman, V., and Eshelman, M. A. (1991). The effect of interface attachment kinetics on solidification interface morphologies, *Metall. Trans. A* **22**, 585–593.

Uwaha, M. (1987). Asymptotic growth shapes developed from two-dimensional nuclei, *J. Crystal Growth* **80**, 84–90.

van Erk, W., van Hoek-Martens, H. J. G. J., and Bartals, G. (1980). The effect of substrate orientation of the growth kinetics of garnet liquid phase epitaxy, *J. Crystal Growth* **48**, 621–634.

van Nostrand, J. E., Chey, S. J., and Cahill, D. G. (1995). Surface roughness and pattern formation during homoepitaxial growth of Ge(001) at low temperatures, *J. Vac. Sci. Technol. B* **13**, 1816–1819.

van Nostrand, J. E., Chey, S. J., Hasan, M.-A., Cahill, D. G., and Greene, J. E. (1995). Surface morphology during multilayer epitaxial growth of Ge(001), *Phys. Rev. Lett.* **74**, 1127–1130.

Wagner, C. (1961). Theorie der Alterung von Niederschlägen durch Umlösen (Ostwald-Reifung), *Z. Elektrochem.* **65**, 581–591.
Wan, G., and Sahm, P. R. (1990). Ostwald ripening in the isothermal rheocasting process, *Acta Metall. Mater.* **38**, 967–972.
Wolf, D. E. (1991). Kinetic roughening of vicinal surfaces, *Phys. Rev. Lett.* **67**, 1783–1786.
Wulff, G. (1901). Zur Frage der Geschwindigkeit des Wachsthums und der Auflösung der Krystallflachen, *Z. Kryst. Min.* **34**, 449–530.
Yeung, C., Rogers, T., Hernandez-Machado, A., and Jasnow, D. (1992). Phase-separation dynamics in driven diffusive systems, *J. Stat. Phys.* **66**, 1071–1088.
Yokoyama, E., and Sekerka, R. F. (1992). A numerical study of the combined effect of anisotropic surface energy and interface kinetics on pattern formation during the growth of two-dimensional crystals, *J. Crystal Growth* **125**, 398–403.
Yoon, M., Mochrie, S. G. J., Tate, M. W., Gruner, S. M., and Eikenberry, E. F. (1998). Anisotropic coarsening of periodic grooves: Time-resolved X-ray scattering, *Phys. Rev. Lett.* **80**, 337–340.
Young, G. W., Davis, S. H., and Brattkus, K. (1987). Anisotropic interface kinetics and tilted cells in unidirectional solidification, *J. Crystal Growth* **83**, 560–571.

6

Disequilibrium

Several processes exist in which a high-power electron or laser beam is focused on the surface of a body. If the beam is stationary with respect to the body, a small, perhaps axisymmetric, pool of liquefied metal is formed, as shown in Figure 6.1(a). This state may be steady, though the liquid may undergo convective motions caused by buoyancy or thermocapillary convection.

If the beam is now translated at some speed V_T, the pool translates as well, as shown in Figure 6.1(b). Now the pool is asymmetric, and melting takes place ahead of the beam whereas solidification occurs behind it; typically $V_T \approx 1 - 10$ m/s. At the rear, one might regard the front as undergoing unidirectional solidification at speed $V_T \sin \theta$, where θ measures the angle between the planer solid surface and the front. At such high solidification rates, new, nonequilibrium microstructures are formed in the solid after the liquid freezes.

Boettinger et al. (1984) observed what seem to be two-dimensional *bands* in Ag–Cu alloys in which layers of cells (or dendrites or eutectics) and segregation-free material alternate in the growth direction. Figure 6.2b is a sketch of the configuration. Bands are not a mode that emerges from Mullins-Sekerka theory. Since this work, bands have been seen in many metallic alloy systems, as discussed by Kurz and Trivedi (1990).

As V_T is increased, the bands disappear and only a segregation-free material is produced, which is consistent with a modified version of the Mullins and Sekerka (1964) theory of morphological instability.

The appearance of new states at high solidification speeds suggests that there is a breakdown of thermodynamic equilibrium in the system, and if such a breakdown exists, it is likely to be important first at the solid–liquid interface, the site of the phase transformation.

The basic model in directional solidification involves interface conditions on heat and solute balances as well as conditions of local thermodynamic

6. Disequilibrium

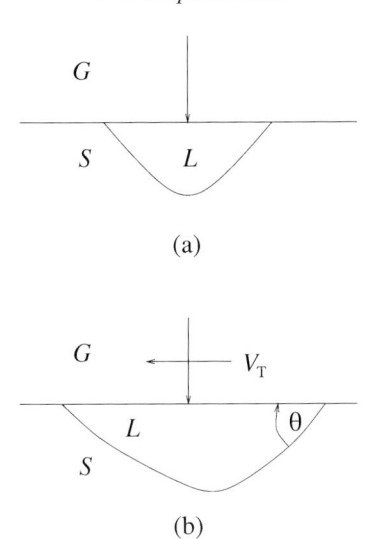

Figure 6.1. A sketch of a liquid pool created by an incident beam (a) with the beam stationary and (b) with the beam translating at speed V_T.

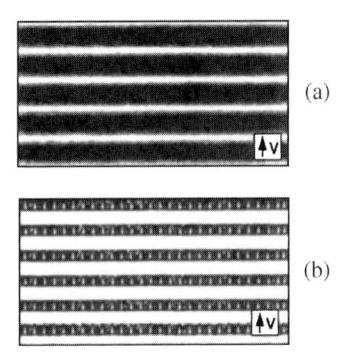

Figure 6.2. Graphical representation of (a) solute bands and (b) bands. For solute bands the concentration of solute, represented by the degree of shading, varies periodically in the pulling direction but does not vary in the lateral direction. For bands the concentration of solute varies periodically in the pulling direction as well, but there is also lateral segregation of solute in the dark regions of the layered structure. These dark regions represent the cellular, dendritic, or eutectic structures that alternate with the lighter-colored, segregation-free structures. From Huntley and Davis (1996). (Reprinted with permission of the American Physical Society.)

equilibrium. In the latter case, there are the Gibbs–Thomson relation

$$T^l = T_e = T_m\left(1 + 2H\frac{\gamma}{L_\nu}\right) + mC \tag{6.1}$$

and the condition defining solute rejection

$$C^s = kC. \tag{6.2}$$

This model is capable of describing, over a wide range of conditions, the formation and evolution of cells and, in principle, the behavior of dendrites. When process conditions are moderate (e.g., V is not too large), one would expect the assumption of local equilibrium to be a good approximation.

In Chapter 2, kinetic undercooling was introduced in situations of pure liquids in which large undercoolings produce large speeds. Thus, additional undercooling

$$T^1 - T_e \propto V_n$$

becomes important at high speeds and, as was seen, in hypercooled liquids. When binary alloys are rapidly solidified, one has to augment the interface conditions on the solute as well. This chapter describes such models and their consequences.

6.1 Model of Rapid Solidification

Mullins and Sekerka (1964) posed local thermodynamic equilibrium at the solid–liquid interface and studied the linear instability of the planar front in directional solidification. They found theoretically that the interface is susceptible to steady cellular instabilities (MS instability) that disappear at sufficiently high pulling speeds at which the instability is suppressed by surface energy at the absolute stability boundary. When the pulling speeds are high enough, the validity of the assumption of local equilibrium at the solid–liquid interface must surely be violated. One would expect here that kinetic undercooling is important.

Because solute rejection is present, one would further expect information derived from the phase diagram to be altered as well. This information incorporated into conditions (6.1) and (6.2) is given by T_m, k, and m. When disequilibrium is present, k and m likely will no longer be take their low-speed values.

Coriell and Sekerka (1983) proposed a model of nonequilibrium interfaces with attachment kinetics in which several functionals of speed, concentration, and temperature replace segregation parameters normally taken from phase disgrams. They posed

$$C^s = k(V_n, C, T^1)C \tag{6.3a}$$

and

$$V_n = f(T^e - T^1, C, T^1). \tag{6.3b}$$

More recently, models have been developed that are simplifications of this general theory in which only the local speed V_n generates disequilibrium; this model will now be discussed.

6.1 Model of rapid solidification

Consider a planar interface moving at speed V and rejecting solute as phase transformation occurs. When V is small enough, the rejected solute has time to diffuse in the liquid, establishing the exponential basic state profile of length scale $\delta_c = D/V$. One describes such a process using the *equilibrium distribution coefficient*, which will now be denoted by k_E.

As V increases, the material rejected has relatively less time to diffuse; if V is comparable to some new speed β_0^{-1}, the interface overtakes a significant amount of rejected solute and incorporates this into the crystal structure of the solid. This *solute trapping*, or *partial solute trapping*, is a nonequilibrium process because the solute cannot "fit" in the crystalline matrix, and a nonequilibrium solid phase is formed.

The process of solute trapping effectively modifies the rejection process by changing k_E to $k(V)$. To be sure, $k(V) \sim k_E$ as $V \to 0$. A form for $k(V)$ was proposed by Aziz (1982) and Jackson, Gilmer, and Leamy (1980) by examining thermodynamics away from equilibrium. They found that

$$k(V_n) = \frac{k_E + \beta_0 V_n}{1 + \beta_0 V_n}, \tag{6.4}$$

where

$$V_n = (V + h_t) \bigg/ \sqrt{1 + |\nabla h|^2}, \tag{6.5}$$

and now the interface is allowed to be nonplanar but not "too" curved.

In Eq. (6.5), as $|V_n| \to \infty$, $k \to 1$, which is called *complete solute trapping*. All the rejected solute is overtaken by the front and incorporated in the solid; there is no solute rejection. Typically, in metals or semiconductors, $\beta_0^{-1} \approx 5$ m/s. Aziz et al. (1986) confirmed form (6.4) quantitatively in silicon alloys for cases in which the planar front is stable. See Table 6.1 for measured values of β_0 in various systems.

Baker and Cahn (1971), Boettinger and Perepezko (1985), and Boettinger and Coriell (1986) argued that if one replaces k_E by $k(V_n)$, then to be self-consistent one should likewise replace the equilibrium liquidus slope m_E by some $m(V_n)$.

Boettinger and Perepezko (1985) and Boettinger and Coriell (1986) derived a thermodynamically consistent expression for $m(V)$ that is represented in terms of the same parameter β_0, namely,

$$m(V_n)/m_E = 1 - \frac{1}{(k_E - 1)} \left\{ k_E - k(V_n) \left[1 - \ell n \frac{k(V_n)}{k_E} \right] \right\}, \tag{6.6}$$

where again V has been replaced by V_n. Here $m \sim m_E$ as $V_n \to 0$ and $m \sim \ell n k_E / k_E - 1$ as $|V_n| \to \infty$.

Table 6.1. *Physical constants for each of the systems considered. The quantities with asterisks attached correspond to values for the pure solvent (no dilute binary component). From Huntley and Davis (1993).*

Parameter	Units	Ag–Cu	Si–Sn	Al–Sn	Al–Fe	Al–Cu
k_E		0.44	0.016	0.0001	0.03	0.10
$\frac{T_M \gamma}{L_v} \times 10^7$	(Km)	1.53	1.5	2.05	1	1
$D \times 10^9$	(m²/s)	1	25	7	1.7	5.0
V_0	(m/s)	1000	58.4	3000	2000	1000
β_0^{-1}	(m/s)	5	17	13	1.7	5.5
m_E	(K/at. %)	−2.8	−4.6	−6.8	−7.29117	−5.37
$G_T \times 10^{-4}$	(K/m)	2	2	4	500	500
$L_v \times 10^{-8}$	(j/s³)	9.7*	41.9	10.7	9.5*	9.5*
$K^\ell \times 10^5$	(m²/s)	13.0*	3.1	3.6	3.7*	3.7*
$K^s \times 10^5$	(m²/s)	17.4*	0.94	7.1	7.0*	7.0*
k_T^ℓ	(J/s Km)	374*	70.0	93	95*	95*
k_T^s	(J/s Km)	429*	22	213	210*	210*

Reprinted with permission of Elsevier Science Ltd.

Given these forms and the existence of linear kinetics, the interface conditions can now be written in what is called the *continuous growth model*. At the interface the temperature is continuous, $T^s = T^\ell$, and

$$T^I = T_m \left[1 + 2H\frac{\gamma}{L_v}\right] + m(V_n)C - \frac{m_E}{k_E - 1}\frac{V_n}{V_0}, \qquad (6.7a)$$

where V_0 is a constant typically about the speed of sound ($\approx 10^3$ m/s in metals); see Table 6.1 for specific values for different materials. Further,

$$C^s = Ck(V_n). \qquad (6.7b)$$

Finally, the Gibbs–Thomson undercooling in Eq. (6.7a) may in fact be altered by disequilibrium (W.J. Boettinger, private communication), though what such changes might be is unknown at present.

When the usual morphological scalings are inserted, two new parameters emerge. These are

$$\beta = \beta_0 V \qquad (6.8a)$$

and

$$\mu^{-1} = V k_E / (k_E - 1)^2 c_\infty V_0. \qquad (6.8b)$$

6.2 Basic State and Linear Stability Theory

Coriell and Sekerka (1983) selected certain forms for the functions in Eqs. (6.3) and for certain parametric values found that the instability of the interface can either be a steady cellular state (a modified MS mode) or an oscillatory state.

For the one-sided *continuous growth model* outlined above and the frozen temperature approximation (FTA) the basic state consists of a planar interface $\bar{h} = 0$, a linear temperature profile $\bar{T} = T_0 + G_T z$, and a concentration profile given by Eq. (3.16a), where now $k = k(V)$. Thus, as V increases from k_E the exponential profile gets flatter, and in the limit $V \to \infty$, the profile becomes flat. These characteristics are consistent with the lowering of the solute rejection with increasing speed.

The temperature T_0 at the interface satisfies Eq. (3.16d), where now both m and k depend on V. It is easy to show, as V increases from zero, that T_0 begins at $T_m + m(0)C_\infty/k(0)$, increases to a maximum, and then decreases owing to the presence of the kinetic undercooling. The difference $T_m - T_0$ is the undercooling of the interface.

Merchant and Davis (1990) used the FTA for the one-sided case and documented the linear stability properties of these two modes. They found an MS mode, steady in time and periodic in space, that is stabilized by disequilibrium and whose neutral curve is unaffected by attachment kinetics. The neutral curve of the cellular mode is shown in Figure 6.3(a). It has a shortwave cutoff, $a = a_s$, and thus if $a > a_s$, surface energy completely stabilizes all disturbances. There is an absolute stability boundary $\Gamma = \Gamma_s$,

$$\Gamma_s = k_E \left[\frac{1 - k_E + (k_E + \beta)\ln\dfrac{k}{k_E}}{(1 - k_E)(\beta + k_E)^2} \right]. \tag{6.9}$$

For low pulling speeds, $\Gamma_s = 1/k_E + O(\beta)$, giving small changes from the MS result. For large pulling speeds, $\Gamma_s \sim k_E \ell n k_E/(k_E - 1)\beta$, which significantly changes the MS boundary. Figure 6.4 shows this in the natural variables $(\mathcal{C}, \mathcal{V})$ defined by Eqs. (3.31). Here, the cellular mode is plotted as a solid curve. Its lower branch has slope -1 corresponding to constitutional undercooling. Its upper branch would have slope $+1$ if disequilibrium were ignored. If disequilibrium is present, the slope becomes $1/2$, showing the large-speed stabilization due to disequilibrium.

The second mode studied by Merchant and Davis (1990) is related to the oscillatory mode found by Coriell and Sekerka (1983) but subject specifically to the continuous growth model. In this FTA limit, the second mode at its minimum has zero (spatial) wave number but is oscillatory in time, as shown

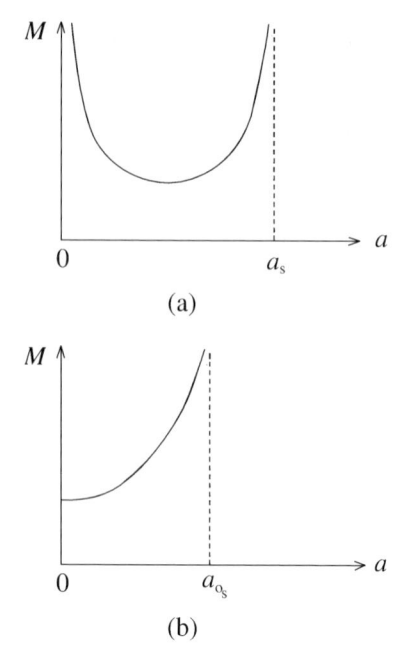

Figure 6.3. Neutral stability curves, M versus a, for (a) the cellular mode with shortwave cutoff a_s and (b) the pulsatile mode with shortwave cutoff a_{0_s}.

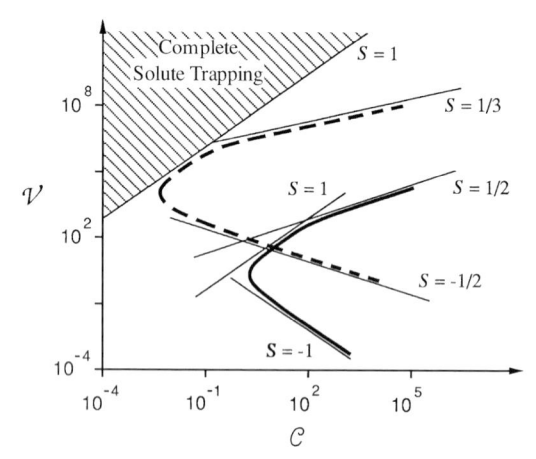

Figure 6.4. Neutral stability curve in the $(\mathcal{C}, \mathcal{V})$ plane for G_T fixed, $k_E = 0.2$, $\beta = 10^{-4}$, and $\mu^{-1} = 10^{-12}$ (such extreme parameters are chosen to highlight the various asymptotic regimes). The dashed line corresponds to the oscillatory instability branch and the solid line to the steady branch. The planar interface is unstable to the right of the curves. The slopes of the asymptotes are denoted by S. The hatches denote the region for which $k(V) \sim 1$. From Merchant and Davis (1990). (Reprinted with permission of Elsevier Science Ltd.)

6.2 Basic state and linear stability theory

in Figure 6.3(b). It has been found that this mode depends on disequilibrium for its existence and is stabilized by kinetic undercooling. There is a shortwave cutoff, $a = a_{0s}$, and an absolute stability criterion for this mode in terms of kinetics; the mode is completely stabilized if $\mu^{-1} \geq \mu_s^{-1}$, where

$$\mu_s^{-1} = \frac{\beta k_E}{(1+\beta)(\beta + k_E)^2}; \tag{6.10}$$

for small β, $\mu_s^{-1} = O(\beta)$ while for large β, $\mu_s^{-1} = O(\beta^{-2})$. For small pulling speeds the neutral curve has the form

$$M \sim \frac{1 + 2k_E}{\beta - k_E \mu^{-1} - (1 + 2k_E)\Gamma a^2}, \quad \beta \to 0 \tag{6.11a}$$

with the corresponding angular frequency ω given by

$$\omega^2 \sim \frac{k_E}{M} - (1 - \Gamma k_E)a^2, \quad \beta \to 0. \tag{6.11b}$$

For large pulling speeds, the neutral curve has the form

$$M \sim \frac{5\beta^2}{2(k_E - \mu^{-1}\beta^2) - 5\Gamma a^2 \beta^2}, \quad \beta \to \infty \tag{6.11c}$$

and

$$\omega^2 \sim \frac{(1 + M\Gamma a^2)(-1 + k_E)\beta}{2Mk_E \ell n k_E}, \quad \beta \to \infty. \tag{6.11d}$$

In Figure 6.5 both modes are simultaneously present with the oscillatory mode terminating on the cellular mode, with the frequency approaching zero at the intersection point. By adjustment of the parametric values, both modes can have the same critical M. As shown, the cellular mode is preferred (is destabilized at lower M) over the oscillatory. Under other conditions, the positions of the relative minima can be reversed, and the oscillatory mode would be preferred.

In Figure 6.4 the upper dashed mode corresponds to the pulsatile case ($a = 0$). When $\mu^{-1} = 0$, the lower and upper branches have slopes $-1/2$ and 1, respectively. However, when kinetics is present, the upper branch is stabilized to slope $1/3$. As μ^{-1} increases, the pulsatile mode moves to the right and disappears as $\mu^{-1} \to \mu_s^{-1}$.

In an experiment in which G_T and C_∞ are fixed and V is increased from zero, Figure 6.4 can be used to interpret transitions. If C_∞ is to the left of both curves, there is no linearized instability. If C_∞ is to the right of the oscillatory nose and to the left of the cellular nose, only pulsatile instabilities are seen. If C_∞ is to the right of both noses but to the left of the intersection of the two curves, the neutral curves, V versus a, look like Figure 6.6.

170   6. *Disequilibrium*

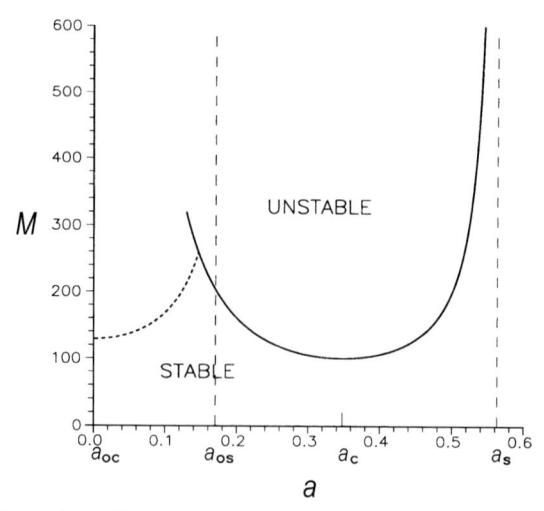

Figure 6.5. Neutral stability curve in the (a, M) plane for $k_E = 0.02$, $\mu^{-1} = 0.001$, $\beta = 0.3$, and $\Gamma = 0.14$. The solid (dotted) curve corresponds to the steady cellular (oscillatory) instability branch. The planar interface is unstable to infinitesimal perturbations above each curve and stable below. The oscillatory and steady wave number cutoffs are denoted by a_{0_s} and a_s, respectively. The critical wave numbers for the onset of the oscillatory and steady instabilities are denoted by a_{0_c} and a_c, respectively. From Huntley and Davis (1993). (Reprinted with permission of Elsevier Science Ltd.)

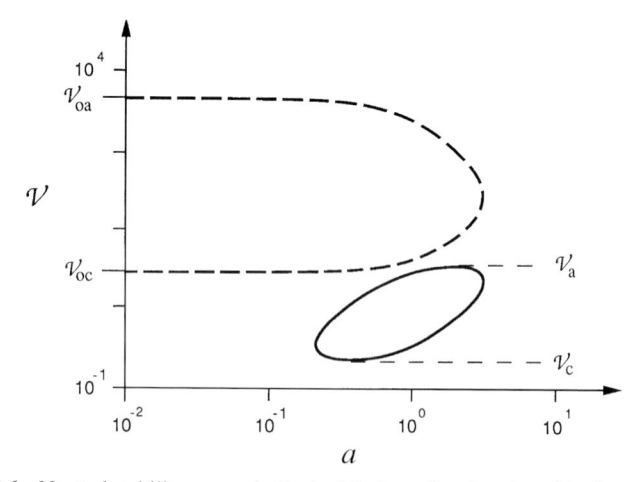

Figure 6.6. Neutral stability curves in the $(a, \mathcal{V})$ plane showing the critical wave number at the onset of instability for $k_E = 0.2$, $\beta = 0.1$, and $\mu^{-1} = 0$. This figure is a cross section of Figure 6.4, and here a denotes the wave number scaled on $\hat{\delta}_T$ of Eq. (3.30). The dashed curve corresponds to the oscillatory branch, and the solid curve to the steady branch. The planar interface is unstable inside the curves. The critical pulling speeds for instability–stability are also marked. From Merchant and Davis (1990). (Reprinted with permission of Elsevier Science Ltd.)

As V crosses V_c, steady cells appear whose wave numbers are near that of the minimum of the solid loop in Figure 6.6, $a_c \approx 0.3$. As V is increased further, the cells should deepen, and a should increase; there may be a cellular–dendritic transition (not covered by the present linear theory), and the cells return as the absolute stability boundary $V = V_A$ is approached. As V increases further, the cells become shallow and finally disappear at $V = V_A$ where the wave number $a_A \approx 2.2$ corresponds to the top of the solid loop in Figure 6.6. Now, since $V_A < V_{0c}$, the planar interface regains stability. The range, $V_A < V < V_{0c}$, where the planar state is stable, is a *window of stability*. As V is increased to $V = V_{0c}$, the oscillatory mode becomes excited with wave number $a_{0c} = 0$. The instability produces *solute bands* shown in Figure 6.2(a), which are solute distributions periodic in the direction of solidification free of lateral segregation. The angular frequency ω_{0c} of the oscillatory instability and the spatial wavelength of the solute band should decrease as V increases. As V is increased further, some nonlinear selection mechanism determines the oscillatory mode in the range $V_{0c} < V < V_{0A}$. It may continue to give solute banding, or some cellular structure may emerge. When $V < V_{0A}$, solute bands are certainly present because the top of the dashed loop in Figure 6.6 has $a_{0c} = 0$. At $V = V_{0A}$ the planar state regains stability. What the system selects away from the linear stability regime is, presumably, determined nonlinear interactions, which are outside the realm of the present theory given that they correspond to parametric regions deep in the unstable regions of the loops.

6.3 Thermal Effects

When the pulling speeds are large, if both D/V and κ/V become very small, and if large wave number instability is important, say, for the oscillatory mode, then the thermal boundary layer likely becomes important and latent heat effects should not be neglected.

Huntley and Davis (1993) generalized the linear stability theory to the case in which the full thermal effects are included subject to the cold boundary condition. In addition to the parameters M, Γ, k_E, m_E, β, and μ, the new parameters that emerge are latent heat

$$\tilde{L} = \frac{L_v V}{k_T^\ell G_T}, \tag{6.12a}$$

the thermal conductivity ratio

$$k_T = \frac{k_T^s}{k_T^\ell}, \tag{6.12b}$$

the temperature gradients

$$\tilde{G}_L = \frac{G_L}{G_T}, \quad \tilde{G}_s = \frac{G_s}{G_T} \quad (6.12c)$$

and the diffusivity ratios

$$D^{(\ell)} = \frac{D}{\kappa^\ell}, \quad D^{(s)} = \frac{D}{\kappa^s}, \quad (6.12d)$$

where the heat balance requires that

$$\tilde{L} = k_T \tilde{G}_S - \tilde{G}_L. \quad (6.13)$$

The FTA is recovered when $k_T = 1$, $\tilde{L} = 0$ and $D^{(s)} = D^{(\ell)} = 0$.
There is a steady basic state solution of the system as follows:

$$C^L = \bar{C}^L(z) = 1 - \frac{k_E}{(k_E + \beta)} \exp(-z) \quad (6.14a)$$

$$C^S = \bar{C}^S(z) = 0 \quad (6.14b)$$

$$T^L = \bar{T}^L(z) = T_0 + \left(\frac{\tilde{G}_L}{D^{(\ell)}}\right)[1 - \exp(-D^{(\ell)}z)] \quad (6.14c)$$

$$T^S = \bar{T}^S(z) = T_0 + \left(\frac{\tilde{L} + \tilde{G}_L}{nD^{(s)}}\right)[1 - \exp(-D^{(s)}z)] \quad (6.14d)$$

$$h = \bar{h} = 0. \quad (6.14e)$$

Here the planar interfacial temperature T_0 is given by

$$T_0 = -M\mu^{-1} + \frac{M}{(k_E - 1)}\left[\frac{\tilde{m}(1)k_E}{k(1)} - 1\right]. \quad (6.14f)$$

This steady state solution will be used as the basic state solution in the linear stability analysis.

6.4 Linear-Stability Theory with Thermal Effects

The linear-stability theory results in the following rather complicated characteristic equation derived by Huntley and Davis (1993):

$$(k_T R_S + R_L)\left\{\frac{k_E}{1+\beta}\left(1 + \frac{\sigma}{k_E + \beta}\right)\tilde{m}(1) - (R + \bar{k} - 1)\right.$$

$$\cdot \left[-\tilde{G}_L M^{-1} - \Gamma a^2 + + \frac{k_E \beta \sigma \ell n(\bar{k}/k_E)}{(k_T + \beta)(k_T - 1)(1 + \beta)} - \mu^{-1}\sigma + \frac{\tilde{m}(1)k_E}{k_E + \beta}\right]\right\}$$

$$+ M^{-1}(R + \bar{k} - 1)\{\tilde{L}\sigma + [\tilde{L} + (1 - k_T)\tilde{G}_L]R_S$$

$$+ \tilde{G}_L(D^{(s)} - D^{(\ell)}) + \tilde{L}D^{(s)}\} = 0, \quad (6.15a)$$

where
$$\bar{k} = \frac{k_E + \beta}{1 + \beta} \tag{6.15b}$$

$$\tilde{m}(1) = 1 - \frac{1}{k_E - 1}\{k_E - \bar{k}[1 - \ell n(\bar{k}/k_E)]\} \tag{6.15c}$$

$$R = \frac{1}{2} + \frac{1}{2}\sqrt{1 + 4\sigma + 4a^2} \tag{6.15d}$$

$$R_S = -\frac{D^{(s)}}{2} + \frac{1}{2}\sqrt{D^{(s)2} + 4\sigma D^{(s)} + 4a^2} \tag{6.15e}$$

$$R_L = \frac{D^{(\ell)}}{2} + \frac{1}{2}\sqrt{D^{(\ell)2} + 4\sigma D^{(\ell)} + 4a^2}. \tag{6.15f}$$

In what follows, some simplifications are made, namely,
$$k_T = 1 \tag{6.16a}$$

and
$$D^{(s)} = D^{(\ell)} \equiv \tilde{D} \tag{6.16b}$$

so that
$$\tilde{G}_L = 1 - \frac{1}{2}\tilde{L}. \tag{6.17}$$

In this simplified case the two phases have equal thermal properties. Then, $\tilde{D}$ measures departures for the temperature field from the linear form (nonthermal steady state), and $\tilde{L}$ measures the latent heat liberated. The aim is to determine how the instabilities found earlier are modified by these thermal effects.

6.4.1 Steady Mode

The steady cellular mode is stabilized by disequilibrium, and the critical conditions are unaffected by kinetics in the FTA. When either $\tilde{D}$ or $\tilde{L}$ is increased from zero, the mode is stabilized, but the shortwave cutoff a_s is independent of the parameters $\tilde{D}$ and $\tilde{L}$, as shown in Figure 6.7. The curve of Figure 6.8 shows for $a = a_c$ the stabilization due to thermal effects and the absence of stabilization in the constitutional undercooling and absolute stability limits.

6.4.2 Oscillatory Mode

In the FTA, the preferred oscillatory mode always has $a \to 0$, and thus it first occurs as a pulsatile mode with frequency $\omega \neq 0$, as shown in Figure 6.5. When

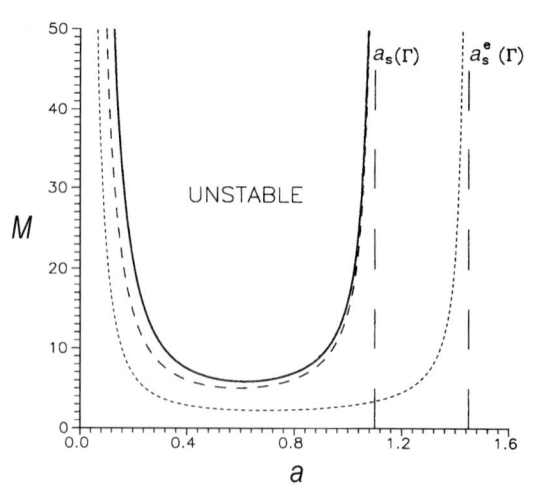

Figure 6.7. The cellular branches of the neutral stability curve, M versus a, for $k_E = 0.2$, $\beta = 0.1$, $\Gamma = 0.4$, $\tilde{D} = 0.2$, and $\tilde{L} = 1.995$. The outermost, short-dotted curve corresponds to the equilibrium case ($\beta = 0$) with the FTA ($\tilde{D} = 0$ and $\tilde{L} = 0$), the long-dashed curve corresponds to the FTA ($\tilde{D} = 0$ and $\tilde{L} = 0$), and the solid curve corresponds to the nonequilibrium model with thermal effects present. The planar interface is linearly unstable above each curve. The steady wave number cutoffs for the equilibrium and nonequilibrium formulations are denoted by a_s^e and a_s, respectively. The critical wave numbers for the different formulations are as follows: local equilibrium $a_c = 0.68$, FTA $a_c = 0.59$, and thermal effects $a_c = 0.61$. From Huntley and Davis (1993). (Reprinted with permission of Elsevier Science Ltd.)

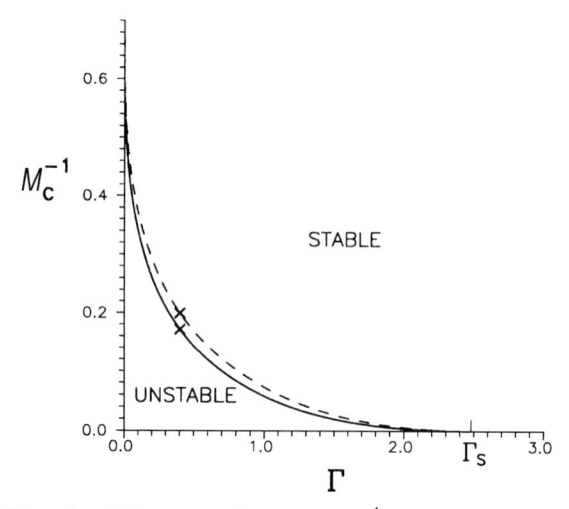

Figure 6.8. Neutral stability curves in the (Γ, M_c^{-1}) plane for $k_E = 0.2$, $\beta = 0.1$, $\tilde{L} = 1.995$, and $\tilde{D} = 0.2$. The solid curve corresponds to the nonequilibrium model with thermal effects present whereas the dashed curve corresponds to the FTA ($\tilde{L} = 0$ and $\tilde{D} = 0$). Note that $M_c^{-1} \to 0$ for both curves as $\Gamma \to \Gamma_s$, the absolute stability boundary. Note also that, as $\Gamma \to 0$, the two curves approach the same intercept. The crosses mark the two points that correspond to Figure 6.7. From Huntley and Davis (1993). (Reprinted with permission of Elsevier Science Ltd.)

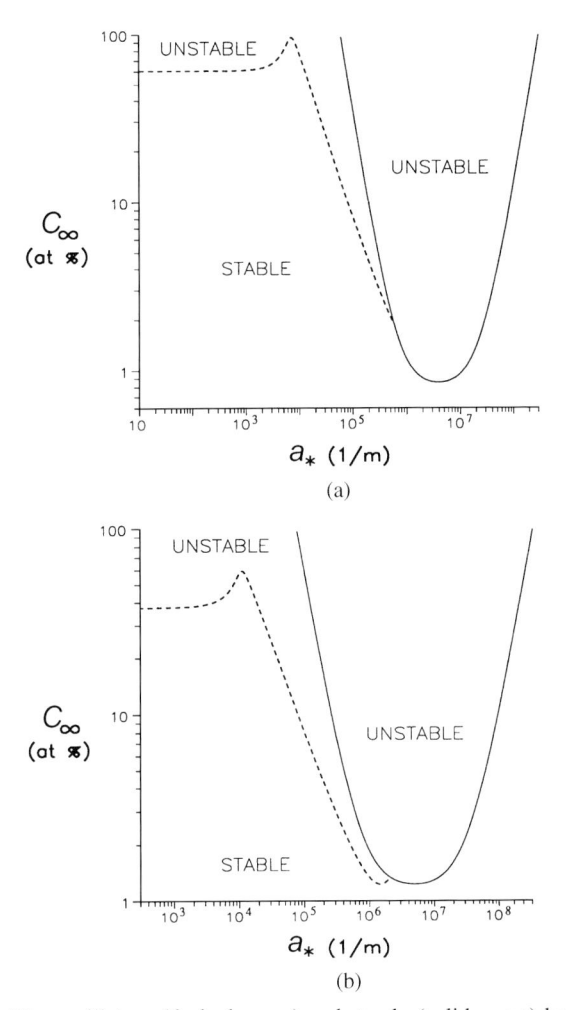

6.4 Linear-stability theory with thermal effects

Figure 6.9. The oscillatory (dashed curve) and steady (solid curve) branches of the neutral stability curve, C_∞ versus a_*, for the Ag–Cu system. The planar interface is linearly stable below these curves and unstable above. In (a) $G_T = 2 \times 10^4$ K/m, $V = 4.1 \times 10^{-2}$ m/s, $a_{*_{0_c}} = 5.8 \times 10^5$ m^{-1}, $\omega^* = 0$, and $a_{*_c} = 3.9 \times 10^{-6}$ m^{-1}. In (b) $G_T = 2 \times 10^4$ K/m, $V = 5.88 \times 10^{-2}$ m/s, and $a_{*_c} = 5 \times 10^6$ m^{-1}; this corresponds to a codimension-two point. From Huntley and Davis (1993). (Reprinted with permission of Elsevier Science Ltd.)

thermal effects are present, there is always a local minimum at $a = 0$, but there can be two other local minima as well. In Figure 6.9(a), the oscillatory mode has an endpoint minimum where its neutral curve intersects that of the cellular mode; this intersection occurs at zero frequency, $\omega = 0$, $a \neq 0$, and constitutes a second cellular mode. Finally, Figure 6.9(b) shows that there can be a local

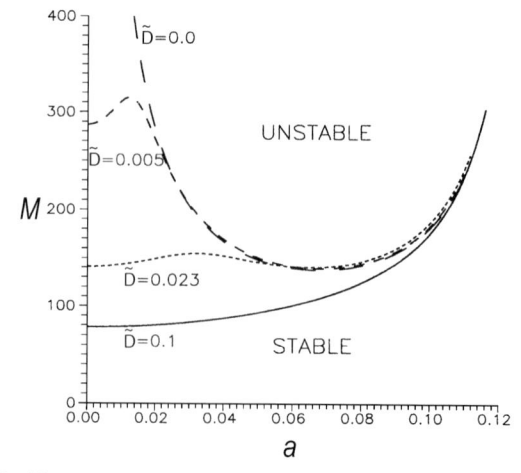

Figure 6.10. Oscillatory branches of the neutral stability curves for $k_E = 0.2$, $\beta = 0.1$, $\Gamma = 2.0$, $\mu^{-1} = 0.001$, $\tilde{L} = 1.995$, and various $\tilde{D}$. From Huntley and Davis (1993). (Reprinted with permission from Elsevier Science Ltd.)

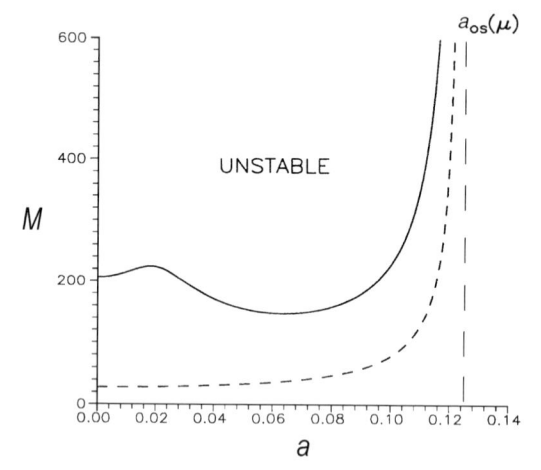

Figure 6.11. Oscillatory branch of the neutral stability curve for $k_E = 0.2$, $\beta = 0.1$, $\Gamma = 2.3$, $\mu^{-1} = 0.001$, $\tilde{L} = 1.995$, and $\tilde{D} = 0.01$. The dashed curve corresponds to the FTA ($\tilde{L} = 0$ and $\tilde{D} = 0$), and the solid curve corresponds to the presence of the full thermal fields. The planar interface is unstable above the curves and stable below. The cutoff wave number is given by $a = a_{0_s}$. From Huntley and Davis (1993). (Reprinted with permission from Elsevier Science Ltd.)

minimum at $a \neq 0$, $\omega \neq 0$, a time-periodic cellular mode. In general, the neutral curve is composed of three pieces.

Figure 6.10 shows how, for $\tilde{L}$ fixed and nonzero, the neutral curve can deform, whereas Figure 6.11 shows how thermal effects stabilize the mode but leave the shortwave cutoff $a_{os}(\mu^{-1})$ unchanged. Figure 6.12 shows the

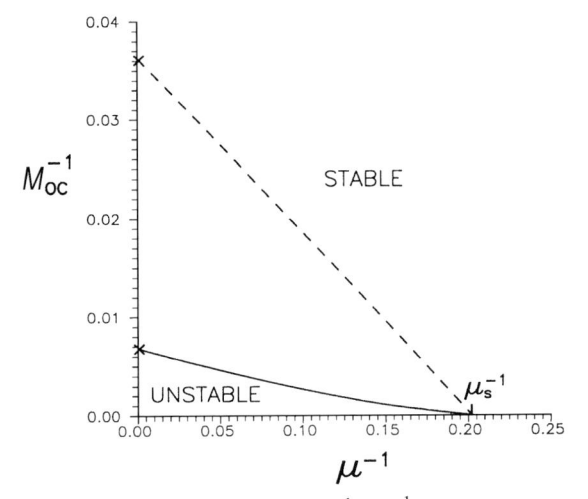

6.4 Linear-stability theory with thermal effects

Figure 6.12. Neutral stability curves in the (μ^{-1}, $M_{0_c}^{-1}$) plane for $k_E = 0.2$, $\beta = 0.1$, $\Gamma = 2.3$, $\tilde{L} = 1.995$, and $\tilde{D} = 0.01$. The solid curve corresponds to the presence of thermal effects, and the dashed curve corresponds to the FTA ($\tilde{L} = 0$ and $\tilde{D} = 0$). Note that $M_{0_c}^{-1} \to 0$ for both curves as $\mu^{-1} \to \mu_s^{-1}$, the absolute stability boundary. The two points that correspond to Figure 6.11 are marked with crosses. From Huntley and Davis (1993). (Reprinted by permission of Elsevier Science Ltd.)

stabilization due to thermal effects and the fact that the absolute stability cutoff does not depend on $\tilde{L}$ and $\tilde{D}$.

6.4.3 The Two Modes

Here, four neutral curves will be overlaid for various materials: two for the cellular mode (FTA and $\tilde{D}, \tilde{L} \neq 0$) and two for the oscillatory mode (FTA and $\tilde{D}, \tilde{L} \neq 0$).

The neutral curves for the Ag–Cu system are represented in Figure 6.13. The oscillatory instability on the neutral curve never has a zero critical wave number, but the associated frequency is zero on the lower branch (short-dotted curve). The triangles correspond to the observed transition (Boettinger et al. 1984) from dendrites to banded structures, and the squares correspond to the observed transition from banded structures to the microsegregation-free structure. Note that there is sizable shift owing to thermal effects of the oscillatory branch versus the FTA oscillatory branch.

The neutral curves for the Si–Sn system are represented in Figure 6.14. The oscillatory instability has zero critical wave number with nonzero associated frequency for smaller pulling speeds (curve denoted by asterisks) and nonzero wave number with zero associated frequency for larger pulling speeds

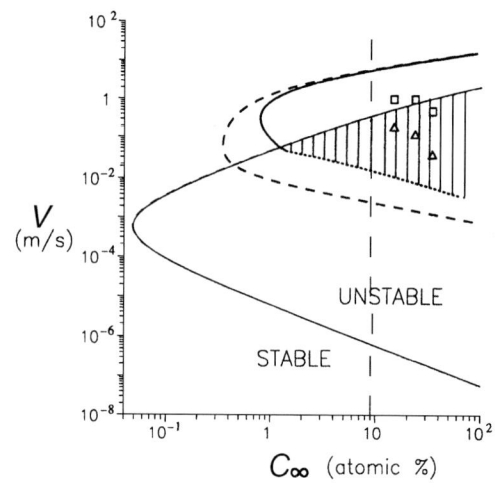

Figure 6.13. The oscillatory and steady branches of the neutral stability curve, V versus C_∞, for the Ag–Cu system with the thermal gradient $G_T = 2 \times 10^4$ K/m. The planar interface is linearly stable to the left of the curves and unstable to the right. The long-dashed curves correspond to the FTA results, and the solid curves correspond to thermal effects present. The two leftmost curves, which here are nearly identical, are the steady branches, whereas the two rightmost curves are the oscillatory branches. When thermal effects are present, the neutral curve for the oscillatory instability is composed of two segments. The lower segment (short-dotted curve) has nonzero critical wave number with an associated frequency of zero (i.e., an endpoint minima), whereas the upper segment (solid curve) has nonzero critical wave number and associated frequency. The vertically hatched region is the coexistence sector. The data points are taken from Boettinger et al. with the triangles corresponding to the observed transition from dendrites to bands and the squares corresponding to the observed transition from bands to microsegregation-free structures. The vertical dashed line denotes the concentration below which no bands were observed. From Huntley and Davis (1993). (Reprinted with permission of Elsevier Science Ltd.)

(short-dotted curve). Notice that the neutral curve for the oscillatory mode is completely contained within the unstable region of the MS mode. The target corresponds to the experiment of Hoglund, Aziz, Stiffler, Thompson, Tsao, and Peercy (1991) in which bands were not seen.

For the cellular mode, the wavelength would be 2500 μm on the lower branch and 0.13 μm on the upper branch. For the oscillatory mode, on the lower branch there is a wavelength of 300 μm with corresponding angular frequency zero, whereas on the upper branch these values are 0.84 μm and zero, respectively.

The neutral curve for the Al–Fe system is represented in Figure 6.15. The oscillatory instability never has a zero critical wave number, but the associated frequency is zero on the lower branch (short-dotted curve). At the initial concentration of 0.17 atom %. The cellular mode has the wavelength 13 μm on the

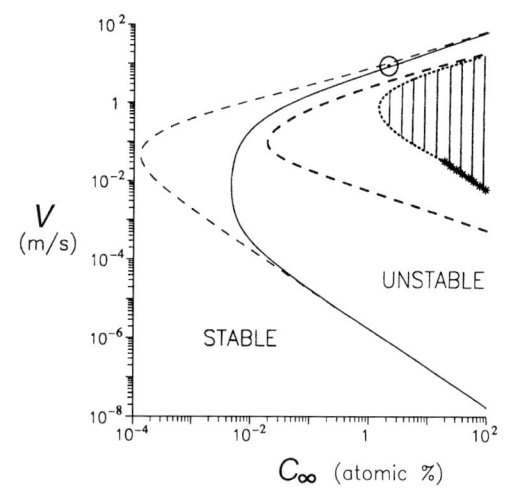

6.4 Linear-stability theory with thermal effects

Figure 6.14. The oscillatory and steady branches of the neutral stability curve for the Si–Sn system with the thermal gradient $G_T = 2 \times 10^4$ K/m. The planar interface is linearly stable to the left of the curves and unstable to the right. The long-dashed curves correspond to the FTA results, and the solid curves correspond to thermal effects present. The two leftmost curves are the steady branches, and the two rightmost curves are the oscillatory branches. When thermal effects are present, the neutral curve for the oscillatory instability is composed of two segments. The lower segment (denoted by asterisks) has zero critical wave number with nonzero associated frequency, whereas the upper segment (short-dotted curve) has nonzero critical wavenumber with an associated frequency of zero (i.e., endpoint minima). The vertically hatched region is the coexistence sector. The target corresponds to the experimental regime investigated by Hoglund et al. (1990). From Huntley and Davis (1993). (Reprinted with permission of Elsevier Science Ltd.)

lower branch and 0.24 μm on the upper. For the oscillatory mode, the lower branch has a wavelength of 18 μm with corresponding frequency zero, whereas on the upper branch the wavelength equals 0.42 μm with a corresponding frequency of $5.3 \times 10^7 \text{s}^{-1}$. The experimental data of Gremaud, Carrard, and Kurz (1990) for the dendrite-banded transition are shown.

The neutral curves for the Al–Cu system are represented in Figure 6.16. The oscillatory instability never has a zero critical wave number, but the associated frequency is zero on the lower branch (short-dotted curve). The noticeable shift in stability for the oscillatory branch occurs when thermal effects are included.

At the representative initial concentration 0.3 atom % the cellular mode has the wavelength 5.7 μm on the lower branch and 0.22 μm on the upper. For the oscillatory mode, the lower branch has a wavelength of 4.5 μm with corresponding angular frequency zero; whereas on the upper branch these values are 0.48 μm and $4.0 \times 10^7 \text{s}^{-1}$. The data of Zimmermann, Carrard, and Kurz (1989) for the dendrite-band transition are shown.

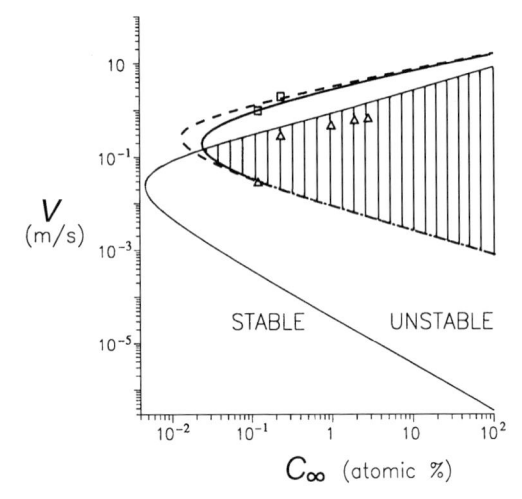

Figure 6.15. The oscillatory and steady branches of the neutral stability curve for the Al–Fe system with the thermal gradient $G_T = 5 \times 10^6$ K/m. The planar interface is linearly stable to the left of the curves and unstable to the right. The long-dashed curves correspond to the FTA results, and the solid curves correspond to thermal effects present. The two leftmost curves, which are nearly identical, are the steady branches, and the two rightmost curves are the oscillatory branches. When thermal effects are present, the neutral curve for the oscillatory instability is composed of two segments. The lower segment (short-dotted curve) has nonzero critical wave number with an associated frequency of zero (i.e., an endpoint minima), whereas the upper segment (solid curve) has nonzero critical wave number and associated frequency. The vertically hatched region is the coexistence sector. The data points are taken from Gremaud et al. with the triangles corresponding to the transition from dendrites to bands and the squares corresponding to the transition from bands to microsegregation-free structures. From Huntley and Davis (1993). (Reprinted with permission of Elsevier Science Ltd.)

The neutral curves for the Al–Sn system are represented in Figure 6.17. The oscillatory instability has zero critical wave number with nonzero associated frequency for smaller pulling speeds (curve denoted by asterisks), nonzero critical wave number with zero associated frequency for an intermediate range in pulling speeds (short-dotted curve), and nonzero critical wave number and associated frequency for larger pulling speeds (solid curve). Note that the shift in stability for the oscillatory branch is over an order of magnitude when thermal effects are included. This gives a value for the cutoff concentration of 0.011 atom %. At the representative initial concentration of 0.30 atom % the cellular mode has the wavelength 1.1×10^4 μm on the lower branch and 0.42 μm on the upper. For the oscillatory mode, the lower branch has a wavelength of 5.4×10^4 μm with corresponding angular frequency zero, whereas on the upper branch these values are 1.2 μm and 4.7×10^6 s^{-1}.

6.5 Cellular modes in the FTA: two-dimensional bifurcation theory

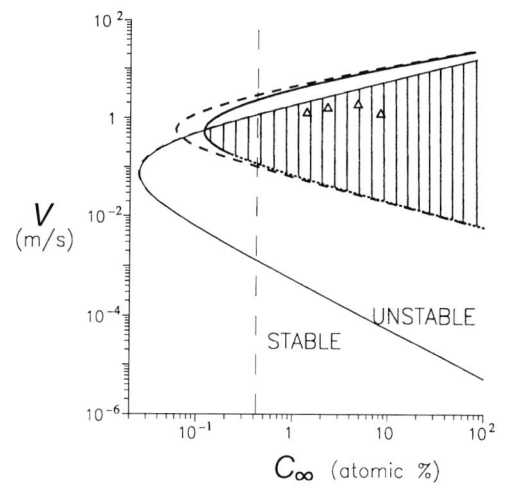

Figure 6.16. The oscillatory and steady branches of the neutral stability curve for the Al–Cu system with thermal gradient $G_T = 5 \times 10^6$ K/m. The planar interface is linearly stable to the left of the curves and unstable to the right. The long-dashed curves correspond to the FTA results, and the solid curves correspond to thermal effects present. The two leftmost curves, which are nearly identical, are the steady branches, and the two rightmost curves are the oscillatory branches. When thermal effects are present, the neutral curve for the oscillatory instability is composed of two segments. The lower segment (short-dotted curve) has nonzero critical wave number with an associated frequency of zero (i.e., an endpoint minima), whereas the upper segment (solid curve) has nonzero critical wave number and associated frequency. The vertically hatched region is the coexistence sector. The data points are taken from Zimmermann et al. (1989) with the triangle corresponding to the transition from dendrites to bands. The vertical dashed line denotes the concentration below which no bands were observed. From Huntley and Davis (1993). (Reprinted with permission of Elsevier Science Ltd.)

6.5 Cellular Modes in the FTA: Two-Dimensional Bifurcation Theory

The discussion in the previous section of the linear stability theory of the *cellular* mode included the facts that an increase in the disequilibrium parameter β results in the stabilization of the high-speed branch of the neutral curve. Further, the neutral curve is independent of the kinetic parameter μ.

Here, weakly nonlinear theory is explored to determine the effects on the Landau constant b_1 of β and μ. It turns out that b_1 depends *strongly* on the value of μ.

The formalism is identical to that given in Chapter 4 and results in an amplitude equation of the type of Eq. (4.3)

$$\frac{dA}{d\tilde{t}} = [\delta - b_1 |A|^2]A, \qquad (6.18)$$

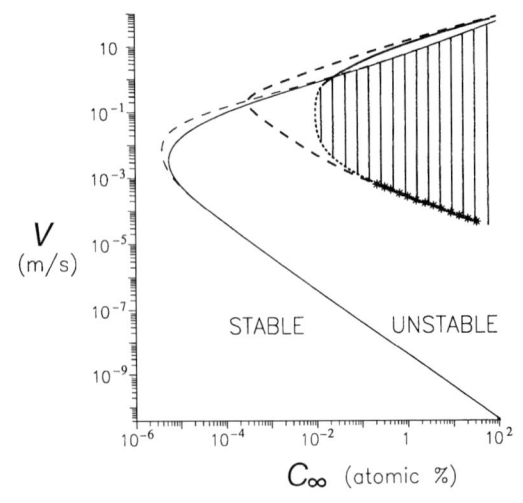

Figure 6.17. The oscillatory and steady branches of the neutral stability curve for the Al–Sn system with thermal gradient $G_T = 4 \times 10^4$ K/m. The planar interface is linearly stable to the left of the curves and unstable to the right. The long-dashed curves correspond to the FTA results, and the solid curves correspond to thermal effects present. The two leftmost curves are the steady branches, and the two rightmost curves are the oscillatory branches. When thermal effects are present, the neutral curve for the oscillatory instability is composed of three segments. The lower segment (denoted by asterisks) has zero critical wave number with nonzero associated frequency, the middle segment (short-dotted curve) has nonzero critical wave number with an associated frequency of zero (i.e., an endpoint minima), whereas the upper segment (solid curve) has nonzero critical wave number and associated frequency. The vertically hatched region is the coexistence sector. From Huntley and Davis (1993). (Reprinted with permission of Elsevier Science Ltd.)

where now b_1 is determined from the governing system, including departures from equilibrium and attachment kinetics, and thus b_1 depends on β and μ as well as Γ and k.

Braun and Davis (1992) found for a given β, k, and Γ that the Landau constant increases strongly with μ^{-1} in a rather complicated manner. Figure 6.18 shows b_1 versus Γ for $k_E = 0.2$ and $\beta = 0$, for increasing μ^{-1}. When $\mu^{-1} = 0$, b_1 is negative for small Γ, which is appropriate near the constitutional undercooling limit, the lower branch of the neutral curve; the bifurcation is subcritical there. When Γ is large enough, b_1 passes through zero and thereafter remains positive, corresponding to supercritical bifurcation.

Figure 6.18 shows the case for $\beta = 0.10$ and various values of μ. There is the tendency for supercriticality to be promoted as μ^{-1} increases. It is, however, apparent that b_1 is singular at a value of Γ equal to $\Gamma_* < \Gamma_s$, that is, before the absolute stability limit. As explained by Braun and Davis (1992), this singularity is created by the presence of the oscillatory mode, which at $\Gamma = \Gamma_*$ increases

6.6 Oscillatory modes in the FTA

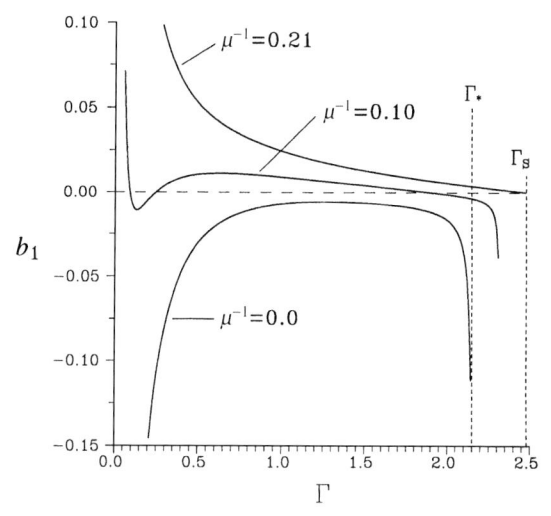

Figure 6.18. The value of the Landau coefficient b_1 for various values of Γ. Here $k_E = 0.2$ and $\beta = 0.10$; μ^{-1} is indicated in the figure. The bifurcation to cells is supercritical (subcritical) for $b_1 > 0$ ($b_1 < 0$). The downward turn in the value of b_1 at $\Gamma = \Gamma_* < \Gamma_s$ is due to a singularity in the Landau constant caused when the minimum of the neutral curve for the cellular mode "moves" inside of the neutral curve for the pulsatile mode. The location of the singularity (Γ_* for the case $\mu^{-1} = 0$) and the absolute stability boundary (Γ_s) are indicated with the short-dashed lines. From Braun and Davis (1992). (Reprinted with permission of Elsevier Science Ltd.)

the dimension of the null space of the linearized stability operator. Figure 6.19 shows the region where $\Gamma < \Gamma_*$ and the minimum of the neutral curves is just to the right of the oscillatory mode. As $\Gamma \to \Gamma_*^-$, the minimum coincides with the new mode, and for $\Gamma > \Gamma_*$ it would lie inside the neutral curve of the oscillatory mode if it existed. Thus, the bifurcation theory employed breaks down near $\Gamma = \Gamma_*$, and a codimension-two point should be analyzed; see Section 6.8. Figure 6.20 shows in the space (C_∞, V) the position of the transition point and the termination of the cellular branch at a finite position.

Hoglund et al. (1991) studied planar to cellular transitions in Si–Sn for V from 2–10 m/s and found good agreement with the linear theory, suggesting that the prediction of smooth transitions is correct.

6.6 Oscillatory Modes in the FTA: Two-Dimensional Bifurcation Theory

The existence of nonequilibrium effects, measured by parameter β, creates the possibility of an oscillatory mode of instability. In the FTA, the preferred mode of linear theory has wave number $a_{c0} = 0$, and thus the mode is pulsatile. This

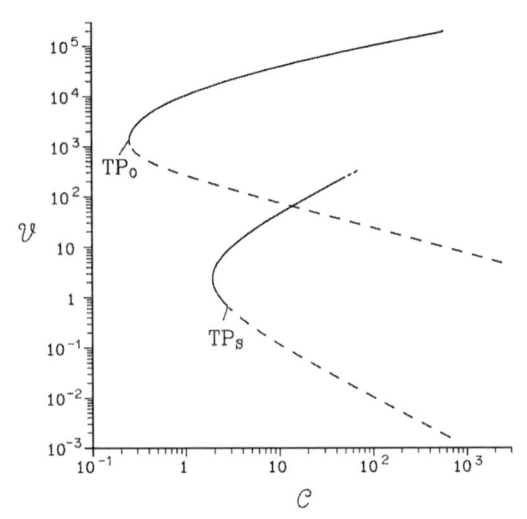

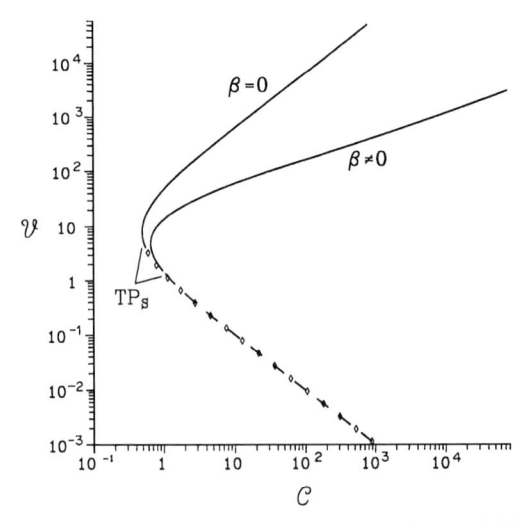

Figure 6.19. Neutral stability curves and bifurcation type in the $(\mathcal{C}, \mathcal{V})$ plane. The upper (lower) paraboloid is the pulsatile (cellular) mode. Bifurcation type is indicated by a solid (dashed) line for supercritical (subcritical) bifurcation. The transition point for the oscillatory mode (TP$_0$) is just below the nose of the curve. The bifurcation to cells appears to be subcritical near the upper end of the cellular branch, but in this region the results are invalid owing to the disappearance of the minimum of the neutral curve. The material parameters are $G_T = 2 \times 10^4$ K/m, $T_M \gamma / L_v = 1.53 \times 10^{-7}$ K/m, $\mu_0^{-1} = 8.57 \times 10^{-3}$ K/m/s, $\beta_0^{-1} = 5$ m/s, $m_E = -4.8$ K/wt%, and $k_E = 0.5$. From Braun and Davis (1992). (Reprinted with permission of Elsevier Science.)

Figure 6.20. Neutral stability curves and bifurcation type in the $(\mathcal{C}, \mathcal{V})$ plane for Si–Sn alloys. The upper (lower) paraboloid is the equilibrium (nonequilibrium) cellular mode. The diamonds indicate subcritical bifurcation for the equilibrium case $\mu_0^{-1} = 8.0$ K/m/s (or $V_0 = 58$ m/s), $\beta_0^{-1} = 17$ m/s. The solid curves denote supercritical bifurcation. The other material parameters are $G_T = 2 \times 10^4$ K/m, $T_M \gamma / L_v = 1.38 \times 10^{-7}$ K/m, $m_E = -4.6$ K/atom %, and $k_E = 0.016$. From Braun and Davis (1992). (Reprinted with permission of Elsevier Science.)

6.6 Oscillatory modes in the FTA

mode is strongly affected in the linear theory by attachment kinetics measured by parameter μ. An absolute stability boundary exists at $\mu^{-1} = \mu_s^{-1}$ above which the oscillatory mode is entirely suppressed.

In this section, weakly nonlinear theory is explored to determine the influence of β and μ. Because the mode has pulsatile or oscillatory behavior, the resulting amplitude equation will have complex coefficients.

Define ϵ by the relation

$$M = M_{c0} + \epsilon^2 \delta, \qquad (6.19a)$$

where M_{c0} is the critical value of the morphological number and $\delta = \text{sgn}(M - M_{c0})$. Given that there is both an oscillation period and a growth rate, two time scales are required, namely,

$$t_0 = \omega(\epsilon)t, \quad t_2 = \epsilon^2 t, \qquad (6.19b)$$

where ω is the angular frequency of the linear theory oscillation and

$$\omega = \omega_0 + \epsilon^3 \omega_3 + \cdots. \qquad (6.19c)$$

Here $\omega_1 = 0$, and ω_2 is taken care of by the slow time.

Because $a_{c0} = 0$, the linear theory gives pulsatile modes, but as M increases from M_{c0}, long spatial structure may emerge. To accommodate this possibility, one allows for two spatial scales, namely x and X,

$$X = \epsilon x. \qquad (6.19d)$$

Now pose an expansion for the dependent variables as follows:

$$[C(x, z, t), h(x, t)] \sim [\bar{C}(z), \bar{h}] + \epsilon\{[A(X, t_2)c_1(z), B(X, t_2)]e^{it_0} + cc\}. \qquad (6.19e)$$

The c_1 is the eigenfunction for the z-structure of C_1 determined by linear theory, and the two complex amplitudes are not independent but are related to each other. Here $\bar{C}$ and $\bar{h}$ are the basic-state quantities identified earlier.

If the forms (6.19) are substituted into the governing system for rapid solidification using the FTA, then a sequence of linear systems is obtained and solved successively. At $O(\epsilon^3)$, an orthogonality equation is obtained. It is a Ginzburg–Landau equation with complex coefficients as follows:

$$\frac{\partial B}{\partial t_2} - d\frac{\partial^2 B}{\partial x^2} = [\lambda\delta - b_1 |B|^2]B. \qquad (6.20)$$

The coefficient d is obtained from the characteristic equation of linear theory and is proportional to $d^2 M/da_1^2$ at $a_1 = a_{0c} = 0$.

This analysis was performed by Braun and Davis (1991), who presented tables of values for the coefficients in Eq. (6.20). All of the coefficients are complex; write

$$\lambda = \lambda' + i\lambda'', \quad b_1 = b' + ib'', \quad \text{and} \quad d = d' + id''.$$

Given that the most dangerous mode has zero wave number, all these coefficients are independent of the surface-energy parameter Γ except d, which provides a means of evolving to nonzero wave numbers. The type of bifurcation is determined by the sign of b'/λ'.

Consider first the pure pulsatile mode by omitting the diffusion term in Eq. (6.20) by formally letting $d = 0$. In this case one can inquire into the type of bifurcation present. Figure 6.21 shows the map of type as functions of μ and β. No oscillations are sustained for $\mu^{-1} > \mu_s^{-1}$, and the dashed line corresponds to the transition point. For small β and μ^{-1} (i.e., for small speeds) there are subcritical bifurcations, and for large β and μ^{-1} supercritical transitions, independent of Γ. The behavior is analogous to the cellular mode, for which small Γ there is subcritical behavior, whereas as $\Gamma \to \Gamma_s^-$, supercritical behavior is expected. Figure 6.22 shows the neutral curve for an Ag–Cu system marked with the transition points.

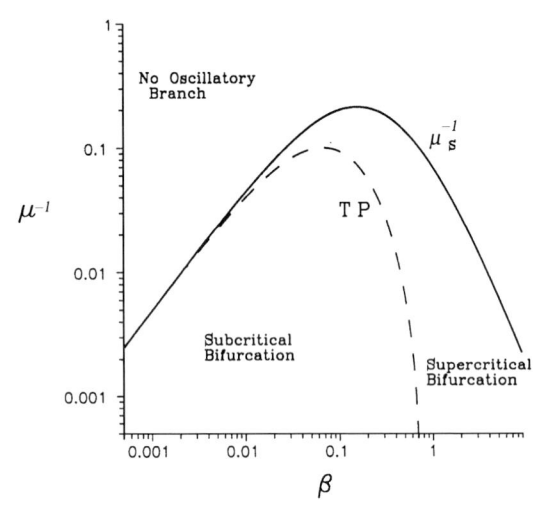

Figure 6.21. A diagram of bifurcation type in the $(\beta, \mu^{-1}0)$ plane for $k_E = 0.2$. The bifurcation of the oscillatory mode for $a = 0$ is unaffected by Γ. The solid curve denotes μ_s^{-1}, and the dashed curve indicates the transition point TP, the zero of g'. The area to the right and above the dashed curve indicates supercritical bifurcation from the planar interface $g' > 0$). The area to the left and below the dashed curve indicates subcritical bifurcation ($g' < 0$); $g' = 0$ for $\mu^{-1} = 0.0$ at $\beta = 0.7265$. From Braun and Davis (1991). (Reprinted with permission of North-Holland.)

6.6 Oscillatory modes in the FTA

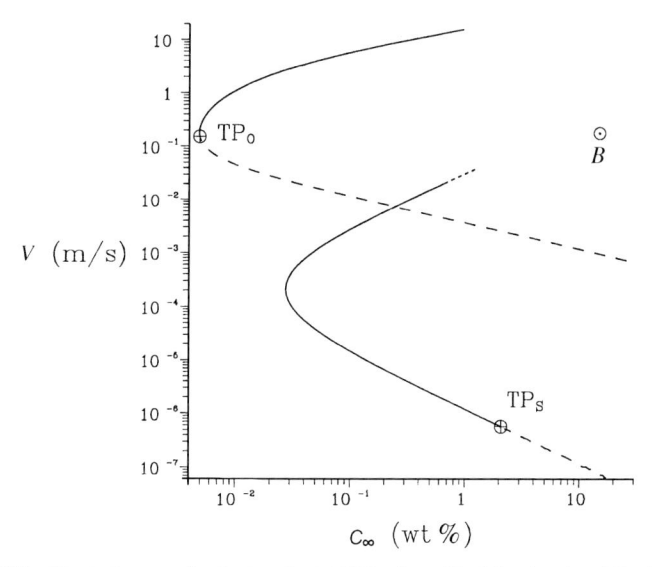

Figure 6.22. Neutral curves for the two instabilities found by Merchant and Davis (1990) for the Ag–Cu system with $m_E = -4.8$ K/wt% and $V_0 = 1000$ m/s. The temperature gradient for the system was taken to be 2×10^4 K/m from the estimate based on heat flow calculations and used by Hoglund et al. (1991). The upper paraboloid corresponds to the pulsatile mode, and the lower paraboloid to the steady mode. The dashed lines denote subcritical bifurcation, and the solid lines denote supercritical bifurcation. The transition point for the steady mode has *not* been calculated herein, but it lies on the lower branch of the neutral curve. The point labeled B denotes the experimental conditions displayed in Figure 11 of Boettinger et al. (1984). From Braun and Davis (1991). (Reprinted with permission of North-Holland.)

Whether or not the pure pulsatile mode can be seen depends on whether it is stable against modulational instabilities, that is, whether or not it survives competition with long waves of finite wavelength. It is the diffusional term $\partial^2 B/\partial x^2$ in Eq. (6.20) that determines this in one space dimension.

Equation (6.20) governs which wave numbers a_1 in a supercritical range are unstable to Eckhaus instabilities. Here, because $a_{0c} = 0$, the analysis determines the range

$$\frac{|a_1 - a_c|}{a_c} > N \tag{6.21}$$

in which there is modulational instability. When one deals with a Ginzburg–Landau equation with real coefficients, $N = 1/\sqrt{3}$ (Eckhaus 1965).

There are two cases for which the present theory delivers useful answers. One is for intermediate values of k_E, β, and μ^{-1} in which the stability region is wider than that for the real coefficient case. Presumably, there is an "island" of parameter values that produces a slightly wider region of stable wave numbers.

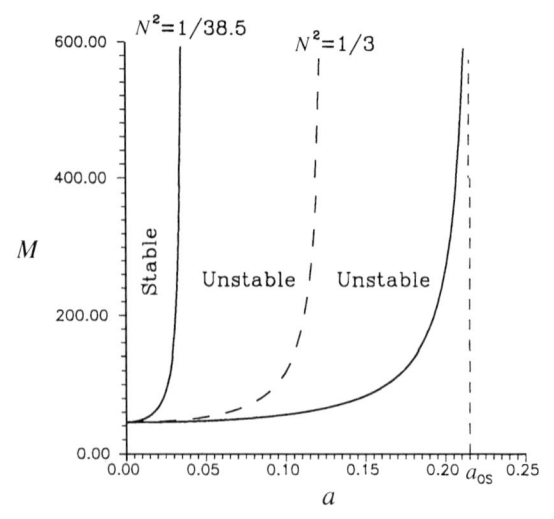

Figure 6.23. Stability of the bifurcated solutions for $k_E = 0.2$, $\Gamma = \Gamma_s$, $\beta = 1.0$, $\mu^{-1} = 0$. The left-hand solid curve indicates the boundary wave numbers within which two-dimensional interfaces are stable against sideband disturbances. From Braun and Davis (1991). (Reprinted with permission of North-Holland.)

All of the other cases result in a narrower region of stable wave numbers; the region is sharply narrowed as β increases. For small β and μ^{-1} near μ_s^{-1}, the stable region approaches that of the real coefficient case, that is, $N^2 \sim 1/3$. For larger β, the narrowing of the region of stable wave numbers indicates a strong preference for the very longest waves. For example, for $\mu^{-1} = 0$ and $\beta \gg 1$,

$$N^2 \sim \frac{\ell n\, k_E}{k_E - 1} \beta^{-2}. \tag{6.22}$$

Also see Figure 6.23. On the basis of this weakly nonlinear theory, one would expect to see approximate solute bands. This is because the very long waves that are selected will have very little lateral segregation, whereas segregation in the direction of growth will occur owing to the pulsations.

Many numerical computations have been made to determine the spatial structure of the solutions to the Ginzburg–Landau equation. The first effort was carried out numerically by Moon, Huerre, and Redekopp (1983). They found a rich variety of solutions close to the onset of instability, including chaos, limit cycles, and quasi-periodic motion. Sirovich and Newton (1986) corrected some of their conclusions and used Floquet theory to find criteria for primary and secondary bifurcation. Perturbation techniques were employed by Bernoff (1988) to analyze the variation of wave trains satisfying Eq. (6.20). In addition, exact solutions have been found by Nozaki and Bekki (1984) and Bekki and Nozaki (1985). These exact solutions are solitary waves, shocklike waves (jumps in wave number), and holes (jumps in phase). The shocklike solutions come about when two regions of different wave number meet (see Bekki and Nozaki 1985 and Bernoff 1988) and a region of rapid transition occurs from one wave number to the other. The hole solutions form when two shocklike solutions meet. For these types of solutions to exist, the wave numbers for the parts of the uniform wave train must be modulationally stable to sideband disturbances according to Stuart and DiPrima (1978). One may expect to see these regions of wave number transition experimentally in stable wave number regimes.

As discussed in Moon et al. (1983) and Kuramoto (1984), for the case when the critical wave number for a Ginzburg–Landau equation is zero, a Kuramoto–Sivashinsky equation may be derived in which the amplitude of the solution remains constant but the phase may vary on the slow time and space scales. If $d' < 0$ in Eq. (6.23), Kuramoto (1984) found that diffusion-induced phase turbulence results. The Ginzburg–Landau equation (6.20) for rapid solidification always has $d' > 0$, and thus one would not expect to see phase turbulence in the present applications.

The form of the evolution equation for the case when a second spatial dimension is present will also be a Ginzburg–Landau equation; the only difference will be that the B_{XX} term becomes $B_{XX} + B_{YY}$ with the same coefficient. The most dangerous mode would still occur at zero wave number. In this case, the analysis of Kuramoto (1984) also applies, but the extra spatial dimension now admits rotating spiral waves as solutions. It is interesting to note that rotating spiral waves have been observed in solidification in the melt-spinning experiments by Thoma et al. (1988). They performed experiments with Ag-19 wt % Cu at wheel speeds of 10–40 m/s and calculated that the average growth rate did not exceed 6 m/s. The wavelength of the spiral far from the core appears to be about 2 μm. The Ginzburg–Landau equation derived above requires that the wavelength of the spiral be long compared with the solute–boundary-layer thickness. The solute diffusivity in the liquid for Ag–Cu is near 10^{-9} m^2/s, whereas the growth rate is near 1 m/s; the solute–boundary-layer thickness is near 10^{-9} m or 10^{-3} μm. Thus, the helical growth observed has a wavelength that is very long compared with the solute boundary layer, and the described model may, indeed, contain the physics necessary to describe this helical growth.

6.7 Strongly Nonlinear Pulsations

In Section 6.6 the weakly nonlinear theory showed that the bifurcation of spatially uniform oscillations from the planar interface is subcritical when nonequilibrium effects are small and supercritical when either β is large or when μ^{-1} is near μ_s^{-1}. When β is small, the nonlinearly stable band of wave numbers has $N = 1/\sqrt{3}$, the Eckhaus limit for the real Ginzburg–Landau equation. As β becomes large, $N \to 0$, and thus very long waves are selected. In both limits of β, the pulsatile mode, $a = 0$, remains stable.

In this section the pulsatile mode will be examined, and it will be demonstrated that *strongly* nonlinear behavior can be described analytically, as shown by Merchant et al. (1992).

6.7.1 Small β

Consider the rapid solidification system with $\partial/\partial x$, $\partial/\partial y = 0$, and let $\beta \ll 1$. From Eqs. (6.11a,b) for $a = 0$

$$M \sim \frac{1 + 2k_E}{\beta - k_E \mu^{-1}}, \quad \omega_0^2 = k_E/M \quad \text{as} \quad \beta \to 0, \tag{6.23}$$

where ω_0 is the angular frequency according to linear theory.

If the two terms in the denominator of Eq. (6.23) are balanced, one can define

$$\mu^{-1} = \frac{\beta}{k_E} - (1 + \bar{\mu}^{-1})\frac{2 + k_E}{k_E^2}\beta^2, \quad M = \bar{M}\beta^{-2}, \quad \bar{M} = O(1) \tag{6.24}$$

consistent with the characteristic equation. Here $\bar{\mu}^{-1}$ is a detuning parameter for μ^{-1} near μ_s^{-1}. An increase in $\bar{\mu}^{-1}$ results in a decrease in μ^{-1}, and thus $\bar{\mu}^{-1}$ measures the deviation from the absolute stability boundary for the pulsatile mode. Here, if $\bar{\mu}^{-1} = 0$, $\mu^{-1} = \mu_s^{-1}$. The restriction of μ^{-1} to $|\mu^{-1} - \mu_s^{-1}| \ll 1$ leads to a uniform description of the bifurcation, even in the subcritical case.

From Eq. (6.23), the critical morphological number corresponds to

$$M \sim \frac{(1 + 2k_E)k_E}{(2 + k_E)\bar{\mu}^{-1}}. \tag{6.25}$$

Asymptotic analysis of the full characteristic equation suggests that there are two time scales

$$t_1 = \beta t \quad \text{and} \quad t_2 = \beta^2 t. \tag{6.26a}$$

The coordinate in the direction of growth remains unchanged, $Z = z$. Map the interface into a fixed domain by letting $\zeta = Z - H(t_1, t_2)$. In the liquid $\zeta > 0$,

$$C_{\zeta\zeta} + C_\zeta - \beta[C_{t_1} - H_{t_1}C_\zeta] - \beta^2[C_{t_1} - H_{t_1}C_\zeta] = 0; \tag{6.26b}$$

6.7 Strongly nonlinear pulsations

on the interface at $\varsigma = 0$,

$$C_\varsigma - (1 + \beta H_{t_1} + \beta^2 H_{t_2})[(k_E - 1)C + 1]\frac{(k-1)}{k_E - 1)} = 0,$$

$$\beta^2 M^{-1} H + \left[\frac{\beta}{k_E} - (\bar{\mu}^{-1} + 1)\frac{2 + k_E}{k_E^2}\beta^2\right](\bar{V} - 1) - C\tilde{m}(\bar{V}) \quad (6.26c)$$

$$- \frac{k_E}{(k_E - 1)}\left[\frac{\tilde{m}(\bar{V})}{k_E} - \frac{\tilde{m}(\bar{V})}{\bar{k}}\right] = 0;$$

and as $\varsigma \to 0$,

$$C \to 1. \quad (6.26d)$$

The domain of the problem is now $0 < \varsigma < \infty$. Also,

$$k(\bar{V}) = \frac{k_E + \beta \bar{V}}{1 + \beta \bar{V}}, \quad (6.26e)$$

$$\bar{V} = 1 + \beta H_{t_1} + \beta^2 H_{t_2}. \quad (6.26f)$$

Pose the following expansions:

$$C(z, t) = C_0 + \beta C_1 + \cdots, \quad (6.26g)$$

$$h(t) = \frac{1}{\beta}(H_0 + \beta H_1 + \cdots). \quad (6.26h)$$

The latter form indicates that the presence of disequilibrium effects drives very large amplitude oscillations compared with the steady-solute boundary-layer thickness δ_c.

At leading order, one obtains, for $\varsigma > 0$

$$C_{0_{\varsigma\varsigma}} + C_{0_\varsigma}(1 + H_{0t_1}) = 0; \quad (6.27a)$$

on $\varsigma = 0$,

$$C_{0_\varsigma} - (1 + H_{0t_1})[1 + C_0(k_E - 1)] = 0, \quad (6.27b)$$

$$C_0 = 0; \quad (6.27c)$$

and, as $\varsigma \to \infty$,

$$C_0 \to 1. \quad (6.27d)$$

The solution to problem (6.27) is

$$C_0 = 1 - \exp[-(1 + H_{0t_1})\varsigma], \quad (6.28)$$

where the interface position H_0 is arbitrary at this order. Note that the solute field is explicitly quasi-steady but depends strongly on time through H_{0t_1}.

Define $W \equiv 1 + H_{0t_1}$ and, at $O(\beta)$ obtain, for $\zeta > 0$,

$$C_{1\zeta\zeta} + C_{1\zeta}W = -(H_{1t_1} + H_{0t_1})C_{0\zeta} + C_{0t_1}; \qquad (6.29a)$$

on $\zeta = 0$

$$C_{1\zeta} - W(k_E - 1)C_1 - H_{1t_1} = H_{0t_2} - W^2, \qquad (6.29b)$$

$$-C_1 + \bar{M}^{-1}H_0 + \frac{1}{k_E}W = 0; \qquad (6.29c)$$

and, as $\zeta \to \infty$,

$$C_1 \to 0. \qquad (6.29d)$$

The Fredholm alternative must be enforced in order to eliminate secular terms that grow algebraically in ζ. This is most easily done by integrating Eq. (6.29a) over the domain in ζ and using the boundary conditions to obtain the compatibility condition

$$\frac{\partial^2 H_0}{\partial t_1^2} + \omega_0^2 H_0 \left(1 + \frac{\partial H_0}{\partial t_1}\right)^3 = 0, \qquad (6.30)$$

where ω_0^2 is given by Eq.(6.23). To leading order, the location of the interface is thus governed by the preceding nonlinear oscillator equation. The first integral may be calculated by multiplying Eq. (6.30) by H_{0t_1} and integrating to obtain

$$\omega_0^2 H_0^2 + \frac{(H_{0t_1})^2}{(1 + H_{0t_1})^2} \equiv A^2(t_2). \qquad (6.31)$$

The phase plane for Eq. (6.31) is shown in Figure 6.24. For amplitudes $A < 1$, the orbits form closed curves, indicating periodic behavior. As the amplitude approaches unity, the velocity becomes unbounded. For $A \geq 1$, the curves are no longer closed, indicating nonperiodic behavior. The energy (or amplitude) evolves on the slow time, and, generally, the frequency is amplitude dependent and thus also evolves on the slow time.

An implicit solution to Eq. (6.31) may be found by integrating once more,

$$H_0 = [A(t_2)/\omega_0] \sin[\omega_0(t_1 + H_0) + \phi(t_2)]. \qquad (6.32)$$

Some results of the numerical calculations of H_0 are shown in Figure 6.25. Note that the velocity profiles become more peaked as the amplitude increases toward unity. The motion becomes more like a relaxation oscillation as the amplitude

6.7 Strongly nonlinear pulsations

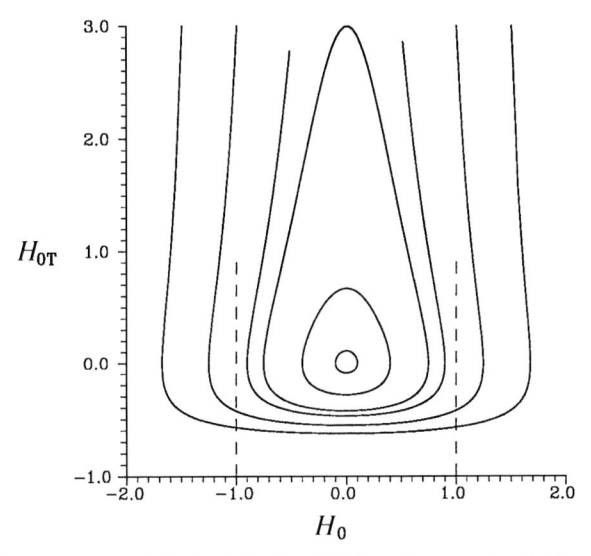

Figure 6.24. Phase plane (H_0, H_{0T}) for Eq. (6.34) with $\omega_0 = 1$ valid for $\beta \ll 1$. The closed curves near the origin are nearly circular, but as $A \to 1$, the maxima tend to infinity. For $A \geq 1$, the curves are no longer closed, and the solutions are no longer periodic. From Merchant et al. (1992). (Reprinted with permission of SIAM.)

increases, and it can also be seen that the period is constant even though the amplitude has changed; the solution (6.32) has the surprising property that the period is independent of the amplitude. This constancy of period may be seen by integrating Eq. (6.31) around a closed contour in the phase plane, leading to

$$\oint \left(\frac{1}{\sqrt{A^2 - \omega_0^2 H_0^2}} - 1 \right) dH_0 = \oint dt_1 \equiv T_1, \quad (6.33)$$

where T_1 is the period, and the left-hand side is easily found to be independent of A. As will be seen, the fact that $\partial T_1 / \partial A = 0$ simplifies the analysis considerably from the general Kuzmak–Luke procedure (see Kevorkian and Cole 1981 and Bourland and Haberman 1988); this simplification occurs only for the case of small β.

The solution of the solute field in the liquid is

$$C_1 = \left[\left(\bar{M}^{-1} H_0 + \frac{1}{k_E} W \right) + \left(H_{1t_1} + H_{0t_2} - \frac{W_{t_1}}{W^2} \right) \zeta - \frac{W_{t_1}}{2W} \zeta^2 \right] \exp[-W\zeta]. \quad (6.34)$$

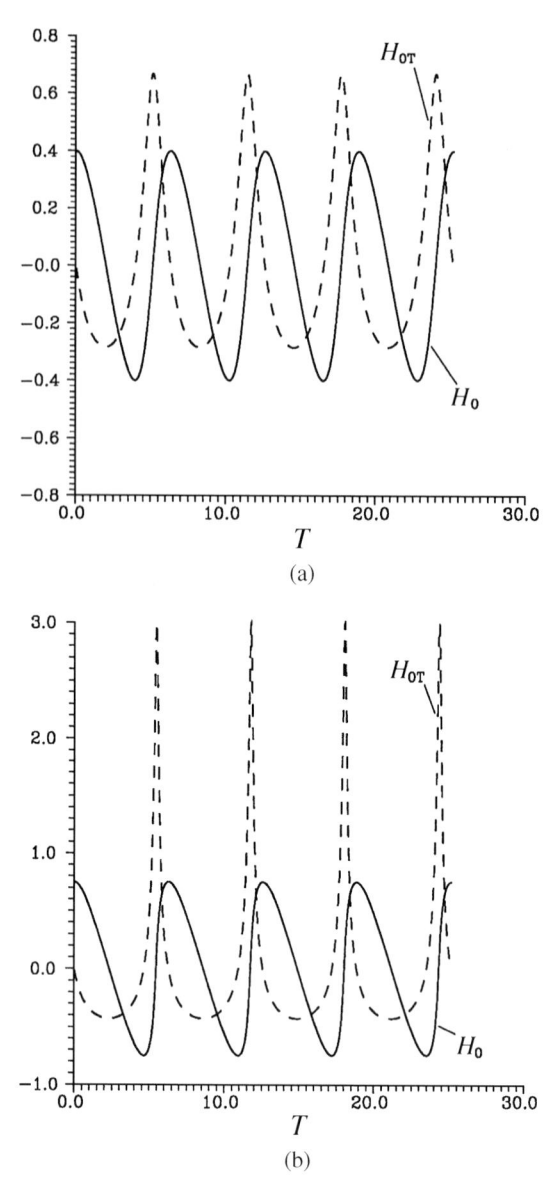

Figure 6.25. Solution (6.35) valued for $\beta \ll 1$. The solid curves give the interface position H_0, and the dashed curves give the interface velocity H_{0T}. Here (a) at $T = 0$, $H_0 = 0.4$, $H_{0T} = 0$, and (b) at $T = 0$, $H_0 = 0.75$, $H_{0T} = 0$. From Merchant et al. (1992). (Reprinted with permission of SIAM.)

6.7 Strongly nonlinear pulsations

At $O(\beta^2)$, the following problem is obtained for $\zeta > 0$:

$$C_{2\zeta\zeta} + C_{2\zeta}W = C_{0t_2} + C_{1t_1} - (C_{1\zeta} + C_{0\zeta})(H_{1t_1} + H_{0t_1}) \qquad (6.35\text{a})$$

with boundary conditions on $\zeta = 0$,

$$C_{2\zeta} - W(k_E - 1)C_2 - H_{2t_1} = -C_1(k_E - 1)(-H_{1t_1} - H_{0t_2} + W^2)$$
$$- 2W(H_{1t_1} + H_{0t_2}) + 3W^2 H_{0t_1} + H_{0t_1}^3 + H_{1t_2}$$
$$- C_2 + \bar{M}^{-1}H_1 + \frac{1}{k_E}H_{t_1}$$
$$= \frac{1}{k_E^2} - \frac{1}{k_E}\left(\frac{1}{2}H_{0t_1}^2 + H_{1t_1} + H_{0t_2}\right)$$
$$- \left(\frac{2 + k_E}{k_E^2}\right)(\bar{\mu}^{-1} + 1)H_{0t_1}, \qquad (6.35\text{b})$$

and, as $\zeta \to 0$

$$C_2 \to 0. \qquad (6.35\text{c})$$

Proceeding as before, one obtains a compatibility condition that is the following forced, damped oscillator equation:

$$\left[\frac{H_{1t_1}}{W^3}\right]_{t_1} + \omega_0^2 H_1 = -\frac{H_{0t_1}}{W^2}\left(\bar{M}^{-1} - 2\omega_0^2\right) - k_E(1 + 2k_E)\bar{M}^2 H_0^2$$
$$- \frac{2H_{0t_1 t_2}}{W^3} - \frac{3\omega_0^2 H_0 H_{0t_2}}{W} + \frac{\bar{\mu}^{-1}(2 + k_E)}{k_E}H_{0t_1}$$
$$- \left(\frac{2 + k_E}{2k_E}\right)H_{0t_1}^2 - \bar{M}^{-1}(k_E - 1)W H_0. \qquad (6.36)$$

The homogeneous solutions to the linearized oscillator operator are H_{0t_1} and $\partial H_0/\partial A$, which has the form $H_0(1 + H_{0t_1})$. Both of these solutions are $2\pi/\omega_0$-periodic in t_1, given the constant period. To preserve periodicity on the fast time (or equivalently, to eliminate secular terms) the forcing function in Eq. (6.36) must be orthogonal to each of the homogeneous solutions. The forcing function is integrated against H_{0T} over a period; because H_{0T} is odd in T, and using parity considerations, the amplitude evolution equation is obtained

$$A_{t_2} = -\frac{(1 + 2k_E)}{4\bar{M}}A + \frac{(2 + k_E)}{2k_E A}\left\{\bar{\mu}\left(\frac{1}{\sqrt{1 - A^2}} - 1\right) - 2 - \frac{(3A^2 - 2)}{(1 - A^2)^{3/2}}\right\}. \qquad (6.37\text{a})$$

Note that this amplitude equation is regular $A \to 0$ and that it is singular as $A \to 1$. The evolution equation (6.37a) is not valid as A approaches unity because of the singularity in H_{0t_1} as $A \to 1$. If the forcing function of Eq. (6.36) is integrated against $H_0(1 + H_{0t_1})$, one obtains the phase evolution equation

$$\phi_{t_2} = \frac{\omega_0(-1 + k_E)}{\bar{M}} \frac{1}{(1 + \sqrt{1 - A^2})}. \qquad (6.37b)$$

These evolution equations are strongly nonlinear and thus allow for the capture of the entire solution branch.

The evolution of the phase is determined by the evolution of the amplitude, and the amplitude is unaffected by the phase. Advantage is taken of this feature to determine the bifurcation structure of the solutions (6.37a), which is of primary interest here.

Some solution branches are pictured in Figure 6.26 for various values of the attachment kinetics parameter $\bar{\mu}^{-1}$. These steady branches (on the t_2 time scale) were found numerically by solving the steady form of Eq. (6.37a). Increasing $\bar{\mu}^{-1}$ corresponds to decreasing μ^{-1}, as can be seen in Eq. (6.24). When $\bar{\mu}^{-1} = 2.0$, the bifurcation switches type near the critical $\bar{M}$ from supercritical for $\bar{\mu}^{-1} < 2.0$ to subcritical for $\bar{\mu}^{-1} > 2.0$. As $\bar{\mu}^{-1}$ becomes large, μ^{-1}

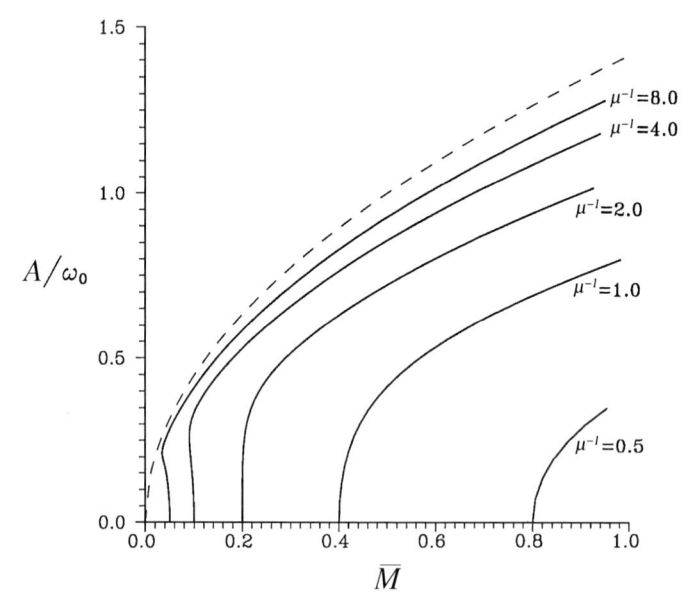

Figure 6.26. Bifurcation diagram from Eq. (6.40a) valid for $\beta \ll 1$ for various values of the kinetics detuning parameter $\bar{\mu}^{-1}$ and for $k_E = 0.5$. As $\bar{\mu}^{-1}$ increases, μ^{-1} decreases. The solid curves are solution branches bifurcating from the planar interface basic state for various values of $\bar{\mu}^{-1}$. The bifurcation is supercritical when $\bar{\mu}^{-1} < 2.0$ and subcritical when $\bar{\mu}^{-1} > 2.0$. The dashed curve corresponds to $A = 1$, and no oscillatory solutions exist for values above this curve. From Merchant et al. (1992). (Reprinted with permission of SIAM.)

6.7 Strongly nonlinear pulsations 197

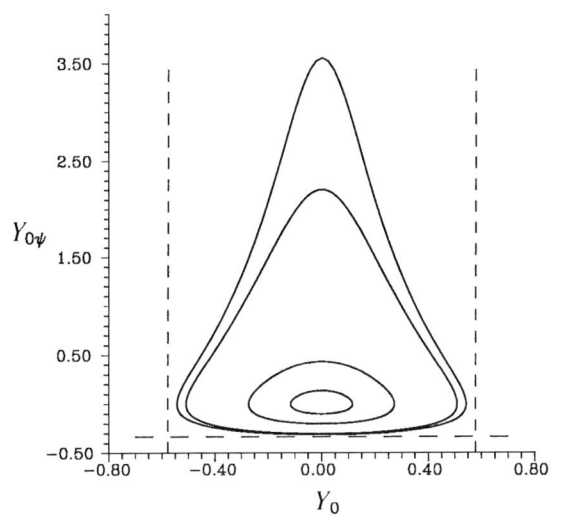

Figure 6.27. Phase plane (Y_0, $Y_{0\psi}$) for Eq. (6.56) valued for $\beta \gg 1$. The closed curves close to the origin are nearly circular (note the different scales on the axes), but as $B \to 1/\sqrt{3}$, the maxima tend to infinity. For $B \geq 1/\sqrt{3}$, the curves are no longer closed, and the solutions are no longer periodic. From Merchant et al. (1992). (Reprinted with permission of SIAM.)

tends toward zero, and the solution branch tends nonuniformly to a subcritical branch with no limit point. The solution branches are stable above the limit point when the bifurcation is locally subcritical, and the entire branch is stable when the bifurcation is supercritical. Note that the frequency does not depend on $\bar{M}$.

Owing to the complicated structure of the evolution equations, the presence of secondary bifurcations would be suspected; however, examination of the Jacobian for the system of evolution equations indicates that no secondary, purely temporal bifurcation occurs. Thus, Figure 6.27 contains the entire bifurcation structure for the amplitude. No bifurcation occurs in the phase either.

The forms of the solution branches suggest that the amplitudes of the oscillations tends to become substantial rather rapidly for bifurcation of either type. This implies that the oscillations that will be seen in an experiment will be more like the oscillations displayed in Figure 6.25(b); that is, they will be relaxation-like. This seems to be consistent with the banded structure observed by Boettinger et al. (1984) in which it is seen that, within a wavelength, the cellular region appears and dies out rather quickly. This is evidence of a relaxation-type modulation of the cells; during the low-velocity portions of the cycles, the segregation coefficient is relatively small and the cells grow. In the relatively short portion of the cycle when the velocity is high, the segregation coefficient rapidly approaches unity, and relatively little solute is rejected.

For this analysis to be physically relevant, one would expect to see the frequency ω observed experimentally, perhaps as the frequency of the cell modulation noted in banding phenomena as studied by Boettinger et al. (1984). On the basis of the material properties and experimental conditions, the cells observed occur at $\beta = 0.034$, and the period of the bands is about 170 diffusion times. The period of the oscillations here is expected to be on the $T = \beta t$ time scale, or a scale of about 30 diffusion times. Using the asymptotic form of the frequency from the linear theory

$$\omega_0^2 = \frac{2 + k_E}{\bar{\mu}^{-1}(1 + 2k_E)}, \qquad (6.38)$$

and *assuming* that $V_0 = 1000$ m/s, one should obtain a period of $2\pi/\beta\omega_0 \approx 80$ diffusion times. The period increases with morphological number to the power $1/2$. The characteristic time for growth or decay of the oscillations is on the $t_2 = \beta^2 t$ time scale and is expected to be on the order of 900 diffusion times. The time scales are of roughly the correct order.

For completeness, the results are discussed in experimentally relevant terms. Experimentally controllable parameters are the speed of the front V and the concentration of the solute C. Typical results of linear stability analysis of the planar interface basic state are shown in Figure 6.4. Small β corresponds to "low" speed, and thus this analysis operates near the lower branch of the oscillatory curve. The experimental conditions under which Boettinger et al. observed bands is just the region where the oscillatory instability may interact with the steady cellular instability (Section 6.8), and this may be the cause of banding phenomena, as suggested by Merchant and Davis (1990). Time scale estimates seem to support this speculation. Furthermore, the period (in dimensional terms) decreases with increasing pulling speed, as observed experimentally by Gremaud et al. (1990).

6.7.2 Large β

As $\beta \to \infty$, the segregation coefficient approaches unity to leading order, yet the pulsatile instability remains. A different nonlinear oscillator governs location of the interface, and the period of the oscillation is amplitude dependent.

The asymptotic forms for $\beta \to \infty$ for the critical morphological number and frequency of linear stability theory, are seen by Eqs. (6.11c,d), which for $a = 0$ are the following:

$$M \sim \frac{5\beta^2}{2(k_E - \mu^{-1}\beta^2)} \quad \text{and} \quad \omega_0^2 = \frac{(k_E - 1)\beta}{2k_E M \ln k_E}. \qquad (6.39)$$

6.7 Strongly nonlinear pulsations

The terms in the denominator of Eq. (6.39) are balanced by choosing μ^{-1} to be $O(\beta^{-2})$ and k_E to be $O(1)$. The following scales emerge:

$$\beta \equiv \frac{1}{\epsilon^2}, \quad M = \frac{\tilde{M}}{\varepsilon^4}, \quad \mu^{-1} = \epsilon^4 \tilde{\mu}^{-1}, \tag{6.40}$$

as $\epsilon \to 0$. The value of μ^{-1} need not be near μ_s^{-1} here because the bifurcation is supercritical (Braun and Davis 1991), and thus a uniform description of the bifurcation is expected even for weak kinetics. Relation (6.39) suggests the existence of a time scale ϵt, and asymptotic analysis of the full dispersion relation suggests a second slow time $\epsilon^2 t$. However, as is often the case for oscillators with slowly varying period, a more general time scale is required (see Kevorkian and Cole 1981 and Bourland and Haberman 1988). In the present case, the appropriate time variables are

$$\psi = \frac{\theta(t_2)}{\epsilon} + \phi(t_2) \tag{6.41a}$$

$$t_2 = \epsilon^2 t \tag{6.41b}$$

These are required because the period in this limit is affected by the amplitude of the oscillator, and perturbations in amplitude affect the functional form of the period. The coordinate in the direction of growth remains unscaled, $Z = z$.

One poses, for $\epsilon \to 0$,

$$C(z, t; \epsilon) = C(\xi, \psi, t_2) = C_0 + \epsilon^2 C_2 + \varepsilon^3 C_3 + \cdots, \tag{6.42a}$$

$$h(t; \epsilon) = H(\psi, t_2) = \frac{1}{\epsilon}(H_0 + \epsilon H_1 + \cdots) \tag{6.42b}$$

and proceeds as before. Again, the presence of disequilibrium effects drives a very large amplitude oscillation.

At leading order,

$$C_0 = 1. \tag{6.43}$$

The solute field is constant to leading order owing to the near absence of solute rejection in the high-speed limit. The interface position H_0 is again arbitrary at leading order. The $O(\epsilon)$ problem is trivial.

At $O(\epsilon^2)$

$$C_2 = -\frac{k_E}{X} \exp[-X\xi], \tag{6.44a}$$

where

$$X = 1 + \Omega H_{0\psi} \quad \text{and} \quad \Omega(t_2) = \theta'(t_2). \tag{6.44b}$$

The correction to the leading order (constant) solute field is strongly dependent on time through $H_{0\psi}$.

At $O(\epsilon^3)$ the integration over the domain yields the compatibility condition

$$\Omega^2 \frac{\partial^2 H_0}{\partial \psi^2} + \Omega_0^2 H_0 \left(1 + \Omega \frac{\partial H_0}{\partial \psi}\right)^4 = 0, \qquad (6.45)$$

where $\Omega_0^2 \equiv (k_E - 1)/(2k_E \tilde{M} \ln k_E)$. As in the previous case, the location of the interface is given to leading order by a nonlinear oscillator equation, now with exponent four. Here the frequency *does* vary with the amplitude of the oscillation. Rescale the amplitude as $Y_{0\psi} \equiv H_0/\Omega_0$ and define $\rho \equiv \Omega/\Omega_0$; Eq. (6.45) then becomes

$$\rho^2 \frac{\partial^2 Y_0}{\partial \psi^2} + Y_0 \left(1 + \rho \frac{\partial Y_0}{\partial \psi}\right)^4 = 0. \qquad (6.46)$$

The first integral may be calculated by multiplying Eq. (6.46) by $\rho Y_{0\psi}$ and integrating; the result is

$$Y_0^2 + \frac{\rho^2 Y_{0\psi}^2 (1 + \rho Y_{0\psi}/3)}{(1 + \rho Y_{0\psi})^3} \equiv B^2(t_2). \qquad (6.47)$$

The phase plane for Eq. (6.47) is shown in Figure 6.27. For amplitudes $B < 1/\sqrt{3}$, the orbits form closed curves, indicating periodic behavior. As the amplitudes approach $1/\sqrt{3}$, the velocity $Y_{0\psi}$ becomes unbounded, and, for $B \geq 1/\sqrt{3}$, the curves are no longer closed, indicating nonperiodic behavior in the same way. The energy B evolves on the slow time; because the frequency depends on the amplitude in this limit, the frequency will also evolve on the slow time.

The first integral (6.47) can not be conveniently integrated again. It is a simple matter, however, to integrate Eq. (6.46) numerically. The solutions are very similar in character to the solutions in the small β case. The velocity profile again becomes very peaked as the amplitude approaches its limiting value, here $1/\sqrt{3}$, and the oscillation becomes more relaxation-like. The period of the oscillation can be normalized to be 2π for a given amplitude by finding the appropriate ρ; this is accomplished by iterating on ρ until the correct value of Y_0 at $\psi = 2\pi$ is obtained; ρ increases monotonically with increasing energy.

6.7 Strongly nonlinear pulsations

The compatibility condition at $O(\epsilon^4)$ gives the following forced, damped oscillator equation:

$$\Omega \left[\frac{\Omega H_{1\psi}}{X^4}\right]_\psi + \Omega_0^2 H_1 = \frac{\Omega}{X^4}(\phi' H_{0\psi\psi} - 2H_{0\psi t_2} + \frac{3\Omega H_{0\psi\psi}}{X^5}(-\phi' + \Omega H_{0t_2})$$

$$+ \frac{3\Omega^3}{2X^6} H_{0\psi\psi\psi} - \frac{15\Omega^4}{2X^7} H_{0\psi\psi}^2 - \Omega_0^2$$

$$\times \left[\frac{k_E \tilde{M}}{2}\left(\frac{1}{X^2} - 1\right) + \Omega \tilde{M} \tilde{\mu}^{-1} H_{0\psi\psi} - \frac{\Omega^2 H_0 H_{0\psi\psi}}{X^3}\right.$$

$$\left. - \frac{1 + H_0 H_{0t_2}}{X} - \frac{1}{X^2} + \frac{\phi' H_0}{\Omega}\left(1 - \frac{1}{X^4}\right)\right].$$

(6.48)

One of the two homogeneous solutions to Eq. (6.48) is periodic on the ψ scale with period 2π, and it is given by $H_{0\psi}$. The other homogeneous solution is not periodic. To preserve periodicity on the fast time, the forcing function in Eq. (6.48) must be orthogonal to the periodic homogeneous solution. Integrate the forcing function against $\Omega H_{0\psi}$ over a period and use the fact that H_0 is odd in ψ, as well as parity considerations, to obtain the (rescaled) amplitude evolution equation

$$\frac{1}{2}\frac{d(B^2)}{dt_2} = \frac{(k_E - 1)\rho^2}{2k_E \ln k_E}$$

$$\cdot \left\{(k_E - \frac{5}{2\tilde{M}})\left[\overline{\frac{Y_{0\psi}^2}{(1 + \rho Y_{0\psi})^2}}\right] - \tilde{\mu}^{-1}\overline{Y_{0\psi}^2} - \frac{1}{2}\left(k_E - \frac{5}{\tilde{M}}\right)\overline{Y_0^2 Y_{0\psi}^2}\right\}.$$

(6.49)

Here

$$\bar{f} \equiv \int_0^{2\pi} f(\hat{\psi})d\hat{\psi}.$$

(6.50)

The procedure used here is identical (except for the choice of the slowest time scale) to a procedure discussed in Bourland and Haberman (1988) for the analysis of weakly dissipative nonlinear oscillators. The form of the amplitude evolution equation (6.49) is an intermediate one for the derivation of the equivalent of Eq. (6.37); there, a closed-form implicit solution is known that allows one to evaluate the mean terms. Here, no such closed-form solution is known, and further simplification is not possible.

It is necessary to proceed to one higher order after completely solving the linear oscillator Eq. (6.48) to obtain the slow-time evolution of ϕ'. The periodic homogeneous solution of Eq. (6.48) is $H_{0\psi}$; the other solution is nonperiodic, and the complete solution may be found as discussed by Bourland and Haberman (1988). The resulting (rather complicated) phase-evolution equation has been found; it is a nonlinear ordinary differential equation for the phase correction (see Braun 1991 for details). Because the amplitude is decoupled from the phase, the phase evolution is forced by the amplitude. There is now the possibility that the phase will undergo bifurcation; however, the leading-order frequency (phase) and the steady amplitude behavior of the nonlinear oscillator are completely determined by Eqs. (6.48) and (6.49). The initial value problem on the slow time t_2 requires that the phase correction be considered to determine the solution; completely here, one is only concerned with the steady amplitude solutions, and thus the phase correction is not required.

The steady solutions are found by first setting $\partial(B^2)/\partial T = 0$ in Eq. (6.49). Then the amplitude is taken as the independent parameter; after choosing an amplitude, one can iterate to find a ρ that keeps the 2π-period on the ψ-scale. Once ρ is known, it is possible to solve for $\tilde{M}$. Choosing a new amplitude and repeating the process allow one to draw the various steady solution branches shown in Figure 6.28. The bifurcation from the steady planar interface for $\beta \gg 1$ is always supercritical, which is consistent with the weakly nonlinear theory

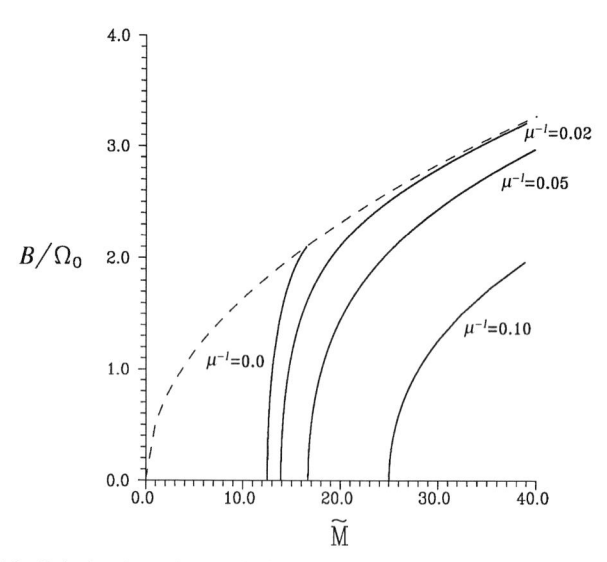

Figure 6.28. Solution branches valid for $\beta \gg 1$ for the steady form of Eq. (6.52) and various values of $\bar{\mu}^{-1}$ and $k_E = 0.2$. The dashed curve corresponds to $B = 1/\sqrt{3}$, and no oscillatory solutions exist for values above this curve. The strongly stabilizing effect of attachment kinetics is easily seen. From Merchant et al. (1992). (Reprinted with permission of SIAM.)

of Braun and Davis (1991). The strongly stabilizing influence of attachment kinetics ($\tilde{\mu}$) is evident in the behavior of the solution branches. For increasing $\tilde{\mu}^{-1}$, the steady-state amplitude of the oscillations decreases, which we would expect if atoms were attaching more sluggishly to the interface. When $\mu^{-1} = 0$, the amplitude rises very quickly to within a very small neighborhood of the limiting amplitude for oscillations to exist. Note that when the solution becomes close enough to the limiting amplitude, the oscillations develop very steep gradients as the velocity $H_{0\psi}$ becomes more sharply peaked, and the multiple scale analysis performed here is no longer valid. A new, faster time scale must be allowed near the limiting amplitude.

One should again expect the amplitude of the oscillations to become substantial very quickly as the morphological number is increased and thus to see relaxation-type oscillations experimentally. The oscillations would be responsible for striations in the solute frozen into the solid.

Attachment kinetics become very important at high speeds, and, for $\beta = 10$, the attachment kinetics for Ag–Cu, one of the best characterized alloys at high solidification rates, are too sluggish to allow the oscillations described here. A material with higher C_∞, smaller k_E, a very large V_0, and a large β_0 is most likely to have experimentally observable solute bands at high speeds; in short, the effect of attachment kinetics μ must be small. Large V_0 implies very rapid attachment of atoms to the solid at the solid–liquid interface, and large β_0 implies that solute rejection at the interface deviates from its equilibrium value at relatively low speeds. Materials without these properties do not exhibit the solute bands analyzed here.

6.7.3 Numerical Simulation

As just seen, for limiting values of β, the pulsatile mode satisfies a strongly nonlinear oscillation equation, whereas instability is determined at higher order. A numerical simulation of the system can be difficult because representations of the interface (boundary integrals, boundary elements, etc.) involve integrals with history-dependent kernels that require cumulative evolution cost. Brattkus and Meiron (1992) have used a method of Greengard and Strain's (1990) to split the kernel into two portions: a regular piece that delivers the history effect and a singular piece. Each portion is uniformly bounded, and the cost of evolution is limited because the singular part is fixed and the remainder can be written as a rapidly convergent series with easily updated coefficients.

Brattkus and Meiron (1992) analyzed the pulsations discussed above. They verified the critical conditions for linearized instability for M_c within 1 percent, ω_0 within a fraction of a percent, and the growth rate.

In the nonlinear range above M_c with instantaneous kinetics, $\mu^{-1} = 0$, amplitudes grow without bound, consistent with the analyses, likely showing the formation of finite-time singularities. When $\mu^{-1} > 0$, the amplitudes equilibrate and give phase portraits for $\beta = 50$ similar to those in Figure 6.27. The properties of the solution for $\beta \to \infty$ show that $\omega_0 \sim \beta^{-1/2}$, and the amplitude $B \sim \beta^{1/2}$, which is consistent with the asymptotics. However, the amplitude itself differs from the asymptotics. Similar conclusions hold for small β. At this time the discrepancy between asymptotic and numerical solutions has not been resolved.

6.8 Mode Coupling

There are parameter values for which the oscillatory and cellular modes have simultaneous onsets of instability (codimension-two points). In Figure 6.4 this occurs at the intersection of the two neutral curves. The equivalent case is shown in Figure 6.5 in which, in nondimensional terms, the two modes have about the same M_c, although the $a_c = 0$ for one and $a_c \approx 0.35$ for the other. Figure 6.9(b) shows a case in which the critical values of C_∞ are equal and $a_c^* \approx 10^6$ and $a_c^* \approx 7 \times 10^6$ for the two modes. A local minimum exists at $a^* = 0$. It is also possible to vary the parameters so that both of these modes have the same critical C_∞ and hence produce a codimension-two interaction between two time-periodic modes, respectively, one with $a_c^* = 0$ and one with $a_c^* \neq 0$.

Given these possibilities, one wishes to determine the possible nonlinear interactions that occur near the codimension-two points and predict, if possible, the conditions for banding.

Bands are alternate strips of cells (or dendrites) and segregation-free material. One can envision these occurring when, say, a cellular mode is modulated by a coexisting pulsation. In the part of the cycle that the speed is maximum, segregation-free states would be stable, whereas in the part of the cycle that the speed is minimum, the cells would be preferred. Such a nonlinear interaction could appear (Braun and Davis 1992) as a tertiary bifurcation from a mixed mode, as shown in Figure 6.29. These possibilities are now considered.

6.8.1 Pulsatile–Cellular Interactions

The oscillatory mode has $a_{0_c} = 0$ in the FTA. In this case, Braun, Merchant, and Davis (1992) wrote

$$\Gamma = \Gamma_0 + \nu\epsilon^2, \quad M \sim M_0 + \epsilon^2 M_2, \quad t_1 = \omega_0 t, \quad t_2 = \epsilon^2 t, \quad (x, z) = (X, Z)$$
(6.50a)

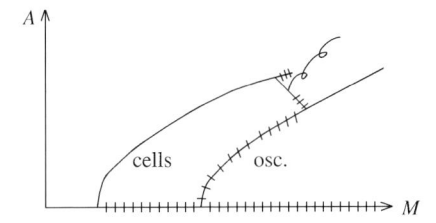

Figure 6.29. A sketch of an initially stable cellular solution and an initially unstable oscillatory solution. The two modes are tied together by a mixed mode that becomes unstable to a Hopf bifurcation as indicated by the curl. The response in time of this Hopf solution would give the characteristics of bands.

and

$$h(x,t) \sim \epsilon h_1(X, t_1, t_2) + \epsilon^2 h_2(X, t_1, t_2). \tag{6.50b}$$

Here $\nu = \text{sgn}(\Gamma - \Gamma_0)$, and Γ_0 is the value of Γ at the codimension-two point. At $O(\epsilon)$, if the linear theory is represented by

$$h_1 = B(t_2)e^{it_1} + F(t_2)e^{ia_cx} + cc, \tag{6.51}$$

the coupled Landau equations have the form

$$\begin{aligned} B_{t_2} &= M_2\sigma_1 B - \left[g_1 |B|^2 + g_2 |F|^2\right] B \\ F_{t_2} &= (M_2\sigma_2 - \nu\sigma_3)F - \left[g_3 |F|^2 + g_4 |B|^2\right] F \end{aligned} \tag{6.52}$$

with σ_1, g_1, and g_2 complex and the other coefficients real. Braun et al. (1992) computed the branches and their stability for many cases. Only the planar state, the pure cellular mode, the pure pulsatile mode, and the mixed mode (cells and pulsations) can be locally state. Bands are not possible with the coefficients calculated from the physical problem (although they are possible in general).

6.8.2 Oscillatory–Cellular Interactions

When the FTA is relaxed, the oscillatory mode has minima other than at $a = 0$. In their study of thermal corrections to mode coupling Huntley and Davis (1996) studied such cases.

They wrote the same scales as above, but now the linear theory solution has the representation

$$h_1 = B^R(t_2)e^{i(t_2+a_{0_c}x)} + B^L(t_2)e^{i(t_2-a_{0_c}x)} + F(t_2)e^{ia_cx} + cc. \tag{6.53}$$

Orthogonality at $O(\epsilon^3)$ gives the complex Landau equations

$$B^L_{t_2} = (M_2\sigma_0 - \sigma_1 v)B^L - \left[g_1 |B^L|^2 + g_2 |F|^2 + g_5 |B^R|^2\right] B^L$$

$$B^R_{t_2} = (M_2\sigma_0 - \sigma_1 v)B^R - \left[g_5 |B^L|^2 + g_2 |F|^2 + g_1 |B^R|^2\right] B^R \quad (6.54)$$

$$F_{t_2} = (M_2\sigma_2 - \sigma_3 v)F - \left[g_3 |F|^2 + g_4 |B^L|^2 + g_4^* |B^R|^2\right] F.$$

Here σ_2, σ_3, and g_3 are real and σ_0, σ_1, g_1, g_2, g_4, and g_5 are complex; an asterisk denotes a complex conjugate.

Huntley and Davis (1996) computed the branches and their stability for the systems Ag–Cu, Al–Cu, Al–Fe, and Al–Sn. Only the planar state, the pure cellular mode, and pure traveling waves can be locally stable. Again, bands are not possible with the physically determined coefficients.

6.8.3 Oscillatory–pulsatile interactions

When the FTA is relaxed, the principal interaction can be between a pulsatile mode, frequency ω_0, and an oscillatory mode, frequency ω_1, with $a_{0_c} \neq 0$. This interaction has been considered by Grimm and Metzener (1998) in two cases: (1) the nonresonant case in which ω_0 and ω_1 are unrelated, and (2) $\omega_0 = 2\omega_1$, a resonant case.

In the nonresonant case, the interface is represented by

$$h(t_1, t_2, x) = A(t_2)e^{i(a_c x + \omega_1 t_1)} + B(t_2)e^{i(a_c x - \omega_1 t_1)} + W(t_2)e^{i\omega_0 t_1} + cc \quad (6.55)$$

and by the usual methods a set of Landau equations is determined as follows:

$$W_{t_2} = \lambda_1 W + W\left[a |W|^2 + b |A|^2 + b |B|^2\right]$$
$$A_{t_2} = \lambda_2 A + A\left[c |W|^2 + d |A|^2 + e |B|^2\right] \quad (6.56)$$
$$B_{t_2} = \lambda_2 B + B\left[c |W|^2 + e |A|^2 + d |B|^2\right],$$

where all coefficients are complex. The cases chosen and the values of the coefficients are given in their tables. The phases are passive in this system, and thus each complex amplitude can be replaced by its real amplitude. These real equations have the usual equilibrium states: the basic state, pure modes, mixed (traveling) modes with W and A or W and B, and mixed (standing) modes with $W \neq 0$ and $A = B$. Time-periodic solutions may arise as Hopf bifurcations from the mixed modes. Grimm and Metzener (1998) give conditions for the existence of this degenerate bifurcation (at the point, an infinity of periodic orbits exists; if the conditions are perturbed, no periodic orbits exist). Higher order corrections are necessary to remove the degeneracy. Periodic solutions

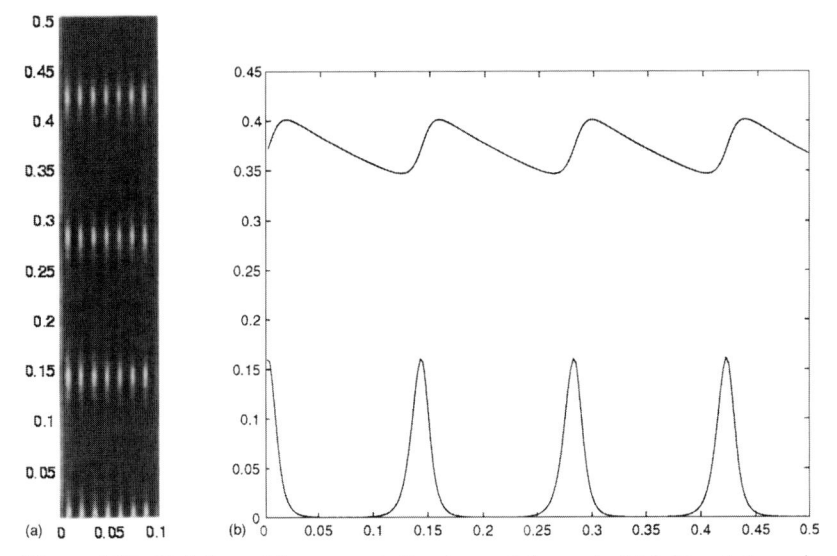

Figure 6.30. Variation of the concentration in a solid sample (at left) and dynamics of the amplitudes of the critical modes for the case of simultaneous criticality of the oscillatory modes in the case of nonresonant interaction. From Grimm and Metzener (1998). (Reprinted with permission of the American Physical Society.)

spend most of a cycle near the origin and the pure-mode fixed points, and thus the physical response has short transients that separate quasi-steady patterns. Figure 6.30 shows the computation in one case illustrating a response that appears to be a banded structure. The interesting point here is that the banded structure can emerge from the interaction of oscillatory states independent of the MS cellular mode.

In the 1:2 resonant case the representation (6.55) still holds, but now the set of Landau equations becomes

$$W_{t_2} = \lambda_1 W + \kappa_1 A^2$$
$$A_{t_2} = \lambda_2 A + \kappa_2 \bar{A} W, \quad (6.57)$$

and $A = B$. Here only quadratic interactions are retained because the cubic terms are negligible when, as is the case here, κ_1 and κ_2 are of unit order. When $\lambda_{1_t} = 2\lambda_{2_t}$, so that there is no detuning, almost all initial conditions lead to finite-time blowup (McDougall and Craik 1991). When the modes are detuned, one can use polar forms to obtain, after rescaling

$$\dot{X} = \hat{\gamma} X + \delta Y - 2Y^2 + Z \cos \Phi$$
$$\dot{Y} = \hat{\gamma} Y - \delta X + 2XY + Z \sin \Phi \quad (6.58)$$
$$\dot{Z} = -2Z(1 + X),$$

where $\hat{\gamma}$ measures $\lambda_{1_R}/\lambda_{2_R}$, δ is the detuning parameter, and Φ is related to the phase difference between the two modes; all the coefficients are real.

Hughes and Proctor (1992) showed that when $X = -1$, there are two fixed points, one stable and one unstable. The state

$$Y = \tfrac{1}{2}\delta + \frac{\tan\Phi + \tfrac{1}{2}}{2 - \delta\tan\Phi}\gamma + O(\hat{\gamma}^2) \tag{6.59}$$

$$Z = \frac{2 + \tfrac{1}{2}\delta^2}{(2 - \delta\tan\Phi)\cos\Phi}\hat{\gamma} + O(\hat{\gamma}^2)$$

is stable if to $O(\hat{\gamma})$,

$$\frac{4 - \delta^2}{12 + \delta^2} < \frac{1}{2}\delta\tan\Phi < 1. \tag{6.60}$$

It ceases to exist at unity and undergoes a Hopf bifurcation at the lower bound. Hughes and Proctor (1992) found that periodic orbits exist beyond the Hopf bifurcation point, and they undergo a cascade of period doublings as the parameters are varied. The dynamics is characterized by the alternation of slow and fast phases. During the slow Z phase, the standing waves have small amplitude, and in the fast phase all variables are important. Again, this suggests the possibility of the prediction of banded structures, as shown in Figure 6.31.

In both of the cases studied involving the interactions of oscillatory modes, the responses suggest banded structures. However, these theories have not yet been tested quantitatively against experiment and hence remain only a possible scenario.

6.9 Phenomenological Models

In Section 6.8 weakly nonlinear interactions of cells and oscillations were discussed for situations near the crossings of the two neutral curves. These codimension-two analyses can in principle give conditions under which band-like structures would emerge. However, to this point these structures are not predicted when the coefficients of the appropriate Landau equations are computed from the material and processing parameters of the system.

An alternative approach is to examine certain physical quantities such as temperature versus speed on interfaces undergoing steady growth and to pose an *ansatz* that, if correct, selects the morphology the system favors.

Gremaud et al. (1991) and Carrard et al. (1992) have posed such a model in the FTA. Figure 6.32 shows plots of the interface temperature for steady growth in two cases. In the first case, as discussed in Section 6.2, there is the interface

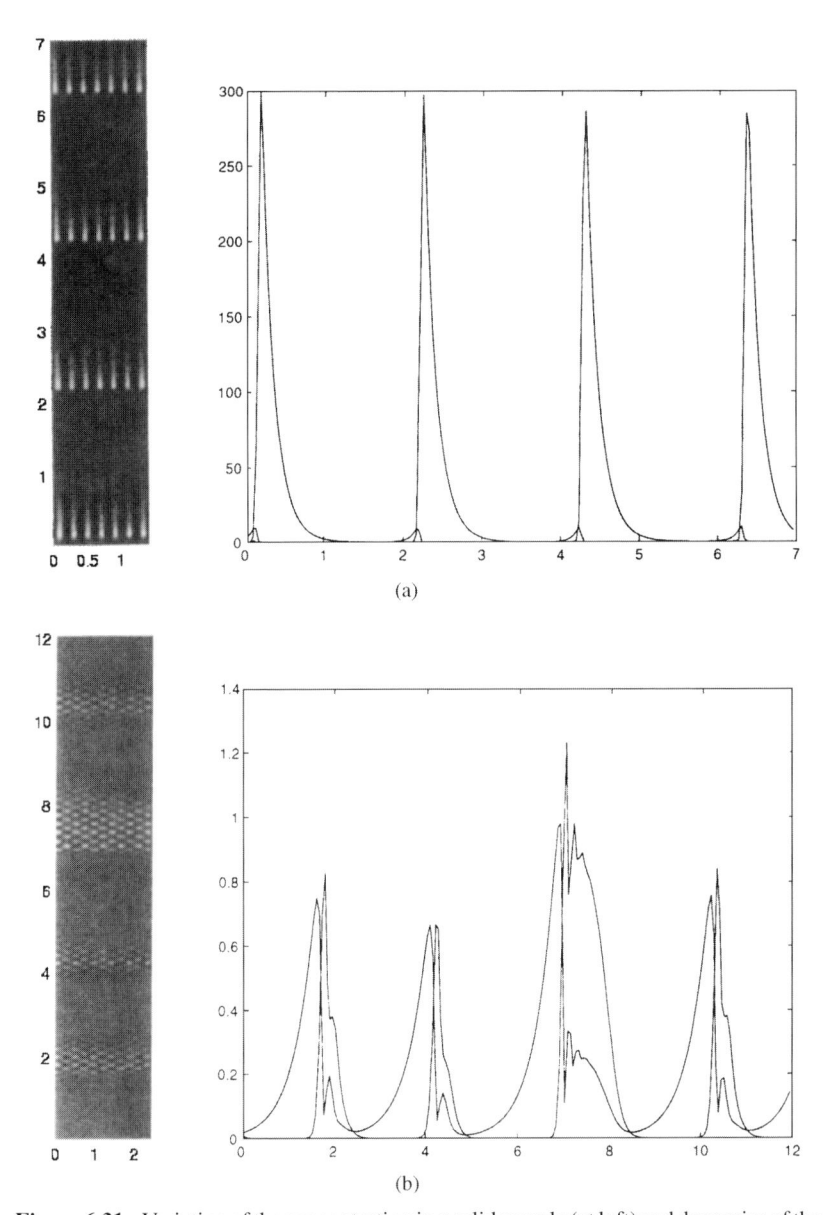

Figure 6.31. Variation of the concentration in a solid sample (at left) and dynamics of the amplitudes of the critical modes for the case of simultaneous criticality of the oscillatory modes. (a) Periodic solutions in the case of the averaged system. (b) Irregular dynamics in the resonant case. From Grimm and Metzener (1998). (Reprinted with permission of the American Physical Society.)

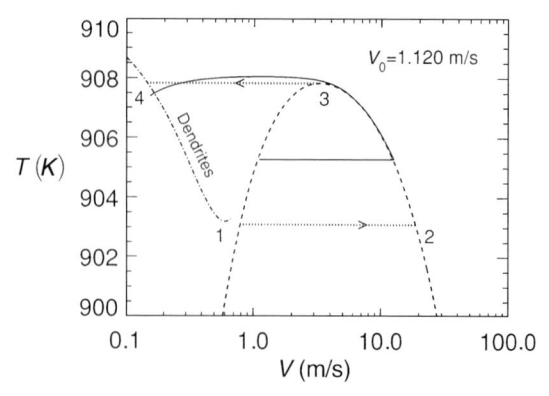

Figure 6.32. The cycle 1–2–3–4–1 corresponds to the Carrard et al. (1992) model in which the steady-state dendrite branch $T_{\text{den}}(V)$ and planar branch $T_{\text{pl}}(V)$ are represented by dash-dotted and dashed lines, respectively. The solid line corresponds to a cycle computed in the FTA. From Karma and Sarkissian (1993).

temperature T_0 for the planar interface subject to disequilibrium and kinetics. It begins at $T_m + [m(0)/k(0)]C_\infty$ for $V = 0$, increases to a maximum, and decreases at large V owing to kinetic undercooling.

In the second case there is the tip temperature of an Ivantsov needle crystal, a paraboloidal, isolated finger of solid discussed in Chapter 7. This crystal has zero surface energy, and its tip temperature as a function of V has been inferred by Kurz, Giovanola, and Trivedi (1986,1988) and is shown in Figure 6.32.

Shown as well in the figure is the *envelope* of these two curves, which has the familiar S-shaped form; for a range of temperatures there are three speeds V for each T. Carrard et al. (1992) supposed that the middle branch having $dT/dV > 0$ is unstable, and thus if V were cycled in time across this figure, the hysteretic path 1–2–3–4 shown would emerge. Part of this cycle is spent on the planar state and part on the cellular–dendritic state. If this occurred, a banded structure would appear with the time spent in each state, and, using the mean pulling speed, one could determine the lengths of each zone.

Carrard et al. (1992) did not directly associate the speed modulation with the pulsatile instability but did use the frequency scale of that oscillation in estimates of the overall wavelength (planar plus cellular) of a band and obtained orders of magnitude that correlated well with observations.

Karma and Sarkissian (1992, 1993) argued correctly that at high pulling speeds the presence of latent heat should be important. They analyzed the system with thermal effects and performed a numerical simulation of the pulsatile planar front. They found, consistent with the FTA analysis of Merchant et al. (1992) and the numerics of Brattkus and Meiron (1992), that relaxation

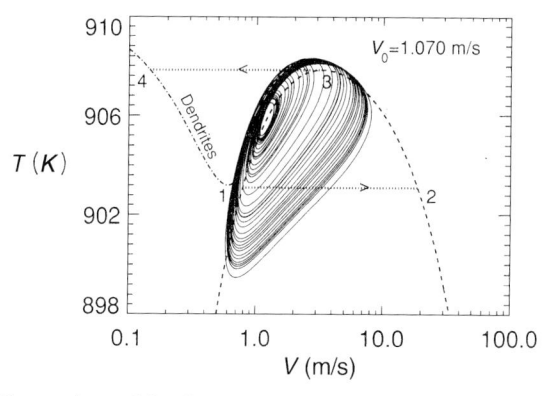

Figure 6.33. Comparison of the Carrard et al. cycle 1–2–3–4–1 and a large-amplitude cycle computed by Karma and Sarkissian. From Karma and Sarkissian (1993).

oscillations are developed and that the thermal effects are quite important for quantitative comparisons.

Karma and Sarkissian (1993) found that the nonlinear, hysteretic trajectories are altered from the FTA and now are as shown in Figure 6.33. The circuit is smaller than before, and the range is distorted. This calculation shows the importance of latent-heat conduction on the cycle period and hence on the overall wavelength; overall wavelengths are predicted within a factor of two of the observations.

This phenomenological theory still has several shortcomings:

1. At the parameter values used, the planar pulsation is *not* preferred; rather a mode with $a_{0_c} \neq 0$. How the planar mode is "selected," if it is, is an open question.
2. The calculation focuses on the plane front only, and there is no interaction with the cellular front. Certainly, such an interaction would determine the fraction of the overall cycle occupied by the cellular or planar states. As noted by Karma and Sarkissian, the theory cannot determine this fraction.

6.10 Remarks

Rapid directional solidification has been discussed using the continuous growth model in which $k = k(V_n)$, $m = m(V_n)$ and a linear model of attachment kinetics has been posed. The presence of these new effects gives rise in the linear stability theory to a stabilization of the high-speed branch of the MS model and the appearance of an oscillatory branch that exists because of disequilibrium and is strongly dependent on attachment kinetics.

These modes have been traced into the weakly nonlinear regime, and their interactions trigger the appearance of nonequilibrium solids called banded structures. As of yet, such microstructures have not been predicted from first principles using measured values of the physical properties of known alloys.

All of the preceding characteristics depend on a particular functional form, Eq. (6.4), for $k(V_n)$, which has a good deal of theoretical and experimental support. However, the case is not closed in this regard. There are other possibilities in the literature. In any case $k(0) = k_E$, $k(\infty) = 1$, and k is monotonic. For all the cases, $m(V_n)$ is a given functional of k, and this functional should still stand.

Likewise, the linear model of kinetics, $V_n = \mu \left(T_m - T^I\right)$, might be more accurately described by a more general $V_n = F\left(T_m - T^I\right)$ or a nonexplicit functional, as suggested by Bates et al. (1997).

References

Aziz, M. J. (1982). Model for solute redistribution during rapid solidification, *J. Appl. Phys.* **53**, 1158–1168.

Aziz, M. J., Tsao, J. Y., Thompson, M. O., Peercy, P. S., and White, C. W. (1986). Solute trapping: Comparison of theory with experiment, *Phys. Rev. Lett.* **56**, 2489–2492.

Baker, J. C., and Cahn, J. W. (1971). Thermodynamics of solidification, in *Solidification*, pages 23–58, American Society for Metals, Metals Park, OH.

Bates, P. W., Fife, P. C., Gardner, R. A., and Jones, C. K. R. T. (1997). Phase field models for hypercooled solidification, *Physica D* **104**, 1–31.

Bekki, N., and Nozaki, K. (1985). Formation of spatial patterns and holes in the generalized Ginzburg-Landau equation, *Phys. Letters* **A110**, 133–135.

Bernoff, A. J. (1988). Slowly varying fully nonlinear wavetrains in the Ginzburg–Landau equations, *Physica D* **30**, 363–381.

Boettinger, W. J., and Coriell, S. R. (1986). Microstructure formation in rapidly solidified alloys, in P. R. Sahm, H. Jones, and C. M. Adam, editors, *Rapid Solidification Materials and Technologies*, pages 81–108, Nijhoff Publishers, Dordrecht, The Netherlands.

Boettinger, W. J., and Perepezko, J. H. (1985). Fundamentals of rapid solidification, in S. K. Das, B. H. Kear, and C. M. Adam, editors, *Rapid Solidified Crystalline Alloys*, Proc. TMS–AIME Northeast Regionional Meeting, Morrison, NJ., pages 21–58, The Metallurgical Society, Warrenton, PA.

Boettinger, W. J., Schechtman, D., Schaefer, R. J., and Biancaneillo, F. S. (1984). The effect of rapid solidification velocity on the microstructure of Ag–Cu alloys, *Metall. Trans.* **A15**, 55–66.

Bourland, F. J., and Haberman, R. (1988). The modulated phase shift for strongly nonlinear, slowly varying, and weakly nonlinear damped oscillators, *SIAM J. Appl. Math.* **48**, 737–748.

Brattkus, K., and Meiron, D. I. (1992). Numerical simulations of unsteady crystal growth, *SIAM J. Appl. Math.* **52**, 1303–1320.

Braun, R. J. (1991). Nonlinear analysis in directional solidification, Ph.D. thesis, Northwestern University, Evanston, IL.

Braun, R. J., and Davis, S. H. (1991). Oscillatory instability in rapid solidification, *J. Crystal Growth* **112**, 670–690.

———. (1992). Cellular instability in rapid directional solidification: Bifurcation theory, *Acta Metall. Mater.* **40**, 2617–2628.
Braun, R. J., Merchant, G. J., and Davis, S. H. (1992). Pulsatile- and cellular-mode interaction in rapid directional solidification, *Phys. Rev. B* **45**, 7002–7016.
Carrard, M., Gremaud, M., Zimmermann, M., and Kurz, W. (1992). About the banded structure in rapidly solidified dendritic and eutectic alloys, *Acta Metall. mater.* **40**, 983–996.
Coriell, S. R., and Sekerka, R. F. (1983). Oscillatory morphological instability due to nonequilibrium segregation, *J. Crystal Growth* **61**, 499–508.
Eckhaus, W. (1965). *Studies in Non-Linear Stability Theory*, Springer-Verlag, New York.
Greengard, L., and Strain, J. (1990). A fast algorithm for the evaluation of heat potentials, *Comm. Pure Appl. Math.* **43**, 949–963.
Gremaud, M., Carrard, M., and Kurz, W. (1990). The microstructure of rapidly solidified Al–Fe alloys subjected to laser surface treatment, *Acta Metall. Mater.* **38**, 2587–2599.
———. (1991). Banding phenomena in Al–Fe alloys subjected to laser surface treatment, *Acta Metall. Mater.* **39**, 1431–1443.
Grimm, H. P., and Metzener, P. (1998). Oscillatory phenomena in directional solidification, *Phys. Rev. B* **58**, 144–155.
Hoglund, D. E., Aziz, M. J., Stiffler, S. R., Thompson, M. O., Tsao, J. Y., and Peercy, P. S. (1991). Effect of nonequilibrium interface kinetics on cellular breakdown of planar interfaces during rapid solidification of Si–Sn, *J. Crystal Growth* **109**, 107–112.
Hughes, D. W., and Proctor, M. R. E. (1992). Nonlinear three-wave interaction with non-conservative coupling, *J. Fluid Mech.* **244**, 583–604.
Huntley, D. A., and Davis, S. H. (1993). Thermal effects in rapid directional solidification: Linear theory, *Acta metall. mater.* **41**, 2025–2043.
Huntley D. A., and Davis, S. H. (1996). Effect of latent heat on oscillatory and cellular mode coupling in rapid directional solidification, *Phys. Rev. B* **53**, 3132–3144.
Jackson, K. A., Gilmer, G. H., and Leamy, H. J. (1980). Solute trapping, in C. W. White and P. S. Peercy, editors, *Laser and Electronic Beam Processing of Materials*, Proceedings of the Symposium of the Materials Research Society, Academic Press, New York, 104–110.
Karma, A., and Sarkissian, A. (1992). Dynamics of banded structure formation in rapid solidification, *Phys. Rev. Lett.* **68**, 2616–2619.
———. (1993). Interface dynamics and banding in rapid solidification, *Phys. Rev. E* **47**, 513–533.
Kevorkian, J., and Cole, J. D. (1981). *Perturbation Methods in Applied Mathematics*, Springer-Verlag, New York.
Kuramoto, Y. (1984). *Chemical Oscillations, Waves, and Turbulence*, Springer-Verlag, New York.
Kurz, W., Giovanola, B., and Trivedi, R. (1986). Theory of microstructural development during rapid solidification, *Acta Metall. Mater.* **34**, 823–830.
———. (1988). Microsegregation in rapidly solidified Ag–15wt% –Cu, *J. Crystal Growth* **91**, 123–125.
Kurz, W., and Trivedi, R. (1990). Solidification microstructures-Recent developments and future directions, *Acta Metall. Mater.* **38**, 1–17.
McDougall, S. R., and Craik, A. D. D. (1991). Blow-up in non-conservative second-harmonic resonance, *Wave Motion* **13**, 155–165.

Merchant, G. J., Braun, R. J., Brattkus, K., and Davis, S. H. (1992). Pulsatile instability in rapid solidification, *SIAM J. Appl. Math.* **52**, 1279–1302.

Merchant, G. J., and Davis, S. H. (1990). Morphological instability in rapid directional solidification, *Acta Metall. Mater.* **38**, 2683–2693.

Moon, H. T., Huerre, P., and Redekopp, L. G. (1983). Transitions to chaos in the Ginzburg-Landau equation, *Physica* **7D**, 135–150.

Mullins, W. W., and Sekerka, R. F. (1964). Stability of a planar interface during solidification of a dilute binary alloy, *J. Appl. Phys.* **35**, 444–451.

Nozaki, K., and Bekki, N. (1984). Exact solutions of the generalized Ginzburg-Landau equation, *J. Phys. Soc. Japan* **53**, 1581–1582.

Sirovich, L., and Newton, P. (1986). Periodic solutions of the Ginzburg-Landau equation, *Physica* **21D**, 115–125.

Stuart, J. T., and DiPrima, R. C. (1978). The Eckhaus and Benjamin–Feir resonance mechanisms, *Proc. Roy. Soc. Lond.* **A362**, 27–41.

Thoma, D. J., Glascow, J. K., Tewari, S. N., Perepezko, J. H., and Jayaraman, N. (1988). Effects of process parameters on melt-spun Ag–Cu, *Mater. Sci. Eng.* **98**, 89–73.

Zimmermann, M., Carrard, M., and Kurz, W. (1989). Rapid solidification of Al–Cu eutectic alloy by laser remelting, *Acta Metall. Mater.* **37**, 3305–3313.

7

Dendrites

Chapter 2 addresses nucleate growth. It was found that a spherical nucleus in an undercooled liquid will melt and disappear if its radius R is smaller than the critical nucleation radius R_*. In this case, the curvature is so large that surface energy effects dominate those of undercooling. When $R > R_*$, the sphere will continue to grow, and, as time increases, the effects of surface energy will decrease. When R reaches R_c, a morphological instability causes the spherical interface to become unstable to spatially periodic disturbances, leading to the growth of "bumps" on the interface. Experimental observation shows that the "bumps" grow, become dendritic, and continue to grow until they impact each other or a system boundary. Figure 7.1 shows a single bump that has become dendritic.

The term dendrite does not seem to have an accepted definition in the literature though it does refer to a treelike structure. Here it will be used to denote a two- or three-dimensional structure with side arms. Cells can, as well, be either two- or three-dimensional.

Dendritic growth is likely the most common form of microstructure, being present in all macroscopic castings. In fact, unless limitations of speed (or undercooling) are taken, a melt will usually freeze dendritically. If a sample of dendritically structured material having coarse microstructure is reprocessed, it will crack or otherwise produce defects. However, if the microstructure were fine enough, the reprocessing could proceed without ill effects. In either case, the "ghost" of the dendrites will remain after reprocessing. One of the key questions in materials science is: How does one predict the microstructure of a solidified sample? Further, how can one control the process to deliver the microstructure that is desired? These questions cannot presently be fully answered, though they do motivate the study of dendritic growth.

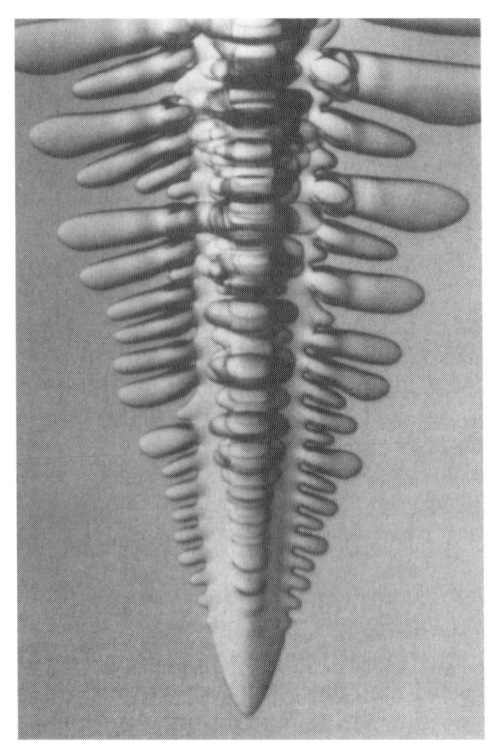

Figure 7.1. A succinonitrile (SCN) dendrite growing downward into its slightly supercooled melt ($\Delta T = 0.1$ K or, equivalently, $S^{-1} = 0.004$). The nearly steady-state tip region evolves rapidly into a train of side branches that form the bulk of the dendritic microstructure. SCN grows in the [100] crystallographic direction and forms four branching sheets, one in each of the four orthogonal [100]-type directions. From Glicksman and Marsh (1993). (Reprinted with permission of Elsevier Science.)

This chapter will begin with the study of an isolated finger of pure solid growing into an undercooled melt. This so-called needle crystal can serve as a basic state of an instability that results in an isolated dendrite. Various stability theories will be discussed aimed at selecting the speed and tip radius of a dendrite that would be seen in experiment. Then, there will be discussion of the appropriateness of studying an isolated dendrite and the advantages of studying alternative models involving dendrite arrays. The study of arrays of needle crystals will begin with one needle in a channel and proceed to the dynamics and instability of the arrays.

7.1 Isolated Needle Crystals

Consider an axisymmetric finger growing into an undercooled melt whose temperature is $T^\ell = T_\infty$, as shown in Figure 7.2. In a coordinate system moving at speed V, one seeks a steady solution of the interface at $\hat{z} = \hat{h}(\hat{x}, \hat{y})$. In the liquid for $\hat{z} < \hat{h}$,

$$VT_z^\ell = \kappa^\ell \nabla^2 T^\ell \tag{7.1a}$$

and in the solid for $\hat{z} > \hat{h}$,

$$VT_z^s = \kappa^s \nabla^2 T^s. \tag{7.1b}$$

For $\hat{x}^2 + \hat{y}^2 = 0$, $\hat{z} > \hat{h}$,

$$|T^s| < \infty. \tag{7.1c}$$

For $\hat{x}^2 + \hat{y}^2 \to \infty$, $\hat{z} < \hat{h}$

$$T^\ell = T_\infty. \tag{7.1d}$$

On the interface $\hat{z} = \hat{h}$,

$$T^\ell = T^s = T_m \left[1 + 2H \frac{\gamma}{L_v} \right] - \mu^{-1} V_n \tag{7.1e}$$

and

$$V_n L_v = (k_T^s \nabla T^s - k_T^\ell \nabla T^\ell) \cdot \mathbf{n}. \tag{7.1f}$$

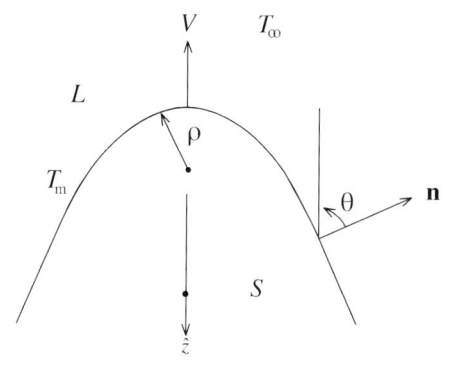

Figure 7.2. A sketch of a parabolic needle growing into an undercooled melt, $T_m > T_\infty$. The tip radius is ρ, and θ measures the angle between the growth direction and the normal $\mathbf{n}$.

Equation (7.1e) is the Gibbs–Thomson equation with kinetic undercooling, Eq. (7.1f) is the interfacial heat balance, and $\mathbf{n} = (\hat{h}_x, \hat{h}_y, -1)(1 + \hat{h}_x^2 + \hat{h}_y^2)^{-1/2}$.

Scale lengths on $\delta_T = \kappa^\ell/V$, and $T - T_\infty$ on $\Delta T = T_m - T_\infty$. The nondimensional system then has the form

$$\theta_z^\ell = \nabla^2 \theta^\ell, \quad z < h \tag{7.2a}$$

$$\theta_z^s = \kappa \nabla^2 \theta^s, \quad z > h \tag{7.2b}$$

$$|\theta^s| < \infty, \quad x^2 + y^2 = 0, \quad z > h \tag{7.2c}$$

$$\theta^\ell = 0, \quad x^2 + y^2 \to \infty, \quad z < h \tag{7.2d}$$

$$\theta^\ell = \theta^s = 1 + H\Gamma - \hat{\mu}^{-1} V_n, \quad z = h \tag{7.2e}$$

$$SV_n = (k_T \nabla \theta^s - \nabla \theta^\ell) \cdot \mathbf{n}, \quad z = h, \tag{7.2f}$$

where

$$\kappa = \kappa^s/\kappa^\ell, \quad k_T = k_T^s/k_T^\ell$$

$$\Gamma = \frac{2\gamma}{\delta_T L_v} \frac{T_m}{\Delta T}, \quad S^{-1} = \frac{c_p^s \Delta T}{L}, \quad \hat{\mu}^{-1} = \frac{\mu^{-1} V}{\Delta T} \tag{7.3}$$

Ivantsov (1947) noticed that a solution of this system could be found when the kinetics is instantaneous, $\hat{\mu}^{-1} = 0$, and when the surface energy is zero, $\Gamma = 0$. In this case Eq. (7.2e) shows that the interface is isothermal, $\theta^I \equiv 1$. Ivantsov then obtained a similarity solution for θ^ℓ and h in the form

$$\theta^\ell = \frac{E_1\left[\frac{1}{2} s(x, y, z)\right]}{E_1[Pe_0]} \tag{7.4a}$$

and

$$h = \frac{1}{4Pe_0}[4\rho^2 - x^2 - y^2]. \tag{7.4b}$$

Here E_1 is the exponential integral

$$E_1(\xi) = \int_\xi^\infty z^{-1} e^{-z} dz, \tag{7.5a}$$

$$s = z + \sqrt{x^2 + y^2 + z^2},$$

and the Peclet number Pe_0 is

$$Pe_0 = \frac{\rho}{2\delta_T}, \tag{7.5b}$$

7.1 Isolated needle crystals

where ρ is the tip radius. Notice that

$$s(x, y, h) = 2Pe_0, \quad (7.5c)$$

and thus $\theta^\ell = 1$ on $z = h$. Finally, when the forms (7.4) are substituted into the interface flux condition, (7.2f), one obtains the characteristic equation

$$S^{-1} = Pe_0 e^{Pe_0} E_1(Pe_0), \quad (7.6)$$

which relates the undercooling to the Peclet number and should be compared with Eq. (2.32) for the freezing of a plane into an undercooled melt. The right-hand side is often called the Ivantsov function $I(Pe_0)$.

Note that

$$I(Pe_0) \sim 1 - \frac{1}{Pe_0} + \cdots \approx \frac{1}{1 + \frac{1}{Pe_0}} = \frac{Pe_0}{Pe_0 + 1}, \quad Pe_0 \to \infty \quad (7.7a)$$

$$I(Pe_0) \sim Pe_0 \ell n \frac{1}{Pe_0}, \quad Pe_0 \to 0 \quad (7.7b)$$

Figure 7.3 gives Pe_0 versus S^{-1}, which shows that solutions exist only for $S^{-1} < 1$, that is, for less than unit undercooling, $\Delta T < L/c_p^s$.

Whenever $S^{-1} < 1$, there is a unique Peclet number for the solution, and a fixed Peclet number implies that the product of ρ and V is known, though each is unknown. The solution is thus a one-parameter family of paraboloids of revolution that lie on the curve Pe_0-is-constant; see Figure 7.4(a).

In order to get a feeling for the result outlined above, consider the simplified system of Kurz and Fisher (1989) shown in Figure 7.5 in which a *cylinder* of solid, having cross-sectional area A, has a hemispherical solid cap that grows at speed V into the melt. The cap of tip radius ρ moves a distance $V \Delta t$ in time Δt.

Assume that the phase transformation occurs only at the hemisphere whose heat flux is $-k_T^\ell A_h \left. \frac{\partial T^\ell}{\partial r} \right|_{r=\rho}$, where A_h is the area of the hemisphere,

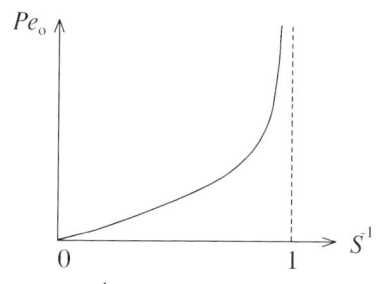

Figure 7.3. The Pe_0 versus S^{-1} for the similarity solution of the parabolic needle.

220   7. Dendrites

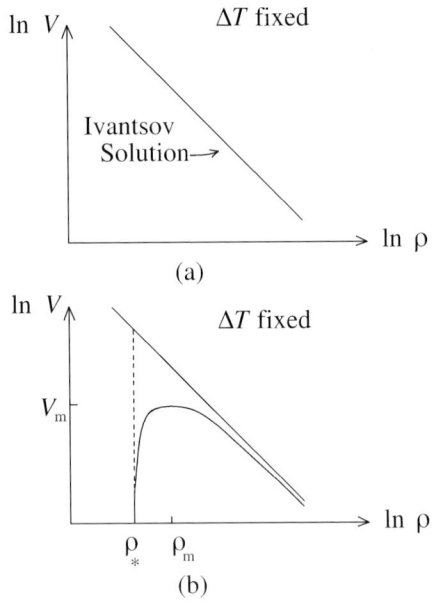

Figure 7.4. The characteristic V versus ρ for (a) the Ivantsov needle crystal and (b) the needle crystal with an ad hoc surface-tension correction.

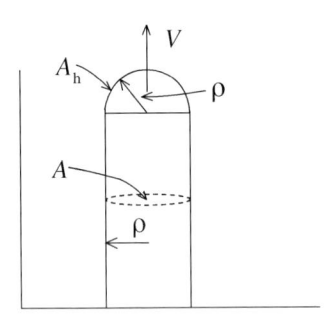

Figure 7.5. The "hemispherical needle" composed of a cylinder with a semi-spherical cap freezing at rate V.

$A_h = 2\pi\rho^2$. Write $-k_T^\ell A_h \left.\frac{\partial T^\ell}{\partial r}\right|_{r=\rho} \approx k_T^\ell A_h \frac{\Delta T}{\rho}$. The latent heat created in time Δt through the creation of the new solid is $AV\frac{L_v}{c_p^s}$, where $A = \pi\rho^2$. These two quantities balance when there is steady growth, giving $S^{-1} = Pe_0$. This model is called the "hemispherical-needle approximation" attributed to Fisher by Chalmers (1966, p. 105).

The axisymmetric parabolic figure described above has a two-dimensional analog in which the front is a parabola and

$$S^{-1} = 2Pe_0 e^{Pe_0} \int_1^\infty e^{-Pe_0 z^2} dz. \tag{7.8}$$

7.2 Approximate selection arguments

The similarity solutions given above (Eqs. 7.4–7.6) can be greatly generalized to general conic sections. Canright and Davis (1989) showed that the conic sections can still be obtained even if any number of components is present and even if cross diffusion is present (i.e., Soret and Dufour effects).

7.2 Approximate Selection Arguments

In Figure 7.4(a) for a given ΔT any point on the line corresponds to an allowable Ivantsov needle. Huang and Glicksman (1981a,b) have shown for a given ΔT, that the Ivantsov prediction for ρV is well verified experimentally; all observed needles lie on the straight line. Much effort has been expended in trying to develop a selection criterion that determines ρ and V individually by means of the introduction of surface energy γ. Because γ multiplies the mean curvature in the Gibbs–Thomson equation, the limit $\gamma \to 0$ is a singular perturbation, which turns out to make the inclusion of surface energy a very subtle business.

When surface energy is present, the needle will no longer have the paraboloidal shape, though when γ is small the departures "may" be small. Temkin (1960) retained the Ivantsov shape and enforced the Gibbs–Thomson condition everywhere but only applied the flux condition at the tip. Glicksman and Schaefer (1967,1968) retained the Ivantsov shape and enforced the flux condition everywhere but only applied the Gibbs–Thomson condition at the tip. None of these attempts solve the full free-boundary problem. In any of these analyses, the monotonic curve of Figure 7.4(a) is altered to give a shape like that in Figure 7.4(b), which has a local maximum of some value ρ_m of ρ.

Let us try to progress in a rather naive way. The tip of the needle has a local radius of curvature ρ, and one would expect from the nucleation theory of Chapter 2 that, if ρ were small enough, surface energy would cause the tip to melt back. Let ρ_* correspond to critical nucleation radius for zero growth speed. In an ad hoc manner, add the capillary undercooling to the undercooling of the Ivantsov equation, that is, let

$$S^{-1} = \mathrm{I}(Pe_0) + \frac{2\delta_{\mathrm{cap}}}{\rho} \qquad (7.9a)$$

where δ_{cap} is the capillary length

$$\delta_{\mathrm{cap}} = \frac{\gamma}{L_\nu} \frac{T_m}{\Delta T}. \qquad (7.9b)$$

Take the hemispherical-tip approximation, $\mathrm{I}(Pe_0) = Pe_0$ and let $Pe_0 = 0$ when $\rho = \rho_*$. Then the new curve, V versus ρ, now has a local maximum due to capillary undercooling, as shown in Figure 7.4(b). The maximum occurs when

$\frac{d}{d\rho}(S^{-1})$ is zero, that is, when $\rho_m = 2\sqrt{\delta_T \delta_{\text{cap}}}$, proportional to the geometrical mean of δ_T and δ_{cap}.

The appearance of the maximum suggested to researchers that the operating point might be at (ρ_m, V_m); this is the "maximum velocity hypothesis." Unfortunately, as Glicksman and Marsh (1993, p. 1088) explained, "the disparity between the predicted and observed dendritic [tip] scales ... [turn out to be] as large as 1–2 orders of magnitude." The predicted tip radii are much too small.

The experiments of Schaefer, Glicksman, and Ayers (1975) and Glicksman, Schaefer, and Ayers (1976) on the transparent organic succinonitrile (SCN) show that, if

$$\Lambda \equiv \frac{\delta_T}{2\delta_{\text{cap}}^{(1)}} \qquad \left(= \frac{1}{2}\Gamma S^{-1}\right), \qquad (7.10a)$$

where

$$\delta_{\text{cap}}^{(1)} = \frac{2T_m c_p^s \gamma}{L_\nu^2},$$

one can only find solutions up to $\Lambda = \Lambda_{\max}$. The Temkin–Trivedi value is

$$\Lambda_{\max} = 0.0254 \, S^{-2.65}. \qquad (7.10b)$$

Nash and Glicksman (1974) were able to convert system (7.2) into a single integral equation using Green's functions and were able to compute a value of $\Lambda_{\max}$, namely,

$$\Lambda_{\max} = 0.064 \, S^{-2.65}. \qquad (7.10c)$$

Both of these results seem to predict the exponent well, though the multiplicative factor is significantly different from experiment, as shown in Figure 7.6. Thus, V of the theory is too large by an order of magnitude. The implication is that the maximum velocity hypothesis is not correct. See Glicksman and Marsh (1993) for a more extensive discussion.

The inclusion of surface energy is thus much more subtle than the preceding "fixes" suggest. Because the ρ selected is orders of magnitude larger than ρ_m, it occurs well down the sloped portion of Figure 7.4(a), where ρ is so large that surface energy seems to have a rather small influence on the shape of the finger.

The selection criterion then involves more than the inclusion of surface energy for steady growth. Oldfield (1973) and Langer and Müller-Krumbhaar (1978a) suggested that even given a *steady* needle with surface energy, a condition of *stability* should be invoked. When surface energy γ is present and kinetic

7.2 Approximate selection arguments

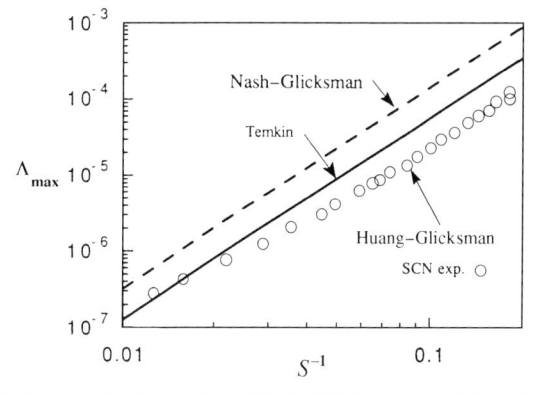

Figure 7.6. Maximum velocity results of Nash, Glicksman, and Temkin are shown for comparison with dimensionless velocity data. All exhibit a slope (exponent 2.6) similar to that of Huang and Glicksman (1981a) and as predicted by the theories but show relatively poor quantitative agreement as to the magnitude of the dendritic growth speed. From Glicksman and Marsh (1993). (Reprinted with permission of Elsevier Science.)

undercooling is neglected, they defined a new nondimensional parameter σ as

$$\sigma = \frac{2\delta_T \delta_{cap}^{(1)}}{\rho^2} \quad \left(= \frac{\Gamma}{S} \left(\frac{\delta_T}{\rho} \right)^2 \right) \tag{7.11}$$

in which surface energy and heat conduction compete.

With surface energy thus included, one would suppose that there is a relation for the marginal stability of a steady growing finger of the form $F(S^{-1}, Pe, \sigma) = 0$, for some function F where Pe is now the Peclet number of the tip in the case $\gamma \neq 0$. Perhaps, this relation could be written

$$\sigma = \sigma(S^{-1}, Pe). \tag{7.12}$$

The Langer and Müller-Krumbhaar argument can be outlined as follows. Using the definitions of δ_T and $\delta_{cap}^{(1)}$, σ can be written as

$$\sigma = \left(\frac{\lambda_c}{2\pi\rho} \right)^2 \tag{7.13a}$$

where λ_c is the cutoff wavelength for the stability of a *planar interface* via the Mullins and Sekerka criterion. Here $\lambda_c = 2\pi\sqrt{2\delta_T \delta_{cap}^{(1)}}$. An estimate for the stability boundary for the finger with the parabolic tip can be obtained if one imposes what is their main *ansatz*, that

$$\rho = \lambda_c. \tag{7.13b}$$

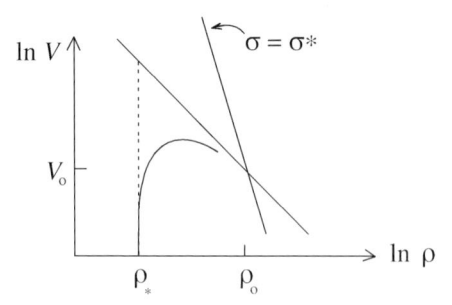

Figure 7.7. A sketch of the V versus ρ plane for the Ivantsov solution with a second line, $\sigma = \sigma^*$, that gives a selection at the point (ρ_0, V_0).

Then there is a critical value σ^* of σ given by

$$\sigma^* = \frac{1}{4\pi^2} \approx 0.0253. \tag{7.13c}$$

The surmised stability boundary should then be given by $\sigma = \sigma^*$ (constant) independent of Pe; see Eq. (7.12). This is a relation of the form given by Oldfield:

$$V\rho^2 = \frac{2\kappa}{\sigma^*} \frac{\gamma}{L_v} \frac{T_m}{\Delta T}. \tag{7.13d}$$

Given the material constants and the undercooling, Eq. (7.13d) yields another relationship between V and ρ that must be solved simultaneously with Pe_0-is-constant. (Here Pe is approximated by the $\gamma = 0$ limit Pe_0.) These two relationships, shown in Figure 7.7, define a unique intersection as shown. This intersection point for marginal stability is the proposed operating point for the system.

With this interpretation, the numerical value of σ^* serves as a measure of the operating point, and it compares reasonably well with the experimental observations of Glicksman, Schaefer, and Ayers (1976), who found $\sigma^* \approx 0.025$ for small ΔT. Huang and Glicksman (1981a) found that σ^* depends on the geometry of the tip; if the tip is taken to be spherical, then $\sigma^* \approx 0.0192$. To understand this *marginal stability hypothesis*, that $\sigma = \sigma^*$ determines the operating point, one must examine the analyses.

Langer and Müller-Krumbhaar (1978b) first considered the stability of the Ivantsov needle crystal (which has $\gamma = \mu^{-1} = 0$), that is, a needle with zero surface energy. They used the one-sided, quasi-steady model in two dimensions and in axisymmetric form both in the asymptotic limit $Pe_0 \to 0$. (They argued that in typical experiments, $Pe_0 \approx 10^{-2}$). A linear stability analysis shows that, when the normal modes are used, the fingers are *always* unstable, their growth rates are real, and their corresponding eigenfunctions are localized near

7.2 Approximate selection arguments

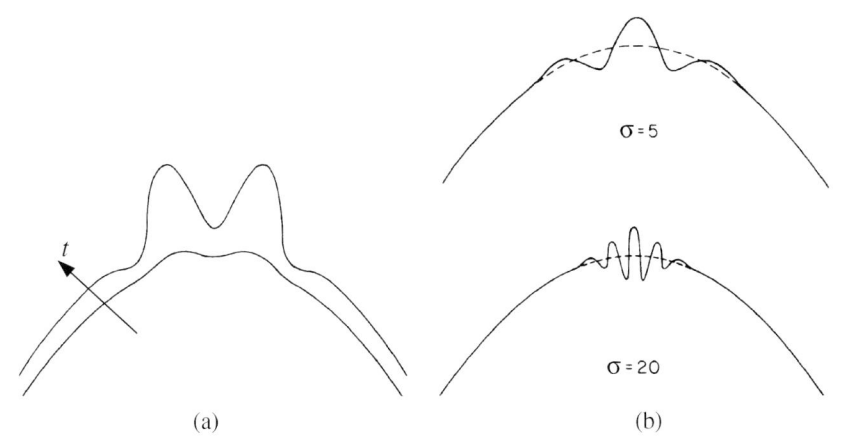

Figure 7.8. The eigenfunction for the tip-splitting instability for zero surface tension for (a) the axisymmetric case with growth rate $\sigma = 5$, and (b) the two-dimensional case with $\sigma = 5$. From Langer and Müller-Krumbhaar (1978b). (Reprinted with permission of Science Press.)

the tip, as shown in Figure 7.8(a). Figures 7.8(a) and (b) show that, either in the axisymmetric or two-dimensional cases, the absence of surface energy renders the crystal unconditionally unstable to a *tip-splitting mode*. Surface tension can in principle stiffen the interface and stabilize against tip splitting.

Müller-Krumbhaar and Langer (1978) then generalized the linear stability analysis to include the presence of surface energy. The linearized disturbance equations now have interface conditions that apply on the underperturbed interface, $h = \bar{h}(x, y; \Gamma)$ with surface energy Γ; $\bar{h}$ is difficult to find. They used the *regular* perturbation argument that one can replace $\bar{h}(x, y; \Gamma)$ by the Ivantsov solution $\bar{h}(x, y; 0)$ and retain surface energy terms in the disturbance equations. When the normal-mode problem was solved, they found two instability modes, one with a real growth rate and one with a complex one. The first of these is a tip-splitting mode whose growth rate approaches zero as $\sigma \to \sigma^*$. When $\sigma > \sigma^*$, so that ρ is small enough, tip splitting is prevented by surface energy.

The second mode has a complex growth rate and corresponds to spatially oscillatory interface deflections that travel away from the tip and increase in amplitude along the crystal. The speed of travel is approximately V, and thus in the laboratory frame, the sidebranches would seem to be stationary. These waves are also unstable for $\sigma < \sigma^*$ and become neutral at $\sigma = \sigma^*$. In the frame of reference moving with the tip, these *sidebranching modes* are stable for $\sigma > \sigma_*$, that is, for a given point on the crystal, the amplitude decreases in time. However, in the frame in which the tip propagates at speed V, the spatial growth down the crystal exceeds the temporal decay, and the sidebranches "seem" to grow.

226

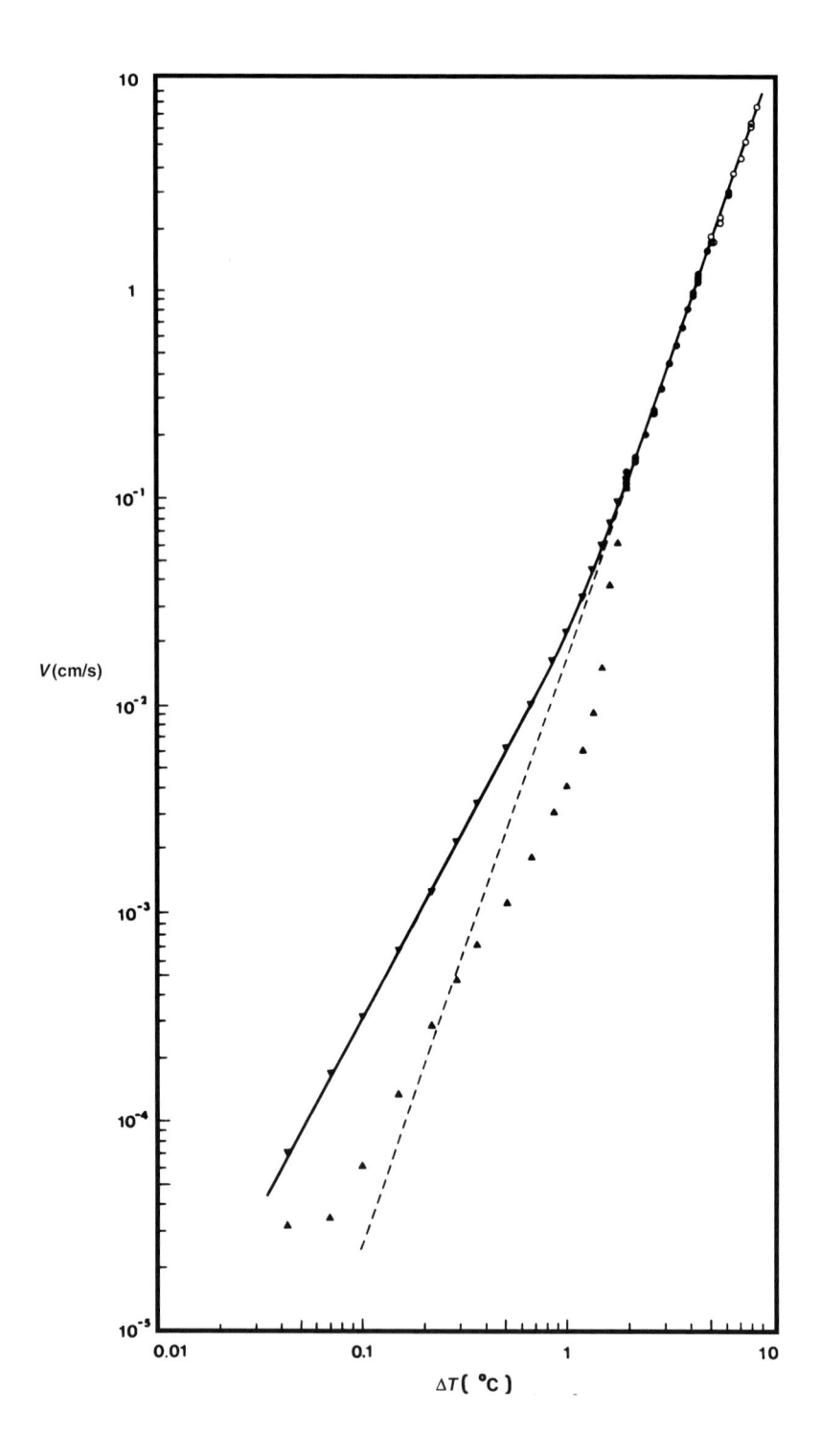

Langer and Müller-Krumbhaar (1978a) argued that the tip region is in this sense stable and that the sidebranches must at *finite amplitude* develop troughs that become deeper. In effect the needle crystal gets thicker, which in turn effectively increases ρ and decreases V, and hence σ. This argument is merely suggestive because it depends on a nonlinear theory that is not available: for $\sigma > \sigma^*$ this sequence drives the system from $\sigma > \sigma^*$ of Figure 7.7 to $\sigma = \sigma^*$. Thus, the theory is called the *marginal stability hypothesis*; the operating point should be $\sigma = \sigma^*$.

Huang and Glicksman (1981a,b) carried out the first comprehensive set of quantitative experiments using 99.9999% pure SCN. The summary of these experiments, outlined by Glicksman and Marsh (1993), can be presented as follows:

1. The tip, locally, is a paraboloid of revolution, as shown in Figure 7.1, with time-independent values of ρ and V at all ΔT examined.
2. For $10^{-3} < S^{-1} < 10^{-1}$, the ρ and V give Peclet numbers consistent with Ivantsov solutions at $T = T_m$.
3. For $10^{-2} < S^{-1} < 10^{-1}$, $\sigma^* = 0.0195$.
4. As S^{-1} decreases from 10^{-2}, σ^* increases to 0.025.
5. The tip radius ρ scales with λ_c over the full range of S^{-1} examined with $\rho/\lambda_c \approx 1.2$; compare this with assumption (7.13b).
6. The ratio of ρ to the critical nucleation radius for the tip, ρ_*, is $\rho/\rho_* \approx 100$ independent of S^{-1}, suggesting that the tip is affected only slightly by surface energy.
7. The sidebranches have a fixed wavelength that scales with ρ uniformly in S^{-1}.

The quite remarkable agreement of marginal stability hypothesis with these small-ΔT observations, notably numbers (1), (2), and (3) or (4), indicates that stability theory is an appropriate ingredient in the selection process. See Figure 7.9 for the comparison with the theory of the data of Glicksman, Schaefer, and Ayers (1976) and the recent experiments of Glicksman and Marsh (1993). However, further inspection shows that the foundations of the theory are suspect.

In order for the stability theory to be valid, obviously, the basic state must exist. Langer and Müller-Krumbhaar assumed that the "true" shape $\bar{h}(x, y; \Gamma)$

Figure 7.9. Measured values of dendritic tip speed V versus the undercooling ΔT in SCN with data from Glicksman, Schaefer, and Ayers (1976) denoted by ● and ○, and Huang and Glicksman (1981a): △ opposite to g and ▽ along g compared with the prediction of MST. From Glicksman and Marsh (1993). (Reprinted with permission of Elsevier Science.)

can be replaced by $\bar{h}(x, y; 0)$, the Ivantsov solution. Certainly, $\bar{h}(x, y; 0)$ is well-defined. As will be seen in a moment, for $\Gamma \neq 0$, $\bar{h}(x, y; \Gamma)$ does not exist if one assumes that, as the roots of the crystal are approached, the behavior is Ivantsov-like! A small amount of surface energy produces a singular perturbation that destroys the existence of shapes uniformly close to the Ivantsov needle crystal.

The question arises then of how to pose the problem with $\Gamma \neq 0$. For convenience, the *two-dimensional case* will be discussed. The approximate solutions of the problem, for example, Nash and Glicksman (1974), are normally sought on half of a symmetric needle, say, from the tip at $s = 0$ to the root at $s = \infty$, where s is the arc length. Among the boundary conditions are that the tip have a horizontal tangent

$$\left.\frac{\partial h}{\partial s}\right|_{s=0} = 0 \tag{7.14a}$$

and that the root be Ivantsov-like,

$$h \to \bar{h}(x; 0) \text{ as } s \to \infty. \tag{7.14b}$$

It was shown by Meiron (1986) and Kessler and Levine (1986) that the steady problem with end conditions (7.14) has *no solutions*. Specifically, if condition (7.14b) holds, then as $\Gamma \to 0$, $\left.\frac{\partial h}{\partial s}\right|_{s=0} \sim \exp(-1/\Gamma)$. The slope at the tip is never zero for any $\Gamma > 0$. This surprising result took many years to establish using analytical continuation and asymptotic methods that retain exponentially small terms; see Langer (1987) for a comprehensive review. Because the problem for $\Gamma \neq 0$ has no solution, the Langer–Müller-Krumbhaar stability analysis of it loses meaning.

In posing the steady needle without kinetics, it was assumed that (1) surface energy is isotropic, (2) the needle is steady, (3) the tip is smooth, and (4) the root is Ivantsov-like. Because solutions of this system do not exist, one or more of these assumptions must be relaxed. Such relaxations have given rise to further attempts at solutions to the isolated needle problem.

If assumptions (2)-(4) are accepted one must include additional physics in the mathematical description. If (1) is relaxed and the surface energy is made anisotropic, then one is led to *microscopic solvability theory* (MST). If (1) and (3), are retained, smooth-tipped fingers can be found that misbehave as $s \to \infty$, and one is led to *interfacial wave theory* (IWT) in which an unstable interface can lead to unsteady tip motions. In Section 7.3, these generalizations will be discussed.

Before continuing with surface–energy-based selection criteria, it should be noted that, if one enforces zero surface energy and anisotropic attachment kinetics, there are solutions for a finger valid for small S^{-1} and β_4. Brener,

Geilikman, and Temkin (1988) found that $V \propto \beta_4^{5/4} S^{-2}$. This suggests that "any" anisotropy may suffice for the existence of steady needles in the case when added physics is allowed.

7.3 Selection Theories

Consider first the possibility that conditions (7.14) hold, but now the surface energy is anisotropic. In this case the capillary undercooling is modified from $2HT_m\gamma/L_v$ to $2H\frac{T_m}{L_v}(\gamma + \gamma_{\theta\theta})$, as discussed in Chapter 5. Here θ is the angle between the normal to the interface and the growth direction. If one consider a material with fourfold symmetry, γ can be written as $\gamma = \gamma_0[1 + \alpha_4 \cos 4\theta]$, and thus $\gamma + \gamma_{\theta\theta} = \gamma_0[1 - 15\alpha_4 \cos 4\theta]$.

It has been shown by Kruskal and Segur (1991) for a model equation with such anisotropy that the steady needle exists, satisfying conditions (7.14) as long as $\alpha_4 \neq 0$. Ben Amar and Pomeau (1986) and Barbieri, Hong, and Langer (1987) have extended the analyses from model equations to the physical model and obtained results consistent with the numerical simulations of Meiron (1986) and Kessler and Levine (1986). Langer (1987) showed for small ΔT that, in two dimensions,

$$\sigma^* \approx \hat{\sigma} \alpha_4^{7/4} \tag{7.15}$$

where $\hat{\sigma}$ is a unit order constant. Again, this is a criterion in which σ^* is a constant, independent of Pe. A countable infinity of steady needle solutions with sufficiently large anisotropy exists (Meiron 1986, Kessler and Levine 1986). Condition (7.15) corresponds to that one having the largest V, which is found to be the one that is not unstable; it is neutrally stable. In order that this neutrally stable needle exhibit sidebranching instabilities, it is necessary that there is persistant noise at the tip.

The prediction of this *microscopic solvability theory* for the symmetric model in two dimensions is given by $Pe \approx (\Delta T/\pi)^2$ so that

$$V \approx \frac{2\kappa}{\bar{\delta}_{\text{cap}}^{(1)}} \sigma^* Pe^2 \approx \sigma^* \frac{2\kappa}{\bar{\ell}_c} \frac{(\Delta T)^4}{\pi^2} \tag{7.16a}$$

and

$$\rho \approx \frac{\bar{\ell}_c}{\sigma^* Pe} \approx \frac{\bar{\ell}_c}{\sigma^*} \frac{\pi}{(\Delta T)^2}, \tag{7.16b}$$

where $\bar{\delta}_{\text{cap}}^{(1)}$ is $\delta_{\text{cap}}^{(1)}$ times a scale factor dependent on the anisotropy. A similar result holds for the axisymmetric case.

In this theory the enforcing of the root condition (7.14b) forbids the existence of the steady needle if $\alpha_4 = 0$, and if $\alpha_4 \neq 0$ the solution at the tip is sensitive to the conditions at the roots. In experiments single needles emerge from capillaries (Glicksman, Schaefer and Ayers 1976), or an array of needles exist side-by-side. The microscopic solvability theory would suggest that the far-field conditions determine the near-tip dynamics, contrary to observation.

Further, result (7.15) suggests that crystalline anisotropy of surface energy is *the* determinant of selection. The experiments of Muschol, Liu, and Cummins (1992) suggest otherwise. When a low α_4-material, $\alpha_4 = (0.55 \pm 0.15)\%$ succinonitrile (SCN), and a high α_4-material, $\alpha_4 = (2.5 \pm 0.2)\%$, pivalic acid (PVA) are formed into needle crystals, the scaling law (7.15) is not seen. The values of σ_{th}^* obtained from the three-dimensional, $Pe = 0$ theory differ from the experimental value, σ_{exp}, by factors of 2 and in opposite directions for the two materials; see Figure 7.10 for the comparisons. Muschol et al. surmise that

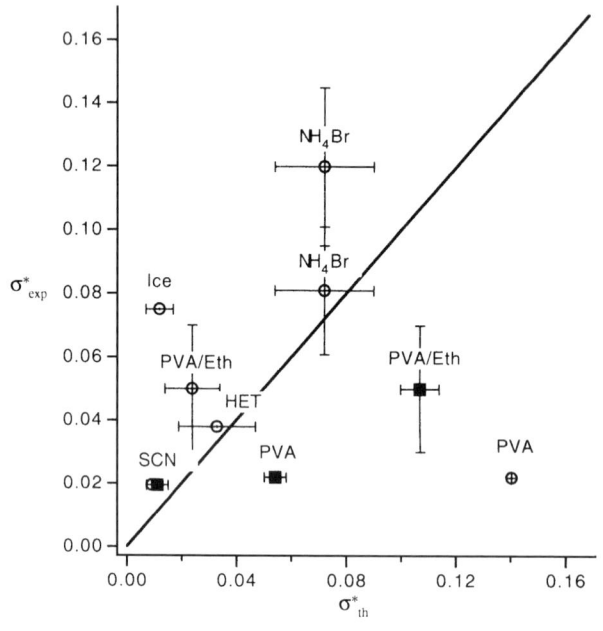

Figure 7.10. Plot of σ_{expt}^* versus σ_{theor}^*. Values of σ_{expt}^* are based on experimental measurements of V and ρ, whereas the results of σ_{theor}^* are derived from measured surface tension anisotropies. Circles are data points from previous anisotropy measurements, and the black squares are based on present data. Whenever available, error bars for both σ_{expt}^* and σ_{theor}^* are included. Errors in σ_{expt}^* are due to uncertainty in V, ρ, and material parameters; errors in σ_{theor}^* are based on uncertainty in surface tension anisotropy. The straight line, indicating perfect agreement with MST, is given as a guide to the eye. From Muschol et al. (1992). (Reprinted with permission of the American Physical Society.)

"... in view of the results for SCN and PVA ..., the correctness of MST in its present form cannot realistically be viewed as being confirmed by experiment."

Consider again the needle crystal with isotropic surface energy. When both conditions (7.14) are imposed, no solution exists for $\Gamma > 0$, whereas infinitely many are present for $\Gamma = 0$. The limit $\Gamma \to 0$ is a singular perturbation. Because the curvature of a needle crystal is significant near the tip, one would think that the reduction of order of the differential system when $\Gamma \to 0$ would require a boundary-layer correction near the tip, whereas the "outer" solution in which $\Gamma \to 0$ is a regular limit would encompass the remainder of the needle, including the roots. With root condition (7.14b) satisfied, no surface-energy correction near the tip would yield a solution; the tip for $\Gamma \neq 0$ is not smooth.

Xu (1997) has taken the point of view that the surface-energy correction should not be at the tip but at the roots. He thus sought an "outer" solution, satisfying condition (7.14a), valid everywhere outside a neighborhood of the roots and an inner surface energy correction at the roots.

His strategy is to express, say in two dimensions, the shape of the needle $\bar{h}(x; \Gamma)$ as an asymptotic series composed of two parts: a regular perturbation series h_R,

$$\bar{h}_R \sim \bar{h}(x, y; 0) + \epsilon^2 h_1 + \epsilon^4 h_2 + \cdots \quad (7.17a)$$

valid for, say, arc lengths in the range $0 \leq s \leq L_0(\epsilon, t)$, and a surface-tension correction obtained by a WKB approximation,

$$\bar{h}_S \sim f(\epsilon) \sin\left(\frac{s}{\epsilon}\right) \left(\frac{\cosh s}{2 \cosh L_0}\right)^{1/2}; \quad (7.17b)$$

$\bar{h}$ is given by

$$\bar{h} \sim \bar{h}_R + \bar{h}_S. \quad (7.17c)$$

A parameter Pe_0 is used and represents the Ivantsov–Peclet number at which the length scale is ρ_0, the tip radius of a needle without surface tension. Here Xu uses a parameter ϵ defined in terms of ρ without surface tension. When $\Delta T \to 0$

$$\epsilon = \frac{1}{Pe_0}\sqrt{\Gamma/S}, \quad (7.17d)$$

which is about $2\sqrt{\sigma}$, where σ is the parameter introduced by Oldfield (1973) and Langer and Müller-Krumbhaar (1978a). When ΔT is not small, ϵ is no longer related directly to σ.

In form (7.17a) $\bar{h}(x, y; 0)$ represents the Ivantsov solution and h_R is a regular series in Γ, which if h_S were omitted, would fail to satisfy root condition (7.14b).

Form (7.17b) represents strongly oscillatory corrections that enable condition (7.14b) to be satisfied but which becomes negligible (exponentially) as the tip is approached. In particular, if s is fixed and $\epsilon \to 0$, then the Ivantsov solution governs the near-tip needle structure independent of near-root conditions. The function f is chosen to satisfy the root conditions.

Finally, the $L_0(\epsilon, t)$ is a slowly varying increasing function of time that approaches infinity as $\epsilon \to 0$; the asymptotic properties can be chosen to validate the approximate solution (7.17c). Xu calls this a generalized steady state, though the terminology "steady" is misleading.

The Xu basic state (7.17) has the attractive feature that the near-tip properties of the needle are very weakly dependent on the root conditions – a characteristic recognized to be present by every experimentalist.

In order to examine linear stability theory, Xu perturbs the basic state $\bar{h}$, symbolically, as

$$h(x, t) = \bar{h} + h'. \tag{7.18}$$

He linearizes in primed quantities and then represents $\bar{h}$ as $\bar{h}(x, y, 0) + O(\epsilon^2)$ and retains only terms of $O(\epsilon)$. Thus, he retains as the basic state the Ivantsov paraboloid without near-tip corrections and without the oscillatory tail near the roots. In effect he returns to the Langer–Müller-Krumbhaar supposition that surface energy is negligible in the basic state but not in the disturbances. Unlike Langer–Müller-Krumbhaar, Xu *derives* this result and applies it only far from the roots. What allows Xu to ignore the root region is his decision to focus on the *spatial growth* of disturbances.

Xu seeks a solution using the WKB approximation in ϵ, where the growth rate $\sigma \sim \sigma_0 + \epsilon \sigma_1$ and the wave number $k \sim k_0 + \epsilon k_1$. At leading order in ϵ, Xu finds a local form of the characteristic equation

$$\sigma_0 = \frac{k_0}{g^2}\left[1 - \frac{2k_0^2}{g}\right] - \frac{isk_0}{g^2}, \tag{7.19}$$

where g is a geometrical scale factor.

Form (7.19) generalizes that of Mullins and Sekerka (1964) to the present situation in which a curved front moves through the melt. At the tip the normal velocity V_n is V. However, as one moves away from the tip, $V_n = V \cos \theta$, and each point experiences a different freezing rate. In a frame moving with the tip, the melt "flows" by like a liquid past a curved body. The result (7.19) shows that $Re\,\sigma_0$ indicates growth for a range of wave numbers k_0. In addition $Im\,\sigma_0$ gives the angular speed of the traveling disturbance, a wave that travels from the tip toward the root.

7.3 Selection theories

Rather than regarding Eq. (7.19) as determining a growth rate as a function of a wave number, Xu takes the view that a disturbance of fixed frequency *grows spatially* (toward the root), and thus Eq. (7.19) gives three complex roots k_0 as functions of the given σ_0. Two of these local solutions have $Re\ k_0 > 0$ and hence are physically relevant, and these correspond to a long- and a shortwave mode. At a fixed time, the amplitude of the shortwave mode first decreases in s and then increases exponentially. The longwave mode first increases in s and then decreases exponentially. Thus, the matching as $s \to \infty$ requires h' to behave like the longwave mode, which is a spatially oscillatory behavior different from the Ivantsov root.

At $O(\epsilon)$, a similar procedure is applied, and the first correction is obtained. Here it is noted that the asymptotics break down at two points: (1) a turning point $\mathbf{r}_c$, which is near the tip but separated from it, and (2) at the tip. The point $\mathbf{r}_c$ is conveniently analyzed in the complex plane and represents a critical layer across which different representations must be matched. To ensure a smooth tip, the near-tip solution must be matched to the outer solution. These subtle analyses yield a characteristic equation for the instability over the entire spatial range excluding a neighborhood of the root. Xu finds a discrete set, $\{\sigma_0^{(n)}, 0, \pm 1, \pm 2, \cdots\}$, of complex growth rates. Figure 7.11 shows both $Re\ \sigma_0$ and $Im\ \sigma_0 \equiv \omega_0$ versus ϵ for a particular case. In Figure 7.11(a) when $\epsilon > \epsilon_*$, all disturbances decay and the first mode to be destabilized is $n = 0$; this occurs $\epsilon = \epsilon_*$. Figure 7.11b, gives the corresponding angular frequency ω_0. At leading order in ϵ, $\omega_0 = -0.21291$ at a critical ϵ, $\epsilon_* = 0.1590$. When $O(\epsilon)$ terms are included, the first corrections of these are $\omega_1 \approx -0.2183$ with $\epsilon^{(1)} \approx 0.1108$, which valid for $Pe \to 0$. Xu finds two classes of solutions: symmetric with respect to the center line and antisymmetric.

At distances well beyond the critical layer, one can use the wave number to convert ω_0 to the phase speed c of an instability wave that could be seen by an observer. Xu finds for the whole range $2 \times 10^{-3} < Pe < 20$ that $c = 1 \pm 0.025$, suggesting that in a laboratory frame the wave crests, the side arms, would appear to be nearly stationary, $c = 1$, consistent with observation. Figure 7.12(a) shows a graphical view of a time sequence of interfacial shapes for a fixed ϵ, which looks very close to the observed measured sequences. Figure 7.12(b) shows the disturbance only for the instability mode. It is concentrated near the critical layer and dies away in both directions.

Xu explained the role played by the critical layer as a "barrier" that takes instability waves arising from the tip, and partially reflects them back toward the tip, effectively trapping waves in the region. The matching across the layer determines the wave that is transmitted and produces the sidebranches.

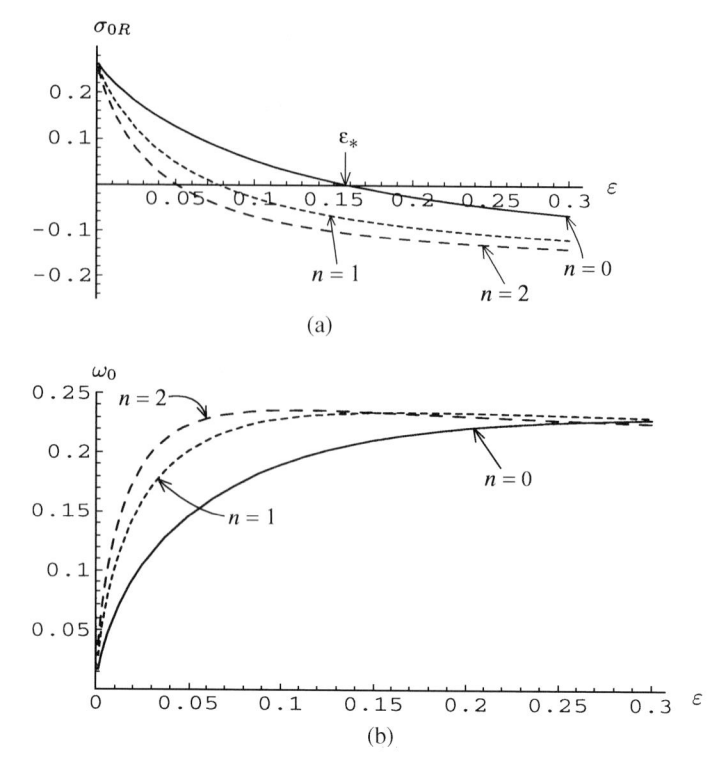

Figure 7.11. The zeroth-order IWT in two dimensions for modes $n = 0, 1, 2$ for (a) σ_{0_R} versus ϵ, and (b) ω_0 versus ϵ. The point $\epsilon = \epsilon_*$ demarks the neutral stability point. From Xu (1998). (Reprinted with permission of Springer-Verlag.)

The preceding theory shows that small disturbances at the tip travel and grow spatially. A weakly nonlinear extension of this, presently unavailable, could show that the system reaches a finite amplitude limit cycle that would result in the oscillation of the whole dendrite in time. This contrasts with the steady mode discussed earlier.

One of the implications of this theory is that the presence of crystalline anisotropy is *not essential* for the existence of the traveling-wave mode. In fact, surface-energy anisotropy can be included as a regular perturbation. When such anisotropy is present, Xu (1997) found in two dimensions the regularly perturbed traveling mode and a second mode that is seemingly steady, a mode directly related to that of the MST. In the remainder of this section, the analysis, including anisotropy, will be discussed.

Consider the case of fourfold symmetries and write the surface energy as

$$\gamma = \hat{\gamma} A_4(\theta), \tag{7.20a}$$

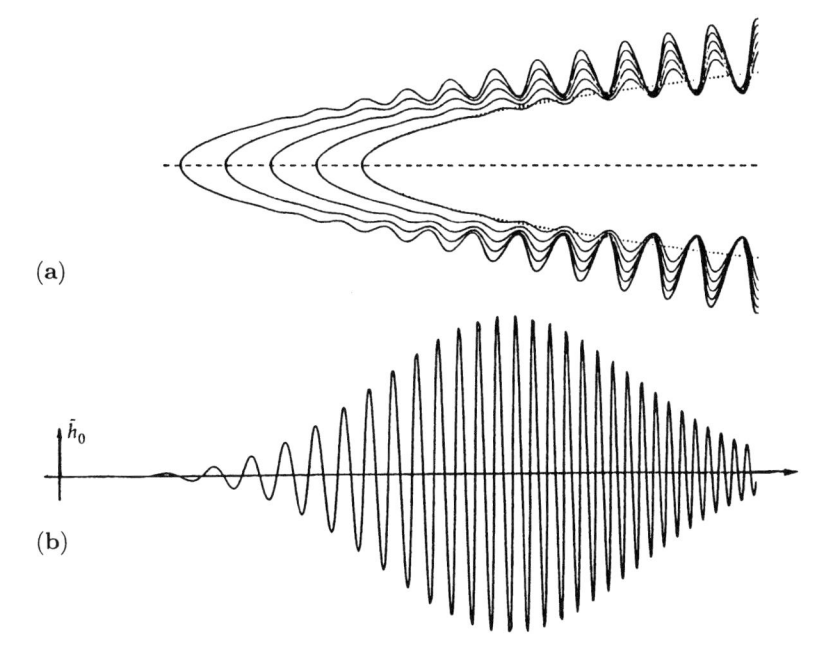

Figure 7.12. The global traveling mode. (a) The superposition of the disturbance and the basic state, the linear theory eigenfunction. From Xu (1997). (Reprinted with permission of Springer-Verlag.)

where

$$A_4(\theta) = 1 + \alpha_4 \cos 4\theta, \qquad (7.20b)$$

and let the small parameter be $\hat{\epsilon}$, where, in the definition of ϵ, $\hat{\gamma}$ replaces γ; also let the new wave number be $\hat{k}$. A similar analysis to that given above yields again three wave numbers for a given σ_0. Two cases arise: (1) $|\sigma_0| = O(1)$, (2) $|\sigma_0| \ll 1$.

When $|\sigma_0| = O(1)$, the regularly perturbed traveling waves already discussed result. When $|\sigma_0| \ll 1$, it turns out that $\sigma_0 \sim \epsilon^{8/7}$, $\alpha_4 \sim \epsilon^{3/7}$, and at leading order the mode is stationary. These are modes symmetric (varicose) with respect to the centerline of the needle and have the characteristic equation for $n = 0, \pm 1, \pm 2, \ldots$

$$c_1 \sigma_0^{\frac{11}{7}} \alpha_4^{\frac{2}{7}} = \frac{c_0}{\sqrt{2}} \alpha_4^{\frac{7}{8}} - \epsilon \left(n + \frac{1}{2} \right) \pi, \qquad (7.21)$$

where $c_0 \approx 1.8025$, $c_1 \approx 3.2886$. These modes correspond to those of the microscopic solvability theory. The preferred mode has $n = 0$, and as $\epsilon \to 0$,

$$\epsilon_* = K_0 \alpha_4^{\frac{7}{4}}, \qquad (7.22)$$

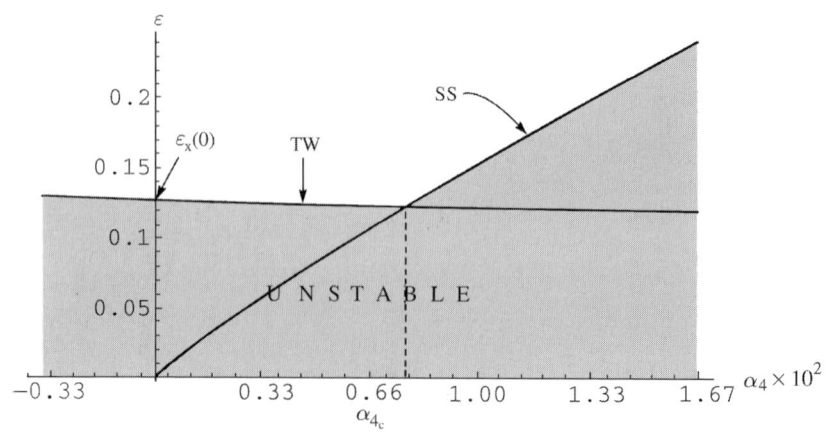

Figure 7.13. The neutral curves, ϵ versus α_4, for IWT containing the traveling-wave mode (TW) and the steady-state mode (SS) for small undercooling. From Xu (1997). (Reprinted with permission of Springer-Verlag.)

where $K_0 \approx 0.81120$. In the microscopic-solvability theory the corresponding $K_0 \approx 1.09$. Kessler and Levine (1986) found this numerically, and Bensimon, Pelce, and Shraiman (1987) found this analytically. As $\epsilon \to 0$, the eigenvalues σ_0 tend to the upper limit $\sigma_{max} \approx 0.5470 \alpha_4^{3/8}$. This steady mode exists only when surface energy anisotropy is present.

When surface energy anisotropy is present, Xu finds two modes that scale differently from each other on α_4. Figure 7.13 shows these neutral curves for the small undercooling case of $Pe = 10^{-3}$. When $\alpha_4 = 0$, only the traveling-wave mode is unstable when $\epsilon < \epsilon_*$. When α_4 is increased, $\epsilon_*(\alpha_4)$ decreases slowly. At the value α_{4_c} of α_4, $\alpha_{4_c} \approx 0.76\%$ for $Pe = 10^{-3}$, the two neutral curves cross, and for $\alpha_4 > \alpha_{4_c}$ the predominant mode becomes the steady one. Notice that when α_4 is greater than the $1/15 = 6.67\%$, the growth should be faceted, not smooth. As Pe increases, α_{4_c} decreases, for example, at $Pe = 1$, $\alpha_{4_c} = 0.57\%$. By these estimates SCN should in the range $\alpha_4 < \alpha_{4_c}$ while PVA should have $\alpha_4 > \alpha_{4_c}$; hence, the two materials should exhibit different dynamical behaviours and for $\Delta T \to 0$ different σ^*. That this is *not* the case may be attributable to PVA having strong anisotropy in μ, not included in the models (Xu, private communication 2001).

The two theories, MST and IWT, give remarkably similar values for ρ and V and would be difficult to distinguish on these bases. However, they should be destinguishable dynamically because one is steady and one is time-periodic, the latter of which could respond resonantly to time-dependent forcing.

7.4 Arrays of Needles

An isolated needle in the absence of surface energy and kinetic undercooling is governed by the Ivantsov analysis, which determines ρV for each undercooling. A selection principle must then be used to determine ρ and V individually. In typical applications needle crystals do not form in infinite universes but in arrays. Needle crystals in one- or two-dimensional arrays will be discussed in this section.

Consider a two-dimensional system in which there is a periodic array of needle crystals growing at constant speed into an undercooled pure melt. One can "isolate" one of these by examining a single needle in a channel of width λ whose sidewalls are perfect insulators. These sidewalls then serve as the lines of symmetry between adjacent needles. As seen in Figure 7.14, the needle shape is asymptotic to lines parallel to the sidewalls as the root is approached; the fractional width W of the needle depends on the undercooling.

Consider the symmetric model with isotropic surface energy. If $P_\lambda \equiv V\lambda/2D$, the limit $P_\lambda \to 0$ is formally equivalent to the Saffman–Taylor problem (Pelce and Pumir 1985) for fluid–fluid material displacement in a Hele–Shaw cell. Kessler, Koplik, and Levine (1986) showed that if $P_\lambda \to 0$, then there are steady-growth solutions only if the undercooling is large enough, that is if $S^{-1} > 1/2$. In this case they found that $V \propto \lambda^{-2}(S^{-1} - 1/2)^{-3/2}$, which is equivalent to the result of Hong and Langer (1986) for the Saffman–Taylor

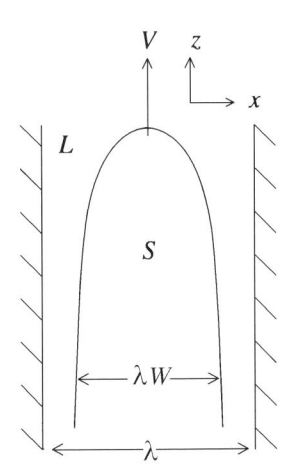

Figure 7.14. A sketch of a two-dimensional finger growing at speed V in a channel. Its width is the fraction W of the channel width.

problem. When surface energy anisotropy is present, then steady needles exist even for $S^{-1} < 1/2$.

The preceding results for $P_\lambda \to 0$ are for a case *distant* from that of the free needle crystal, which would have $P_\lambda \to \infty$. To understand how the needle passes to this limit, Brener, et al. (1988) considered the case of P_λ arbitrary. They defined Λ to be

$$\Lambda = \lambda/d_0, \qquad (7.23a)$$

where

$$d_0 = \mathcal{F}(\theta)\frac{T_m c_p}{L_v^2}, \qquad (7.23b)$$

$$\mathcal{F}(\theta) = \gamma + \gamma_{\theta\theta} = \bar{\gamma} A_4(\theta), \qquad (7.23c)$$

$$A_4 = 1 + \alpha_4 \cos 4\theta. \qquad (7.23d)$$

Brener, et al. (1998) solved the problem directly using the techniques of microscopic solvability theory. For isotropic surface energy they, too, found solutions only for $S^{-1} > 1/2$, which translates into solutions existing only for Λ *greater* than a critical value. Figure 7.15(a) shows the case $S^{-1} = 0.6$ in which there are two branches of solutions. The lower branch is equivalent to that found by Kessler et al. (1986), whereas the upper branch is new. Figure 7.15(b) shows V versus S^{-1} for a pair of separations Λ. Solutions exist only for $S^{-1} > S_*^{-1}(\Lambda)$, where S_*^{-1} decreases with increasing Λ. The *upper* branch corresponds to the width W of a finger increasing with undercooling S^{-1}, and hence it is this branch that corresponds to the isolated needle as $\Lambda \to \infty$. The symbol $\otimes$ denotes points calculated by Karma (1986) for $S^{-1} = 1$.

When the surface energy is anisotropic, $\alpha_4 \neq 0$, solutions exist below $S^{-1} = 1/2$. Figure 7.16 shows for $\alpha_4 = 10^{-2}$ that there are again two branches, and as $\Lambda \to \infty$ a single branch emerges from the origin, which consistent with both microscopic solrability theory (MST) and interfacial wave theory (IWT).

In the preceding the solutions are symmetric with respect to the centerline of the channel. It turns out that in addition to center-symmetric solutions, there exist pairs of solutions that have this symmetry broken; see Brener et al. (1993) and Ihle and Müller-Krumbhaar (1994). Kupferman, Kessler, and Ben-Jacob (1995) reexamined the two-dimensional problem for the one-sided model and found solutions for both $\alpha_4 = 0$ and $\alpha_4 \neq 0$, as shown in Figure 7.17, but now they found symmetry-breaking solutions as well that bifurcate from the

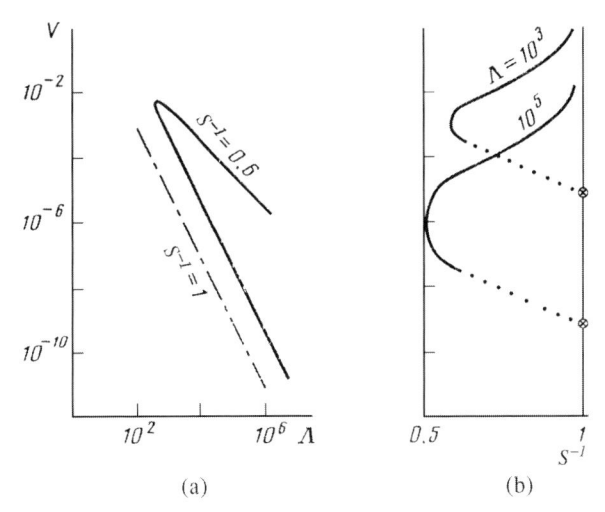

Figure 7.15. Dependences of the growth rate V on (a) the channel width Λ and (b) on the supercooling S^{-1} in the absence of the surface tension anisotropy ($\alpha_4 = 0$). The crosses enclosed by circles are the numerical results from Karma (1986); the dotted curves in Figure 7.15(b) are the proposed interpolation of the curves in the range $S^{-1} > 1/2$, which is outside the framework of the adopted approximations. From Brener et al. (1988). (Reprinted with permission of the American Institute of Physics.)

symmetric ones, as also shown. Each bifurcation represents a solution pair, one bending left and one bending right. As α_4 increases from zero, the bifurcation structure evolves in a rather complex way, as shown.

Bechhoeffer, Guido, and Libshaber (1988) have performed experiments on SCN in a channel having rectangular cross section with aspect ratio 10 and found smooth needles forming, which is consistent with the upper branch of solutions because the width W decreases with V. When the same experiment was performed with pivalic acid, $\alpha_4 \approx 5.5\%$, the anisotropy sharpens the tip, as shown in Figure 7.18.

These bifurcations of asymmetric solutions in a channel are suggestive of instabilities in arrays of needles. However, the constraint that the sidewalls are nondeformable is inconsistent with the constraints provided by neighbors in arrays. In an array, the line of symmetry between needles can deform, allowing the needles to meander; either varicose or sinuous disturbances may occur. As a result, one would suppose that, in an array, the configuration would become unstable *before* the aforementioned bifurcations appear. Further, all the analyses of array calculated so far have concentrated on steady solutions.

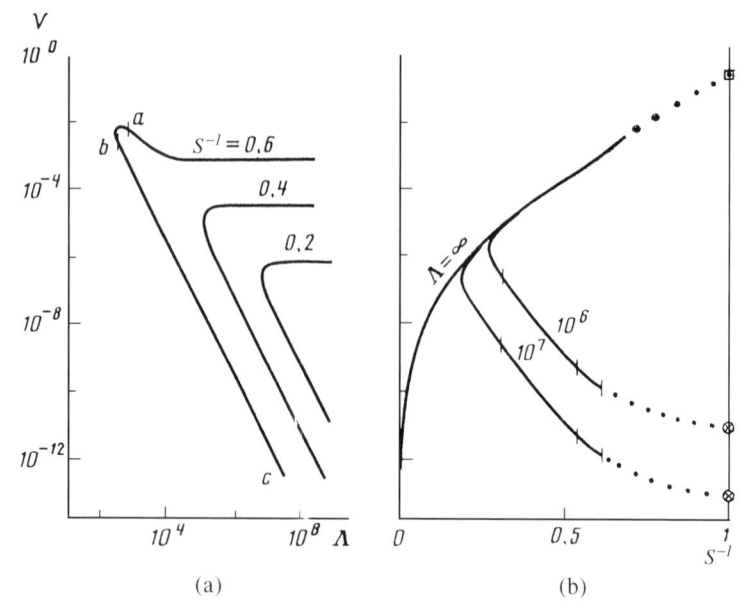

Figure 7.16. Dependences of the growth rate V on (a) the channel width Λ and (b) the undercooling S^{-1} for $\alpha_4 = 0.01$. The continuous curves are from the theory; the points corresponding to $S^{-1} = 1$ are from Karma (1986) and from Hong and Langer (1986). The dotted curves represent interpolations from Brener, et al. (1988). (Reprinted with permission of the American Institute of Physics.)

When each needle is three-dimensional, the arrays are necessarily two-dimensional. Spencer and Huppert (1995) examined such arrays in the absence of surface energy and attachment kinetics. Their analysis addressed an Ivantsov needle in the presence of neighbors in the small undercooling limit.

Consider the array sketched in Figure 7.19, where λ is the needle spacing, ρ is the tip radius, a is the needle radius, and $\delta_T = \kappa/V$ is the thermal-boundary-layer thickness. Spencer and Huppert considered the situation in which the undercooling is very small so that the tips strongly interact thermally. They examined asymptotically the situation in which $\rho \ll a \ll \lambda \ll \delta_T$ and $S^{-1} \to 0$. In this limit four regions are present; see Figure 7.20. There is the inner region near the crystal and away from the tips and roots. There is the outer region between, and in front of, the needles. There is a near-tip region and, finally, a near-root region, not shown, in which the adjacent roots are nearly parallel and the diffusion field is nearly one-dimensional, which is sometimes called the Scheil region. Spencer and Huppert used slender-body theory, found the local solutions, matched them together, and obtained a uniformly valid representation for the temperature in the liquid; the

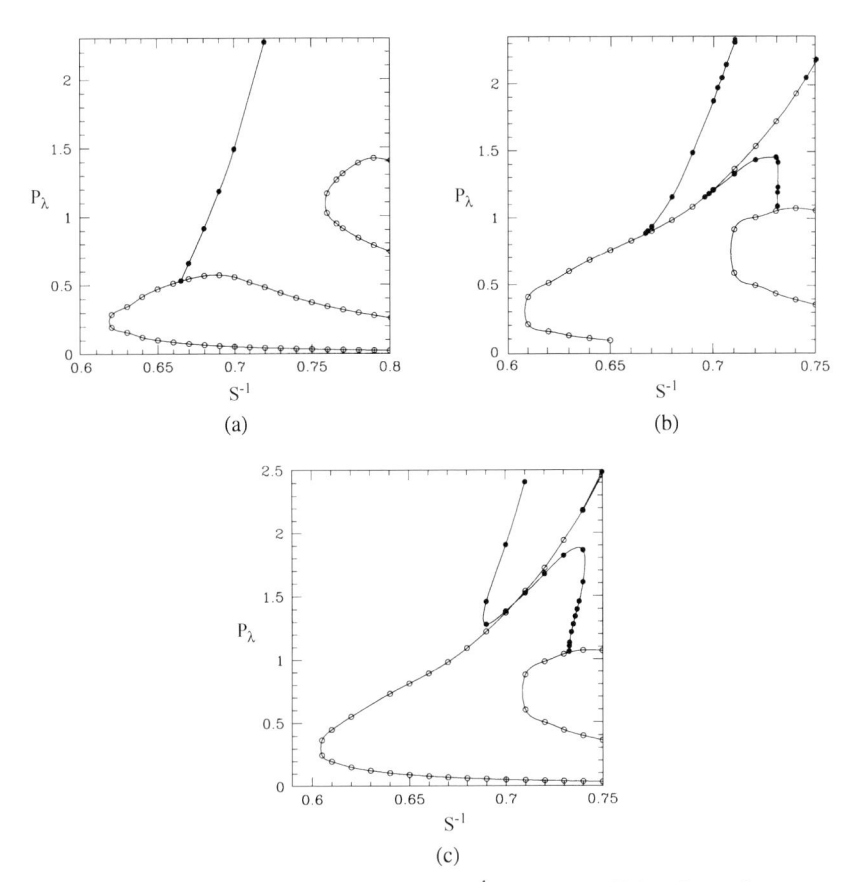

Figure 7.17. Bifurcation maps, P_λ versus S^{-1}, for a solidifying finger in a two-dimensional channel for (a) $\alpha_4 = 0$, (b) $\alpha_4 = 0.09$, and (c) $\alpha_4 = 0.1$. The filled (open) dots denote parity-broken (symmetric) fingers. From Kupferman et al. (1995). (Reprinted with permission of the American Physical Society.)

temperature in the solid is constant. In addition, they obtained a characteristic relation among the parameters consistent with a solution steady in the moving frame.

The scaling is explained as follows. Let Pe be the Peclet number based on the tip radius. When $S^{-1} \to 0$, so does Pe, and from the Ivantsov solution $S^{-1} \sim Pe \ln(1/Pe)$, which is supposed to be still correct for λ finite but not too small. Spencer and Huppert defined ϵ by $S^{-1} = \epsilon^2 \ln(1/\epsilon)$ and $\delta = 1/(\ln 1/\epsilon)$ and supposed that as $\epsilon \to 0$, and $Pe = \epsilon^2 P$, $P = O(1)$. For strong tip interactions $\lambda \ll \delta_T$ and thus $\lambda/\delta_T = \hat{\lambda}\delta^{1/2}$ and $\hat{\lambda} = O(1)$, which is obtained from the outer solution that sees the needle as a semi-infinite line source of heat moving at speed V; the source strength is $\phi(z)$. The needle radius is small,

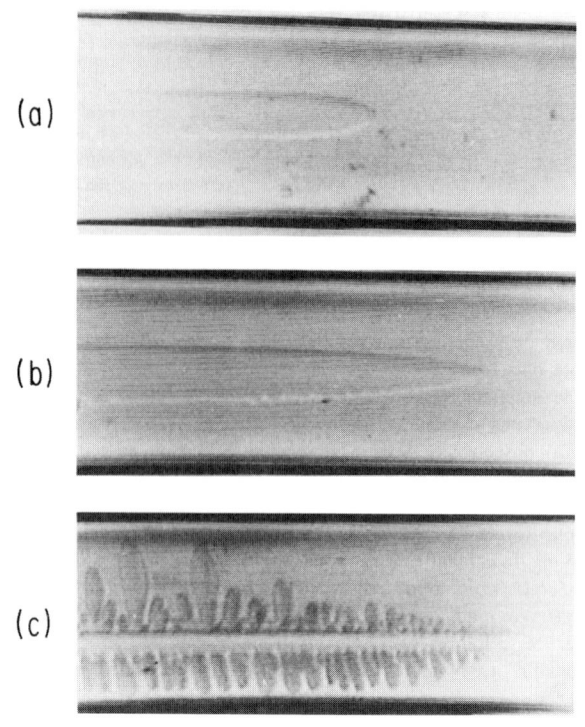

Figure 7.18. The solidifying of a finger of pivalic acid in a channel with (a) $\lambda = 0.25 \pm 0.02$, $V = 1.3\,\mu$m/s, (b) $V = 1.5\,\mu$m/s (slightly dendritic), and (c) $V = 61\,\mu$m/s (dendritic). From Bechhoeffer et al. (1998). (Reprinted with permission of Éditions Elsevier.)

$\tilde{a} = \epsilon \hat{a}$, $\hat{a} = O(1)$, and thus from an overall heat balance $\pi \hat{a}^2 = \hat{\lambda}^2$. The asymptotic analysis in the inner region is in powers of δ.

The results at leading order involve the parameter ν,

$$\nu = \frac{1}{2}\left[-1 + \sqrt{1 + 8\pi/\hat{\lambda}^2}\right], \tag{7.24a}$$

and the leading-order needle shape is given by

$$R_0(z) = \frac{\hat{\lambda}}{\sqrt{\pi}}\sqrt{1 - e - \nu z}, \tag{7.24b}$$

which is matched near the tip. The matching yields the key result

$$Pe = \frac{1}{2(1+\nu)}, \tag{7.25}$$

7.4 Arrays of needles

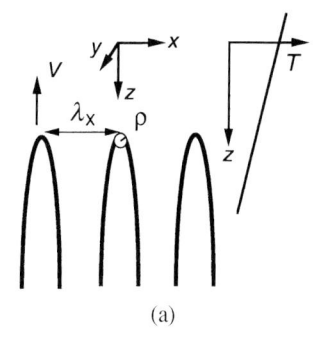

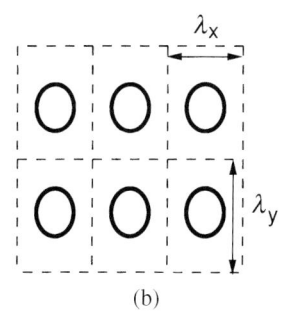

Figure 7.19. Schematic of an array of fingers. (a) A side view. (b) A top view. From Spencer and Huppert (1997). (Reprinted with permission of Elsevier Science Ltd.)

that is, the scaled Peclet number $Pe = 1/2$, for as $\hat{\lambda} \to \infty$ and as the spacings decrease, Pe decreases. Because Pe is determined by the undercooling, Eq. (7.24a) states that the "array-modified" undercooling is smaller than that for the isolated needle. In particular, in nondimensional terms using scale length $\delta_{\text{cap}}^{(1)}$,

$$\rho'V' = 1 - 2\pi \left(\frac{\rho'}{\lambda'}\right)^2. \quad (7.26)$$

Thus, for a given speed V', the tip radius is decreased by tip–tip thermal interactions.

Spencer and Huppert noted that corrections that make the individual needles nonaxisymmetric only appear at $O(\epsilon^4)$, and thus all the above results are valid for needle arrays of any shape such as squares, rectangles, and hexagons.

The array interactions found by Spencer and Huppert suggest that the neighbors in an array of needle crystals growing into an undercooled melt should be important. For one thing, their presence completely alters the roots of the needles; Ivantsov needles spread to infinity in width as $|z| \to \infty$, whereas in an array they asymptotically approach finite widths.

If such array interactions are important here, they should also be important in directional solidification of a binary alloy. Spencer and Huppert (1997) have

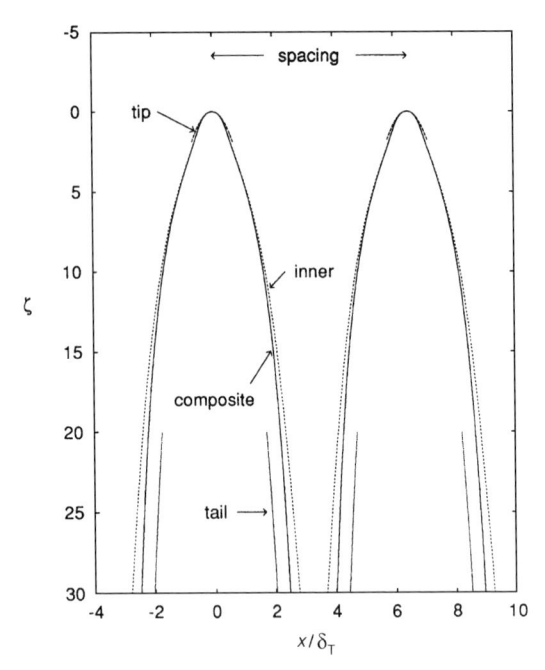

Figure 7.20. The composite solution consists of appropriately matched tip, inner, and tail solutions. The horizontal scale is expanded to show the details of the slender shape. The tail solution is taken to be axisymmetric. Here $k = 0.1$, $\epsilon = 0.0116$, $\Lambda_x = 6.47$, $\Lambda_y = 7.75$. This solution has a scaled-tip radius of 10.8 and a scaled-tip position of 2.24. From Spencer and Huppert (1998). (Reprinted with permission of Elsevier Science Ltd.)

investigated this case as well. For the freezing range $\Delta T_0 = m(k-1)C_\infty/k$, they defined a thermal length $\delta_T^{(1)} = k\Delta T_0/G_T$, and a capillary length $\delta_{\text{cap}} = \frac{\gamma}{L_v}/k\Delta T_0$. They examined the experiments of Somboonsuk, Mason, and Trivedi (1984) and concluded that, in the dendritic range (at speeds higher than the cellular-dendritic transition), the ordering

$$\delta_{\text{cap}} \ll \rho \ll \delta_c \ll \delta_T^{(1)} \tag{7.27}$$

should hold and that the array spacing should satisfy

$$\lambda_x, \lambda_y \sim \delta_c. \tag{7.28a}$$

From Trivedi and Kurz (1994) they deduced that

$$\rho \propto (\delta_c \delta_{\text{cap}})^{\frac{1}{2}} \quad \text{and} \quad \lambda \propto \left[\delta_T^{(1)}\right]^{\frac{1}{2}} (\delta_c \delta_{\text{cap}})^{\frac{1}{4}}. \tag{7.28b}$$

They defined a new parameter ϵ as

$$\epsilon = \frac{\delta_c}{\delta_T^{(1)}}, \tag{7.28c}$$

which satisfies $\epsilon = k^{-1}M^{-1}$, and sought solutions for $\epsilon \to 0$ (large speed),

$$Pe = \frac{\rho}{\delta_c} \equiv \epsilon \hat{P}, \quad \hat{P} = O(1), \tag{7.28d}$$

and

$$\frac{\delta_{cap}}{\delta_c} = O(\epsilon^2). \tag{7.28e}$$

Typically, in the experiments of Somboonsuk et al. (1984), $\epsilon < 0.1$, except near the cellular–dendritic transition.

In scaled form they solved the following system: In the liquid

$$\nabla^2 C - C_\zeta = 0 \tag{7.29a}$$

on the interface

$$C = \epsilon \hat{\Theta} + \epsilon \zeta \tag{7.29b}$$

$$\mathbf{n} \cdot \mathbf{k}[1 + (1-k)C] = -C_n, \tag{7.29c}$$

where

$$\Theta = \frac{T_0 - T_{tip}}{k \Delta T_0} = \frac{c_{tip} - c_\infty}{c_\infty (1-k)} \equiv \epsilon \hat{\Theta} \tag{7.30}$$

is the solute undercooling of the tip relative to the liquidus, $\zeta = z - z_{tip}$, $\mathbf{n}$ and $\mathbf{k}$ are unit vectors normal to the interface and in the ζ-direction, respectively, and the $O(\epsilon^2)$ capillary undercooling has been omitted.

They decomposed the solution into four regions; see Figure 7.20. In the *outer* region, $r \sim 1$, $\zeta \sim 1$, and the needle appears to be a line source of strength $Q(\zeta) \sim \epsilon$ from $0 < \zeta < \infty$. The *tip* solution is given by the Ivantsov needle, $\zeta \sim \epsilon$, $r = R_{tip}(\zeta)$, where

$$R_{tip}^2 = 2\hat{P}\zeta. \tag{7.31}$$

The *root* solution corresponds to radial diffusion with $r \sim 1$, and $\zeta \sim 1/\epsilon (\zeta > 0)$. The "unit cell" encompasses a nonaxisymmetric needle whose cross-sectional area A_R is

$$A_R = \Lambda_x \Lambda_y \{1 - [1 + (1-k)\epsilon\zeta]\}^{-\frac{1}{1-k}}. \tag{7.32}$$

The *inner* solution $r \sim \epsilon^{1/2}$ and $\zeta \sim 1$ ties all of these together because it contains effects of the tip, the neighbors, and the roots. For the case of $\Lambda_x = \Lambda_y$, R_{in} is axisymmetric at leading order and satisfies

$$\pi R_{in}^2 = \int_0^\zeta Q(s) ds. \tag{7.33}$$

7. Dendrites

Finally, the strength $Q(\zeta)$ of the line source is determined by local equilibrium on the surface of the slender needle; it requires that the C generated by Q vary linearly in z as given by the temperature field. Thus, after matching, Somboonsuk et al. found that

$$\hat{\Theta} + \delta\zeta = \frac{Q(\zeta)}{4\pi\delta} - \frac{1}{4\pi}Q(\zeta)\left[\ln\left(\frac{R_{in}^2}{4\zeta}\right) + \gamma_E\right]$$
$$+ \int_0^\infty [Q(s) - Q(\zeta)]G_{00}(\zeta;s)ds + \sum_{\substack{i=-\infty \\ i^2+j^2\neq 0}}^\infty \sum_{j=-\infty}^\infty \int_0^\infty Q(s)G_{ij}(\zeta,s)ds,$$

(7.34a)

where

$$\delta = \frac{1}{\ell n\left(\frac{1}{\epsilon}\right)},$$
(7.34b)

γ_E is Euler's constant, and

$$G_{ij}(\zeta,s) = \exp\left\{-\frac{1}{2}[d_{ij}+s-\zeta]\right\}/4\pi d_{ij},$$
(7.34c)

$$d_{ij} = \left\{(i\Lambda_x)^2 + (j\Lambda_y)^2 + (s-\zeta)^2\right\}^{1/2}.$$

For $\epsilon \ll \zeta \ll 1$, the inner solution must match the Ivantsov tip,

$$Q(0) = 2\pi\hat{P},$$
(7.35a)

and for $1 \ll \zeta \ll 1/\epsilon$, the inner solution must match the (quasi–one-dimensional) Scheil-like solution

$$Q(\infty) = \Lambda_x\Lambda_y.$$
(7.35b)

Given ϵ and (Λ_x, Λ_y), the unknown $\hat{\Theta}$ and $Q(\zeta)$ are simultaneously determined by a nonlinear solvability condition. The constraint (7.35b) is satisfied automatically, and the tip radius through $\hat{P}$ is determined from Eq. (7.35a).

Spencer and Huppert (1998) subsequently solved this nonlinear integral system. They found that the parabolic tip extends back at a distance δ_c, after which the cross section becomes thinner than the Ivantsov needle. They also found that the inner region extends back from the tip a distance of $10\delta_c$. Figure 7.20 shows the computed solutions in each region and the composite solution. Again, note that the Ivantsov tip occupies a very small region. For small ϵ, the tip radius agrees well with experiment. One can convert this to the tip undercooling Θ and hence relate the results to the single isothermal solutal dendrite (Figure 7.21).

Spencer and Huppert suggested that there is a type of selection in the steady growth problem. Two parameters are present, say the spacing λ and either ρ

7.4 Arrays of needles

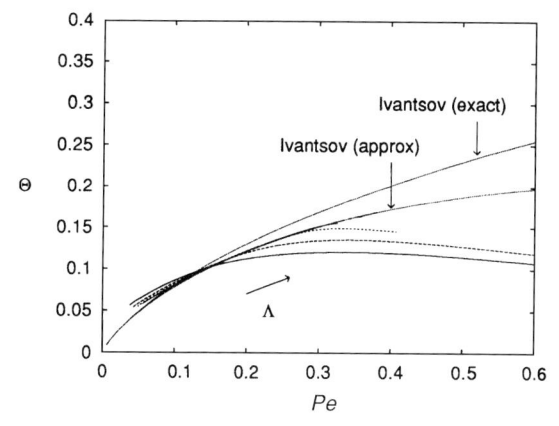

Figure 7.21. Theoretical calculations of the nondimensional tip undercooling Θ as a function of the nondimensional tip radius Pe for different values of the control parameter ϵ. Along each curve, the spacing Λ increases as Pe increases. From Spencer and Huppert (1998). (Reprinted with permission of Elsevier Science Ltd.)

or V that need to be determined. When $\lambda < \infty$ is fixed, both ρ and V are determined. The case $\lambda \to \infty$, the Ivantsov needle is a singular limit and only ρV is determined, not each individually.

The rationale of Spencer and Huppert (1997) is as follows: The tip region is Ivantsov-like with corrections to the undercooling owing to the presence of neighbors. By themselves, these constitute a one-parameter family of tips. However, the root region yields a unique solution determined by the spacing because global solute concentration determines the cross-sectional area of the needle. Via the matching of tip and root though the inner region, a single-tip solution is selected, and hence both ρ and V are determined. The lesson here is that there is really a double indeterminacy in the selection problem, λ and ρ (or V). In the free-needle case, one specifies $\lambda = \infty$ and requires an extra selection theory to give ρ (or V). In the array case, $\lambda < \infty$, one finds both ρ or V. The question that remains unanswered is, What determines λ?

These results represent a self-consistent solution of the free-boundary problem. There were previous attempts at the problem that used "guesses" as to the tip shape and inferred behavior of the array by using integral balances of heat or solute. These require the use of selection criteria such as minimum undercooling or marginal stability. Hunt (1991) and Trivedi (1984) used a spherical tip and a Scheil approximation. Makkonen (1991) used a parabolic tip. Figure 7.22 is the comparison of these partial theories with the present case, as taken from Spencer and Huppert (1999). Apart from the Makkonen result, all the others

248   7. Dendrites

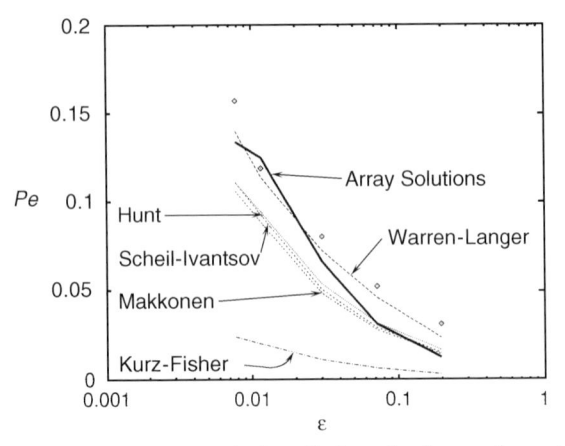

Figure 7.22. Comparison of theoretical predictions for the nondimensional tip radius Pe to data from Somboosuk, Mason, and Trivedi (1984) on SCN–acetone (indicated by diamonds). From Spencer and Huppert (1999). (Reprinted with permission of North-Holland.)

are grouped near each other, and for small ϵ, the present result agrees well with experiment.

The Spencer–Huppert theory of steady growth ignores the presence of surface energy. As they acknowledge, surface energy must enter in at least two ways:

1. The tip is predicted to be an Ivantsov paraboloid at least for a distance of one δ_c. Without surface energy, the tip is unconditionally unstable (Langer and Müller-Krumbhaar 1978b), and thus surface energy must be present to stabilize the needle tip against splitting.
2. Spencer and Huppert (1998) showed that, given $(\Lambda_x, \Lambda_y) = \Lambda$ and the temperature gradient ϵ, a unique tip scale is determined. However, nothing is said about what determines Λ. There is a one-parameter family of solutions relating Λ, ρ, and V. The Spencer and Huppert view is that the knowing of Λ determines ρ.

What likely resolves the one parameter in this family is a stability criterion; only a single or a range of Λ is stable to changes of wavelength. The full stability problem is a daunting one, indeed. The basic state would be represented by the array of needles, and the linear stability problem would involve the solving of a system of partial differential equations with spatially periodic coefficients – a Floquet problem.

The beginning of such an analysis has been made by Warren and Langer (1990), who treated the dendrite array as a set of tips that interact through the concentration field. These tips are local sources of rejected solute that depend on the local speeds V_i and local radii of curvature ρ_i. The tips are well separated so that $\rho_i \ll \lambda, d$, where d is the gap width of the Hele–Shaw cell if one considers a single row of needles. The position z_i of each tip is on the liquidus, $C_i = -G_c/m z_i$ so that capillary undercooling is neglected. Warren and Langer presumed that the tips each behave as isolated needles growing slowly in a homogeneous melt.

Because the work of Warren and Langer was done before that of Spencer and Huppert, they were not aware that each V_i and ρ_i can be determined by array interactions. By examining tips only they needed an extra condition to determine these. They used the condition given by MST (dependent on surface energy), namely,

$$\frac{2Dd_0}{V_i \rho_i^2} = \sigma^*,$$

where for this one-sided model they take $\sigma^* = 0.04$. The result is a linear dynamical system for the z_i, and they found conditions for marginal stability. In the experiments of Somboonsuk et al. (1984), $d < \lambda$, and thus no instability modes "fit" across the gap. The result of Warren and Langer is that the most dangerous instability is spatially subharmonic, that is, alternate needles grow at the expense of the nearest neighbors. Figure 7.22 compares these results for large $M \propto \epsilon^{-1}$. The tip radius given compares quite well for $\epsilon \ll 1$ with the Spencer–Huppert theory and for ϵ not so small is better than theirs.

A possible explanation for the fine comparison of this theory is that the array interactions do determine ρ_i and V_i for fixed λ but that the instability of the steady growth is determined by effects only local to the tips. The selection question awaits a full stability theory.

The results of Warren and Langer allow for the existence of a whole range of spacings that correspond to stable needle arrays. The situation that seems to exist is that "all" spacings λ are allowable for the *existence* of steady growth; see Figure 7.23. When λ is small enough, the front becomes unstable to an overgrowth effect in which alternate needles run ahead, suppressing their neighbors and effectively increasing λ. This is the Warren–Langer subharmonic instability. When λ is large enough, so that ρ is large, the "flattened" tip becomes susceptible to a morphological instability locally, which results in the splitting of the tip. The arrays would then be stable in between these limits. In experiments, however, the tip-splitting limit is never reached. Instead, another instability intervenes. Han and Trivedi (1994) showed that the needles develop sidearms that develop orthogonally to the main needles, and these side branches themselves grow side branches, called tertiary arms, that grow in the pulling

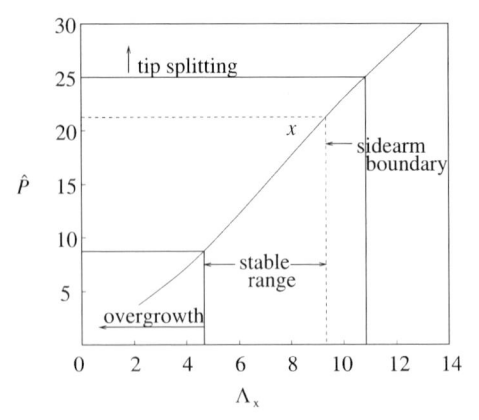

Figure 7.23. A sketch of the range of stable spacings. The solid curve shows the variation of the tip radius parameter $\hat{P}$ with the dendrite spacing parameter Λ_x in the theory. Also shown is a lower bound for array stability given by the Warren–Langer theory (1990) for array overgrowth, an upper bound for the tip splitting according to the marginal stability theory, and a broken line suggesting the (guessed) position of the boundary for sidearm instability. The maximum stable-tip radius translates to an upper bound for stable array spacings. The single experimental data point (x) is indicated within the stability bounds and is close to the theoretical curve. From Spencer and Huppert (1998). (Reprinted with permission of Elsevier Science Ltd.)

directions and compete with the main needles; they can even overgrow them. This narrowed range is indicated in Figure 7.23.

The agreement between the Warren–Langer stability analysis and the experiments is poorer when one compares wavelengths. The theory predicts wavelengths substantially greater than those seen in experiment. Because wavelengths should scale directly with δ_c, which varies inversely with V, it would seem that the experimentally observed instability is occurring when V is smaller than one supposed, which is a situation that would be present if the pulling started from rest and accelerated to the value V. If the instability occurred before the basic state were fully developed, it might explain the observations. To test this idea, Warren and Langer (1993) considered the basic state with the usual exponential profile replaced by one with a time-dependent boundary-layer thickness $\ell(t)$, where $\ell(t) \to \delta_c$ as $t \to \infty$. They performed a quasi-steady linear stability analysis (the basic state is frozen in time and normal modes in time are used) that predicts instability wavelengths that are closer to those observed.

To test the Warren–Langer stability results, Losert, Shi, and Cummins (1998) used SCN doped with 0.43 wt% Coumarin 152 in a rectangular glass cell of volume $0.1 \times 2.0 \times 150$ mm pulled through a temperature gradient $G_T = 11.7\,\text{K/cm}$ at $V = 49.82\,\mu\text{m/s}$ until a one-dimensional, steady-state dendritic array was established. As V was stepped down $1.25\,\mu\text{m/s}$ every 30 s, the separation length

λ remained at $170\,\mu$m until $V = 17.4\,\mu$m/s, when the array became unstable through alternate dendrites that began to overgrow their neighbors. The Warren and Langer (1990) critical speed is given by $V = 14.0\,\mu$m/s.

Losert, Mesquita, Figeiredo, and Cummins (1998) reconsidered the same system in the same channel. Coumarin 152 is a laser dye that has fluorescence absorption in the ultraviolet range so that if one illuminates the sample along the interface with a mercury ultraviolet lamp, the sample will be heated locally and the interface will melt back. This allows one to select a uniform wavelength for study. Here $V = 19.8\,\mu$m/s initially, and a row of ultraviolet spots is applied continuously until a uniform array of dendrites is created with $\lambda \approx 150\,\mu$m. The ultraviolet forcing is then turned off. In order to test for subharmonic instability, the ultraviolet spots are focused on the tips of alternate dendrites, creating an amplitude modulation equal to 2λ, which decays at a measured rate. This process is repeated at several different V values between 19.8 and $6.6\,\mu$m/s, and the decay rates are calculated. When these are extrapolated to zero, the apparent critical V for stability is $V \approx 4\,\mu$m/s. (When no ultraviolet modulation is introduced, the array seems to become unstable at the same critical V). The measured decay rates are those predicted by Warren and Langer occur at pulling speeds about three times greater than those in the experiments.

Losert, Mesquita, Figeiredo, and Cummins (1998) found a quantitative fit for this instability in the weakly nonlinear range to a Landau equation for the amplitude of the modulation,

$$\dot{A} = \sigma_0 A - b_1 |A|^2 A \tag{7.36a}$$

with

$$\sigma_0 = -0.090 S^{-1} \text{ and } b_1 = -1.1 \times 10^{-5} \mu\text{m}^{-2} S^{-1}. \tag{7.36b}$$

As V is decreased, there is a bifurcation that is subcritical with a deep range of subcriticality.

This remarkable study documenting the overgrowth instability shows how overgrowth occurs at the left-hand boundary of Figure 7.23. There is no comparable study for the instability on the right-hand boundary. The significance of Figure 7.23 is that no single spacing is globally preferred; there is an interval of λ that corresponds to stable, regular needle arrays, and these presumably can be accessed by the system through varying the history of the process.

7.5 Remarks

The Ivantsov solution for zero surface energy and zero attachment kinetics presents a parabola (two-dimensional) or paraboloid (three-dimensional) valid for steady growth. This deceptively simple solution is difficult to extend to situations with small surface tension. The implied singular perturbation destroys the uniform validity of the Ivantsov solution. If one preserves the Ivantsov root, no solution is present unless one makes the system anisotropic, say, in the surface tension. If one abandons the Ivantsov-root behavior, a solution with a smooth tip and isotropic surface energy can be found.

The stability of the isolated needle is summarized in Figure 7.13. When $0 \leq \alpha_4 < \alpha_{4_c}$, the traveling wave predominates, and when α_4 exceeds α_{4_c}, the steady-needle predominates. Every numerical simulation to date applies to the latter range, and the MST seems well tested. In some of these simulations, when $\alpha_4 < \alpha_{4_c}$, no MST solution could be found in anecdotal agreement with the picture shown. No comprehensive numerical test of IWT has yet appeared.

It seems to be the case that all the theories give tip radii and tip speeds of remarkably similar magnitudes, and thus by these measures it would be very difficult to distinguish these experimentally. Because the theories predict intrinsically different behaviors, steady versus time-periodic, and persistent noise required versus persistent noise not required, more dynamically based experiments are needed. Further, experiments, physical and numerical, cannot approach the small undercooling limit most accessible to analysis since in this vicinity V is very small and hence δ_T is very large, and the presence of "distant" boundaries affects the thermal field.

An array of needles possesses root regions significantly different from those of isolated needles, and their stability behavior differs markedly. In this case, the presence of neighbors induces collective instabilities as summarized in Figure 7.23. Regular arrays with separation distances λ too small are susceptible to overgrowth instabilities, whereas those with λ too large are made unstable by tip splitting or tertiary-sidearm overgrowth. The large λ instabilities are not well understood theoretically. When λ is between these critical values, the regular array is seemingly stable, and any value in this range is in principle accessible. Cummins et al. have addressed overgrowth instabilities, but much more needs to be done experimentally to demark the range of λ that must be accessed. No theory exists in the literature that addresses the conceivable appearance of irregular arrays.

References

Barbieri, A., Hong, D. C., and Langer, J. S. (1987). Velocity selection in the symmetrical models of dendritic crystal growth, *Phys. Rev. A* **35**, 1802–1808.

Bechhoeffer, J., Guido, H., and Libshaber, A. (1988). Solidification dans un capillaire rectangulaire, *Comptes Rendus* **306**, 619–625.

Ben Amar, M., and Pomeau, Y. (1986). "Theory of dendritic growth in a weakly under cooled melt." *Europhys. Lett.* **2**, 307–314.

Bensimon, D., Pelce, P., and Shraiman, B. I. (1987). Dynamics of curved fronts and pattern selection, *J. Physique* (Paris) **48**, 2081–2087.

Brener, E. A., Geilikman, M. B., and Temkin, D. E. (1988). Growth of a needle-shaped crystal in a channel, *Sov. Phys.* JETP **67**, 1002–1009.
Brener, E. A., Müller-Krumbhaar, H., Saito, Y., and Temkin, D. E. (1993). Crystal growth in a channel: Numerical study of the one-sided model, *Phys. Rev. E* **47**, 1151–1155.
Canright, D., and Davis, S. H. (1989). Similarity solutions to phase-change problems, *Metall. Trans.* **20A**, 225–235.
Chalmers, B. (1966). *Principles of Solidification*, Wiley, New York.
Glicksman, M. E., and Marsh, S. P. (1993). The dendrite, in D. T. J. Hurle, editor, *Handbook of Crystal Growth*, Chap.15, North-Holland, Amsterdam.
Glicksman, M. E., and Schaefer, R. J. (1967). Investigation of solid-liquid interface temperature via isenthalpic solidification, *J. Crystal Growth* **1**, 297–310.
———· (1968). Comments on theoretical analyses of isenthalpic solidification, *J. Crystal Growth* **2**, 239–242.
Glicksman, M. E., Schaefer, R. J., and Ayers, J. D. (1976). Dendritic growth – a test of theory, *Metall. Trans.* **A7**, 1747–1759.
Han, S. H., and Trivedi, R. (1994). Primary spacing selection in directionally solidified alloys, *Acta Metall. Mater.* **42**, 25–41.
Hong, D. C., and Langer, J. S. (1986). Analytic theory of the selection mechanism in the Saffman–Taylor problem, *Phys. Rev. Lett.* **56**, 2032–2035.
Huang, S.-C., and Glicksman, M. E. (1981a). Fundamentals of dendritic solidification: I. Steady-state tip-growth, *Acta Metall. Mater.* **29**, 701–716.
———· (1981b). Fundamentals of dendritic solidification: Development of sidebranch structure, *Acta Metall. Mater.* **29**, 717–731.
Hunt, J. D. (1991). A numerical analysis of dendritic and cellular growth of a pure material investigating the transition from array to isolated growth, *Acta Metall. Mater.* **39**, 2117–2133.
Ihle, T., and Müller-Krumbhaar, H. (1994). Diffusion-limited fractal growth morphology in thermodynamical 2-phase systems, *Phys. Rev. Lett.* **70**, 3083–3086.
Ivantsov, G. P. (1947). Temperature field around spherical, cylindrical, and needle-shaped crystals which grow in supercooled melt, *Dokl. Akad. Nauk SSSR* **558**, 567–569.
Karma, A. (1986). "Solidification cells at low velocity – The moving symmetric model" *Phys. Rev. A* **34**, 4353–4362.
Kessler, D. A., and Levine, H. (1986). Stability of dendritic crystals, *Phys. Rev. Lett.* **57**, 3069–3072.
Kessler, D. A., Koplik, J., and Levine, H. (1986). Dendritic growth in a channel, *Phys. Rev. A* **34**, 4980–4987.
Kruskal, H., and Segur, H. (1991). Asymptotics beyond all orders in a. model of crystal growth, *Stud. Appl. Math.* **85**, 129–181.
Kupferman, R., Kessler, D. A., and Ben-Jacob, E. (1995). Coexistence of symmetric and parity-broken dendrites in a channel, *Physics A* **213**, 451–464.
Kurz, W., and Fisher, D. J. (1989). *Fundamentals of Solidification*, Trans Tech, Aedermannsdorf, Switzerland.
Langer, J. S. (1987). Lectures in the theory of pattern formation, in, J. Souletie, J. Vannimeus, and R. Stora, editors *Le hasard et la matiére, Les Houches session XLVI*, pages 629–712, North-Holland, Amsterdam.
Langer, J. S., and Müller-Krumbhaar, H. (1978a). Theory of dendritic growth – I. Elements of a stability analysis, *Acta Metall. Mater.* **26**, 1681–1688.
Langer, J. S., and Müller-Krumbhaar, H. (1978b). Theory of dendritic growth – II. Instabilities in the limit of vanishing surface tension, *Acta Metall. Mater.* **26**, 1689–1696.

Losert, W., Mesquita, O. N., Figeiredo, J. M. A., and Cummins, H. Z. (1998). Direct measurement of dendritic array stability, *Phys. Rev. Lett.* **81**, 409–412.

Losert, W., Shi, B. Q., and Cummins, H. Z. (1998). Evolution of dendritic patterns during alloy solidification: Onset of the initial instability, *Proc. Nat. Acad. Sci.* **95**, 431–438.

Makkonen, L. (1991). Primary dendrite spacing in constrained solidification, *Mat. Sci. Eng. A-Struct.* **148**, 141–143.

Meiron, D. I. (1986). Selection of steady states in the two-dimensional symmetric model of dendritic growth, *Phys. Rev. A* **33**, 2704–2715.

Müller-Krumbhaar, H., and Langer, J. S. (1978). Theory of dendritic growth – III. Effects of surface tension, *Acta Metall. Mater.* **26**, 1697–1708.

Mullins, W. W., and Sekerka, R. F. (1964). Stability of a planar interface during solidification of a dilute binary alloy, *J. Appl. Phys.* **35**, 444–451.

Muschol, M., Liu, D., and Cummins, H. Z. (1992). Surface-tension-anisotropy measurements of succinonitrile and pivalic acid: Comparison with microscopic solvability theory, *Phys. Rev. A* **46**, 1038–1050.

Nash, G. E., and Glicksman, M. E. (1974). Capillary-limited steady-state dendritic growth – I. Theoretical developments, *Acta Metall. mater.* **22**, 1283–1290.

Oldfield, W. (1973). "Computer model studies of dendritic growth." *Mater. Sci. Eng.* **11**, 211–218.

Pelce, P., and Pumir, A. (1985). Cell shape in directional solidification in the small Peclét limit, *J. Crystal Growth* **73**, 337.

Schaefer, R. J., Glicksman, M. E., and Ayers, J. D. (1975). High-confidence measurements of solid–liquid surface energy in a pure metal, *Philos. Mag.* **32**, 725–743.

Somboonsuk, K., Mason, J. T., and R. Trivedi (1984). Interdendritic spacing: Part I. Experimental studies, *Metall. Trans. A* **15A**, 967.

Spencer, B. J., and Huppert, H. E. (1995). State–state solutions for an array of strongly-interacting needle crystals in the limit of small undercooling, *J. Crystal Growth* **148**, 305–323.

——— (1997). On the solidification of dendritic arrays: An asymptotic theory for the directional solidification of slender needle crystals, *Acta Metall. Mater* **45**, 1535–1549.

——— (1998). On the solidification of dendritic arrays: Selection of the tip characteristics of slender needle crystals by array interactions, *Acta Metall. Mater* **46**, 2645–2662.

——— (1999). The relationship between dendrite tip characteristics and dendrite spacings in alloy directional solidification, *J. Crystal Growth* **200**, 287–296.

Temkin, D. E. (1960). Growth rate of a needle crystal formed in a supercooled melt, *Dokl. Akad. Nauk SSSR* **132**, 609–613.

Trivedi, R. (1984). Interdendritic spacingz. A comparison of theory and experiment, *Metall Trans A* **15**, 977–982.

Trivedi, R., and Kurz, W. (1994). Solidification microstructures – A conceptual approach, *Acta Metall. Mater.* **42**, 15–23.

Warren, J. A., and Langer, J. S. (1990). Stability of dendritic arrays, *Phys. Rev. A* **42**, 3518–3525.

———(1993). Prediction of dendritic spacings in a directional-solidification experiment, *Phys. Rev. E* **47**, 2702–2712.

Xu, J.-J. (1997). *Interfacial Wave Theory of Pattern Formation*, Springer-Verlag, New York.

8

Eutectics

In all the previous discussions of the freezing of binary liquids it was assumed that the alloy is dilute so that the relevant parts of the phase diagrams governing equilibrium are neighborhoods of their extreme left-hand or right-hand sides. Figure 3.1 shows the phase diagram for Pb–Sn, as one can see, much additional structure is displayed. Two equilibrium melting temperatures T_m^A and T_m^B are located at points A and F. At points B and G the solidi intersect the eutectic temperature $T_E \approx 183°C$. The point E is the eutectic point with coordinates (C_E, T_E). Here Pb is called component A, and Sn, the solute, is called component B.

Above the liquidi, the material is always liquid. If C is the composition of Sn and the alloy is cooled in the region $C < C_E$, the first phase change will result in a mixture of alloy α and liquid. When the solidus is crossed, and $C < 18.3$ wt%, only alloy α is formed. At still lower temperatures below the solvus, two solid alloys, α and β, are formed. If $18.3\% < C < C_E$, then the cooling of $\alpha + L$ results in the formation of solids α and β. When $C = C_E$, the alloy is called eutectic, and when $C < C_E (C > C_E)$, it is called hypoeutectic (hypereutectic). A simplified diagram is given in Figure 8.1

The eutectic temperature T_E represents the lowest temperature at which the alloy in thermodynamic equilibrium can persist as a liquid. As the temperature of an alloy at the eutectic concentration C_E is lowered past T_E, there is a spontaneous formation of solid composed of two alloys, α and β, that have concentrations C^α and C^β, respectively, the endpoints of the horizontal line at $T = T_E$, as shown in Figure 8.1.

Figure 8.2 shows a lamellar eutectic of CBr_4–$C_2C\ell_6$; this is a two-dimensional configuration composed of alternate plates of alloys α and β. The second common type is the rod eutectic in which solid cylinders of, say, alloy α grow in a frozen sea of alloy β. Kurz and Fisher (1985, p. 98) have given

256   8. Eutectics

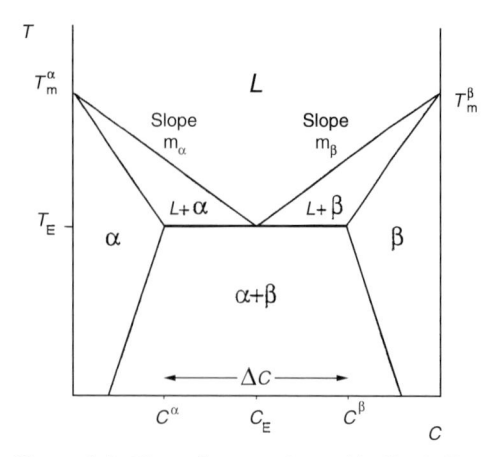

Figure 8.1. Phase diagrams for an idealized alloy.

the rule of thumb that when C_E is between 25 and 75%, the lamellar eutectic will form, whereas, outside these limits, the rod ecutectic is to be expected. In the present discussion, only the lamellar case will be discussed; consequently, roughly symmetric phase diagrams will be considered.

Eutectic alloys are commercially very important for two distinct reasons. They can be reprocessed very easily without defect formation because the typical wave length λ of a pair of α, β plates is an order of magnitude smaller than a dendrite produced under similar conditions (Kurz and Fisher 1985, p. 98). Further, eutectic alloys represent composites produced in situ. If the α and β have vastly different properties (say, in an extreme case, that one is a good electrical conductor and the other an insulator), the lamellar eutectic will be a directional device that conducts electricity in one direction but not in the orthogonal direction.

8.1 Formulation

Consider the one-sided FTA model that gives the temperature T permanently by

$$T = T_E + G_T z, \tag{8.1}$$

where the reference temperature is now taken to be T_E.

If the concentrations are measured in volume percent, then, in the liquid, C_A^ℓ and C_B^ℓ are related by $C_A^\ell + C_B^\ell = 1$. The concentration that will be used here is that the concentration C_B^ℓ of component B will be denoted by C so that

$$C_A^\ell = 1 - C \tag{8.2}$$

8.1 Formulation

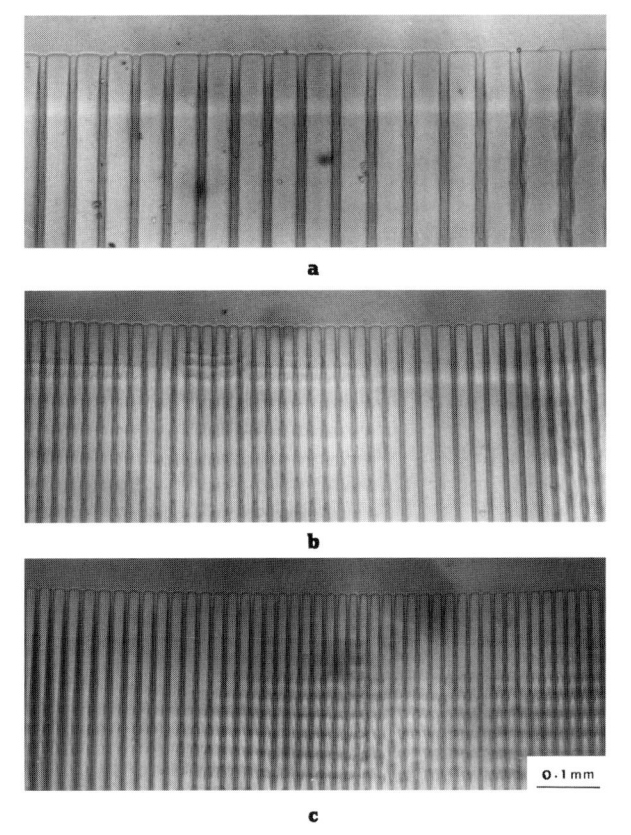

Figure 8.2. Steady-state eutectic microstructures of directionally solidified CBr$_4$–8.4 wt % C$_2$Cℓ$_6$ alloys at different pulling speeds: (a) $V = 0.2$ μm/s, (b) $V = 0.5$ μm/s, and (c) $V = 1.0$ μm/s. From Seetharaman and Trivedi (1988). (Reprinted with permission of Metall. mater. Trans.)

In the liquid there is the diffusion of component B,

$$C_t - VC_z = D\nabla^2 C, \tag{8.3}$$

where D represents the interdiffusion coefficient because the concentration of B is not small.

At large distances from the interface, the concentration is given as

$$C \to C_E + C_\infty \quad \text{as} \quad z \to \infty, \tag{8.4}$$

where C_∞ is the far-field departure from C_E of the concentration of B; $C_\infty = 0$ corresponds to the eutectic composition.

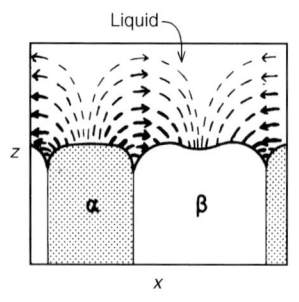

Figure 8.3. A sketch of one wavelength of a lamellar eutectic with the diffusion field of component B indicated. From Kurz and Fisher (1985). (Reprinted with permission of Trans Tech Sa.)

Figure 8.3 shows a detail of the front of an α–β–liquid interface. As α is being created, this A-rich alloy rejects component B. As β is being created, the B-rich alloy rejects component A. These simultaneous rejections create concentration gradients in the liquid, and hence lateral diffusion, even in the steady growth of eutectic alloys.

At the interface in two dimensions, $z = h(x, t)$, there are two rejection coefficients. At phase α,

$$C^s/C^\ell = k_\alpha < 1, \qquad (8.5\text{a})$$

and at phase β,

$$k_\beta = \frac{1 - C^s}{1 - C^\ell} < 1, \qquad (8.5\text{b})$$

and thus across the interface

$$C^\ell - C^s = \begin{cases} (1 - k_\alpha)C, & \text{phase } \alpha \\ (1 - k_\beta)(C - 1), & \text{phase } \beta. \end{cases} \qquad (8.5\text{c})$$

At the interface for each phase,

$$(C^s - C^\ell)V_n = D\nabla C^\ell \cdot \mathbf{n};$$

consequently,

$$\begin{aligned} (V + h_t)(k_\alpha - 1)C &= D(C_z - C_x h_x), \quad \text{phase } \alpha \\ (V + h_t)(k_\beta - 1)(C - 1) &= D(C_z - C_x h_x), \quad \text{phase } \beta. \end{aligned} \qquad (8.6)$$

Likewise, there are two Gibbs–Thomson equations

$$\begin{aligned} T &= T_\text{E}\left(1 + 2H\frac{\gamma_\alpha}{L_v^\alpha}\right) + m_\alpha(C - C_\text{E}), \quad \text{phase } \alpha \\ T &= T_\text{E}\left(1 + 2H\frac{\gamma_\beta}{L_v^\beta}\right) + m_\beta(C - C_\text{E}), \quad \text{phase } \beta, \end{aligned} \qquad (8.7)$$

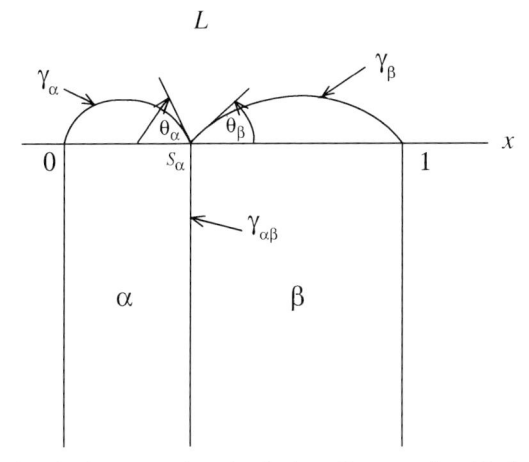

Figure 8.4. A sketch of one wavelength of a lamellar eutectic with the contact angles θ_α and θ_β and interfacial energies γ_α, γ_β, and, $\gamma_{\alpha\beta}$.

where $m_\alpha < 0$ and $m_\beta > 0$ are the slopes of the respective liquidi, γ_α and γ_β are the surface energies, and L_v^α and L_v^β are the latent heats per unit volume.

If one combines relations (8.7) with the given temperature field (8.1), the following are obtained:

$$G_T h = 2H \frac{T_E \gamma_\alpha}{L_v^\alpha} + m_\alpha (C - C_E), \text{ phase } \alpha$$

$$G_T h = 2H \frac{T_E \gamma_\beta}{L_v^\beta} + m_\beta (C - C_E), \text{ phase } \beta.$$

(8.8)

Finally, at the contact lines there are trijunction conditions obtained by minimizing energies, as discussed in Chapter 2. These thermodynamic–equilibrium conditions apply at a contact line located at $x = x'$, $z = z'$. As shown in Figure 8.4,

$$\gamma_\alpha \cos\theta_\alpha = \gamma_\beta \cos\theta_\beta$$
$$\gamma_\alpha \sin\theta_\alpha + \gamma_\beta \sin\theta_\beta = \gamma_{\alpha\beta},$$

(8.9)

where $\gamma_{\alpha\beta}$ is the interfacial energy between phases α and β, and θ_α and θ_β are the equilibrium contact angles, as shown.

Let us consider now growth periodic in x when an (α, β) pair has length λ; call S_α the volume fraction of phase α, and thus phase α lies in $0 < x < S_\alpha \lambda$ and phase β lies in $S_\alpha \lambda < x < \lambda$. Scale z on δ_c, x on λ and time on δ_c/V and define the nondimensional concentration $\hat{C}$ by $\hat{C} = (C - C_E)/\Delta C$ where

$\Delta C = C_\beta - C_\alpha$, as shown in Figure 8.1. If the carets are dropped, system (8.3)–(8.9) takes the form

$$C_t - C_z = C_{zz} + \Lambda^{-2} C_{xx}, \quad z > h(x,t). \tag{8.10a}$$

On $z = h(x,t)$, $0 < x < S_\alpha$,

$$(1 + h_t)[(1 - k_\alpha)C + \Delta] = -(C_z - \Lambda^{-2} h_x C_x), \tag{8.10b}$$

$$M_\alpha^{-1} h = -C + 2H \Gamma_\alpha. \tag{8.10c}$$

On $z = h(x,t)$, $S_\alpha < x < 1$,

$$(1 + h_t)\left[(1 - k_\beta)C + \Delta - 1\right] = -(C_z - \Lambda^2 h_x C_z), \tag{8.10d}$$

$$M_\beta^{-1} h = C + 2H \Gamma_\beta. \tag{8.10e}$$

At the trijunctions,

$$m \Gamma_\alpha \cos \theta_\alpha = \Gamma_\beta \cos \theta_\beta, \tag{8.10f}$$

$$m \Gamma_\alpha \sin \theta_\alpha + \Gamma_\beta \sin \theta_\beta = \Gamma_{\alpha\beta}, \tag{8.10g}$$

and as $z \to \infty$,

$$C \to C^\infty, \tag{8.10h}$$

where

$$M_\alpha = -\frac{m_\alpha \Delta C}{G_T \delta_c} \qquad M_\beta = \frac{m_\beta \Delta C}{G_T \delta_c}$$

$$\Gamma_\alpha = -\frac{\gamma_\alpha T_E}{m_\alpha (\Delta C) L_\nu^\alpha \delta_c} \qquad \Gamma_\beta = -\frac{\gamma_\beta T_E}{m_\beta (\Delta C) L_\nu^\beta \delta_c}$$

$$\Gamma_{\alpha\beta} = \frac{\gamma_{\alpha\beta} T_E}{m_\beta (\Delta C) L_\nu^\beta \delta_c} \qquad m = -\frac{m_\alpha}{m_\beta} \frac{L_\nu^\alpha}{L_\nu^\beta} \tag{8.11}$$

$$\Delta = \frac{C_E - C_\alpha^s}{\Delta C} \qquad C^\infty = \frac{C_\infty - C_E}{\Delta C}$$

$$\Lambda = \frac{\lambda}{\delta_c}$$

8.2 Approximate Theories for Steady Growth and Selection

The central analysis of steady growth was undertaken by Jackson and Hunt (1966), who solved the steady growth problem with rather severe approximations.

(1) They sought solutions for the concentration C with the interface replaced by a plane. In effect they obtained an "outer" solution valid where the interfacial corrugations are negligible. To do this, (2) they approximated the left-hand side of the conditions (8.10b,d) with constants obtained by presuming that the concentration departs from C_E by a negligibly small amount. Because $C = 0$ corresponds to the eutectic concentration, the new flux balances become

$$\Delta = C_z, \quad 0 \leq x < S_\alpha \tag{8.12a}$$

and

$$\Delta - 1 = C_z, \quad S_\alpha < x \leq 1. \tag{8.12b}$$

On the right-hand sides of balances (8.10b,d), h_x is replaced by zero.

Jackson and Hunt solved (8.10a) for the concentration by writing

$$C(x, g) = C^\infty + B_0 e^{-z} + \sum_{n=1}^{\infty} B_n e^{-\left[\frac{1}{2}+\sqrt{\frac{1}{4}+n^2\pi^2\Lambda^2}\right]z} \cos n\pi \left(x - \frac{1}{2}S_\alpha\right). \tag{8.13}$$

For purposes of simplicity, (3) they utilized the condition that for $n = 1, 2, \ldots$, $n\pi \gg (1/2)\Lambda$, which means that $\lambda \ll \delta_c$ is valid for all cases in which the solidification is not too "rapid." As a result

$$C(x, z) \approx C^\infty + B_0 e^{-z} + \sum_{n=1}^{\infty} B_n e^{-\frac{n\pi z}{\Lambda}} \cos n\pi \left(x - \frac{1}{2}S_\alpha\right). \tag{8.14}$$

They then obtained the coefficients B_n using the orthogonality of the Fourier modes to find that

$$-B_0 = S_\alpha \Delta + (1 - S_\alpha)(\Delta - 1), \tag{8.15a}$$

which is a scaled version of the lever law, and

$$B_n = -\frac{2\Lambda}{n^2\pi^2} \sin n\pi S_\alpha, \tag{8.15b}$$

which represents the ratio of Δ to ΔC. Thus,

$$C(x, z) \approx C^\infty - [S_\alpha \Delta + (1 - S_\alpha)(\Delta - 1)] e^{-z}$$
$$- 2\Lambda \sum_{n=1}^{\infty} \frac{1}{n^2\pi^2} e^{-\frac{n\pi z}{\Lambda}} \sin n\pi S_\alpha \cos n\pi \left(x - \frac{1}{2}S_\alpha\right). \tag{8.16}$$

This approximated distribution of solute represents a concentration boundary layer of thickness δ_c (the first two terms), similar to that for directional solidification of a dilute alloy, corrected by x-periodic distributions that reflect the plate-like structure of eutectic growth. The boundary layer has constant thickness, but the coefficient B_0 varies with concentration; it is zero at the eutectic concentration that reflects the fact in this case that there is no net rejection over one wavelength. What α rejects, β incorporates.

Given the approximate $C(x, z)$, Eq. (8.16), one can now turn to shape of the interface. From Eq. (8.10c),

$$2H\Gamma_\alpha - M_\alpha^{-1} h = C(x, 0)$$
$$= C^\infty + [S_\alpha \Delta + (1 - S_\alpha)(\Delta - 1)]$$
$$- 2\Lambda \sum_{n=1}^{\infty} \frac{1}{n^2 \pi^2} \sin n\pi S_\alpha \sin n\pi x$$
$$\equiv f(x). \tag{8.17}$$

Jackson and Hunt (1966) ignored the term $M^{-1}h$ and solved Eq. (8.17) subject to the edge conditions. For present purposes, it is sufficient to replace $2H$ by its linearization h_{xx} appropriate to small contact angles θ_α and θ_β. Thus,

$$\Gamma_\alpha h_{xx} = f(x), \tag{8.18a}$$

and thus

$$\Gamma_\alpha h_x = \int_0^x f(s)\,ds + \theta_\alpha \tag{8.18b}$$

($\tan \theta_\alpha \approx \theta_\alpha$), and

$$\Gamma_\alpha h = \int_0^x \int_0^\xi f(s)\,ds\,d\xi + \theta_\alpha x, \tag{8.18c}$$

which satisfies $h(0) = 0$. Likewise, from Eq. (8.10e) with similar approximations,

$$\Gamma_\beta h_{xx} = -f(x) \tag{8.19a}$$

so that

$$\Gamma_\beta h_x = -\int_{S_a}^x f(s)\,ds + \theta_\beta \tag{8.19b}$$

and

$$\Gamma_\beta h = -\int_{S_a}^x \int_{S_a}^\xi f(s)\,ds\,d\xi + \theta_\beta (x - S_\alpha), \tag{8.19c}$$

which satisfies $h(S_\alpha) = 0$. The shapes (8.18c) and (8.19c) are concave downward segments that meet at the trijunctions at prescribed angles.

8.2 Approximate theories for steady growth and selection

The validity of the approximations made, and hence the predictions for steady growth, will be discussed shortly. Before this, let us review a selection argument that Jackson and Hunt suggested.

From Eqs. (8.7) the dimensional undercoolings $T_E - T^1$ on each phase have the forms

$$\Delta T_\alpha = -\frac{T_E \gamma_\alpha}{L_v^\alpha} 2H - m_\alpha(C - C_E) \tag{8.20a}$$

and

$$\Delta T_\beta = -\frac{T_E \gamma_\beta}{L_v^\alpha} 2H - m_\beta(C - C_E). \tag{8.20b}$$

Jackson and Hunt then averaged each of these over its respective phase to obtain $\overline{\Delta T_\alpha}$ and $\overline{\Delta T_\beta}$ and then set them equal, $\overline{\Delta T_\alpha} = \overline{\Delta T_\beta} \equiv (\Delta T)_m$, and evaluated C using Eq. (8.16). In *dimensional* variables, the undercooling has the form

$$\frac{1}{\bar{m}}(\Delta T)_m = V\lambda Q_1 + \frac{Q_2}{\lambda}, \tag{8.21}$$

where $1/\bar{m} = 1/|m_\alpha| + 1/|m_\beta|$ and Q_1 and Q_2 are positive functionals of the material properties. Equation (8.21) has a minimum value at $\lambda_m^2 = \frac{Q_2}{VQ_1}$, as shown in Figure 8.5. One can then "suppose" that the operating point is this minimum (Zener 1946), giving

$$V\lambda^2 = c_1 \text{ (constant)}. \tag{8.22}$$

In the calculation of $C(x, z)$, the wavelength λ was taken to be fixed but unknown. The preceding argument claims that λ is selected by the minimum of the curve $(\Delta T)_m$ versus λ.

Independent of the selection criterion, when the Peclet number $P_\lambda = \frac{\lambda V}{2D} \to 0$, the similarity principal for a given material

$$\lambda \propto \sqrt{\delta_{\text{cap}} \delta_c} f(M)$$

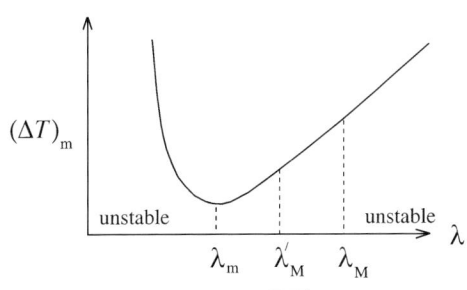

Figure 8.5. A sketch of average undercooling $\overline{\Delta T}$ versus interlamellar spacing λ for a given pulling speed V. Steady eutectic growth is allowable in $\lambda_m < \lambda < \lambda'_M$.

is shown to be the case by Kassner and Misbah (1991a) through the invariance of the governing system to the transformation $\lambda \to q\lambda$, $(G_T, V) \to q^{-2}(G_T, V)$ for any $q > 0$. Here the parameters δ_{cap} and M are taken for either material α or β. Thus, if M is fixed in a set of experiments, Eq. (8.22) is necessarily true (for $P_\lambda \to 0$).

Note that relation (8.22) would then imply two others, namely,

$$[(\Delta T)_m]^2 / V = c_2 \text{ (constant)} \tag{8.23a}$$

and

$$\lambda (\Delta T)_m = c_3 \text{ (constant)}. \tag{8.23b}$$

The minimum point, which has λ_m^2 proportional to the product of the capillary length and a diffusion length (itself proportional to ΔT), here serves in a sense the same purpose as the marginal stability boundary in dendritic growth. Jackson and Hunt argued that all $\lambda < \lambda_m$ correspond to unstable states because overgrowth will always occur. This was shown to be the case for eutectics by Langer (1980) though by means of a much simplified analysis.

As λ increases from λ_m, the interface of the phase with the larger volume fraction tends to dimple, and at some $\lambda \equiv \lambda_M$ the interface becomes vertical at two points. Jackson and Hunt argued that beyond λ_M the steady growth, as described, could not persist without tip splitting and that every steady growth morphology must have $\lambda \in [\lambda_m, \lambda_M]$, as shown in Figure 8.5.

The rather drastic approximations in the Jackson and Hunt theory make one wonder under what conditions the results would be reliable. Brattkus and Davis (1990) suggested that when G_T is very large in magnitude that the interface deformations would be small if θ_α and θ_β were both small. Brattkus et al. (1990) used the large G_T limit, with more general θ_α and θ_β, and employed a bounding technique to show that the Jackson and Hunt theory is recovered. In both analyses, when M_α and M_β are vanishingly small, the Jackson and Hunt theory is sufficient for the description of $C(x, z)$ and $h(x)$, though it turns out that $\overline{\Delta T_\alpha} \neq \overline{\Delta T_\beta}$.

Kassner and Misbah (1991a) used the boundary-integral method to solve the two-dimensional problem in the one-sided FTA numerically and posed an improved Jackson and Hunt theory. This IJH theory still uses the concentration field for the flat interface but now fully solves Eqs. (8.10c,e) with the proper trijunction conditions. Kassner and Misbah showed that the undercoolings on the two phases are not equal and that, if the *correct* average undercooling $\overline{\Delta T}$ is calculated, it departs $(\Delta T)_m$ of the Jackson and Hunt theory, especially at large λ; see Figure 8.6. Surprisingly, except at large λ, the interface shapes given by the IJH theory are quite accurate even for $M_\alpha, M_\beta \approx 2$.

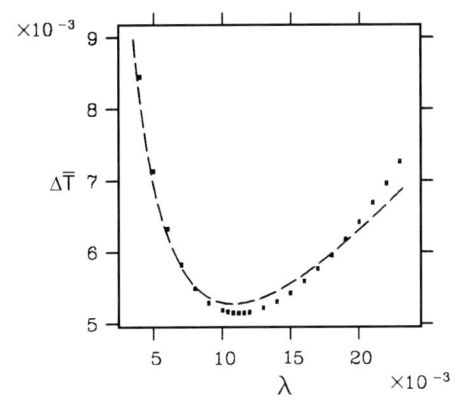

8.2 Approximate theories for steady growth and selection 265

Figure 8.6. $\overline{\Delta T}$ versus λ where the dashed line is a least-squares fit of a hyperbola $z = a\lambda + b/\lambda$ to the data points. From Kassner and Misbah (1991a). (Reprinted with permission of the American Physical Society.)

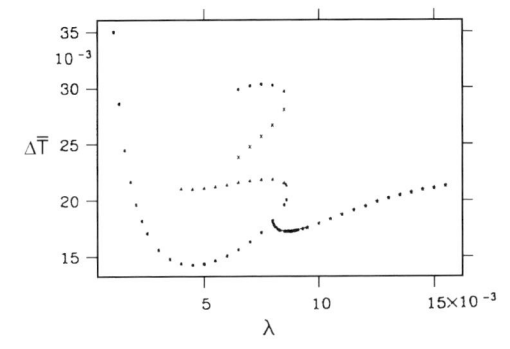

Figure 8.7. $\overline{\Delta T}$ versus λ for four branches of symmetric solutions and one branch of tilted solutions. The four branches form two pairs whose members coalesce at a limit point at $\lambda \approx 0.0086$. Beyond this λ value, no symmetric solutions could be found. The asterisks correspond to the tilted solution. From Kassner and Misbah (1991b). (Reprinted with permission of the American Physical Society.)

When the full free-boundary problem in two dimensions is solved, Kassner and Misbah (1991b) found that the steady growth solutions exhibit a bifurcation. Figure 8.7 for, say, $\lambda = 6 \times 10^{-3}$ shows four solutions, and it is likely there are others as well for larger $\overline{\Delta T}$. Interestingly, when λ increases enough, the solutions develop a limit point *before* which another solution bifurcates. This is shown only on the lowest curve and represents the creation of a tilted eutectic at an angle ϕ to the temperature gradient. This is a supercritical bifurcation in which ϕ increases monotonically from its value of zero at the bifurcation point.

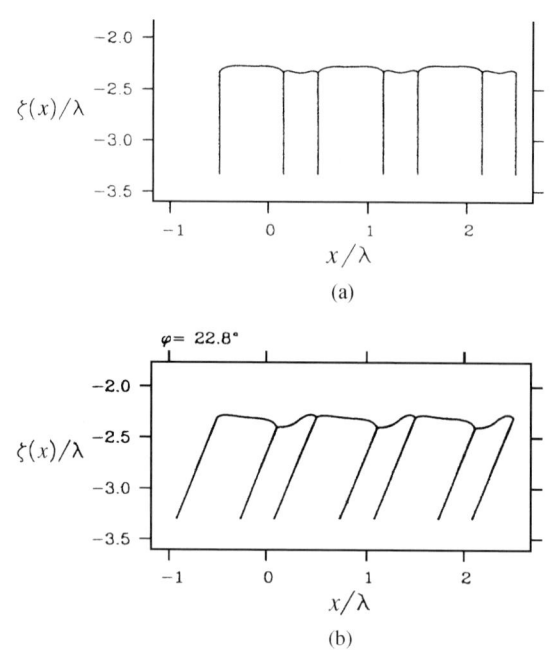

Figure 8.8. The calculated eutectic front for (a) $\lambda = 0.008$ and (b) $\lambda = 0.012$. From Kassner and Misbah (1991b). (Reprinted with permission of the American Physical Society.)

Figure 8.8 shows the untilted and tilted modes, respectively, on the two sides of the bifurcation.

The bifurcation always occurs before tip splitting appears, and the bifurcation site is always on the lower branch. The $\overline{\Delta T}$-versus-λ curve for the tilted mode often has a local minimum as well, as shown in Figure 8.7, which itself might indicate a selection site. The ratio of the λ for the two modal minima has approximately the value 2 in all the simulations.

Finally, the bifurcations to tilting cells are predicted for $-1.4 < C_\infty/\Delta C < 0.1$ for the parameters shown, indicating their existence for hypo- and hypereutectic alloys as well as eutectic alloys.

A comprehensive set of experiments on CB_4–$C_2C\ell_6$ at $C = C_E$ was performed by Seetharaman and Trivedi (1988). They saw no unique selection (consistent with earlier observations of Jordan and Hunt 1972) with the preferred λ always lying in an interval from λ_m to λ'_M, a value well below λ_M, as shown in Figure 8.5. They increased V impulsively from zero or from a nonzero steady state and found that the magnitude of the jump affects the selected λ. Finally, they found that the scaling law $V\lambda^2$-is-constant holds, but only for small enough P_λ.

8.3 Instabilities

Datye and Langer (1981) used the Jackson and Hunt model to examine linearized instabilities. They assumed that the solid–solid interfaces always remain perpendicular to the local average solid–liquid front and found that the regular lamellar eutectic is Eckhaus unstable for $\lambda < \lambda_m$. They also found for large M an oscillatory instability of wavelength 2λ for systems sufficiently off-eutectic; see Figure 8.9. Caroli, Caroli, and Roulet (1990) tested the steady growth for instability with respect to quasi-uniform variations in mass fraction of the phases and $M_\alpha, M_\beta \to 0$. They found that the instability occurs only for alloys with $C - C_E$ on the side of the phase diagram with the larger liquid-solid temperature gap. All patterns with $\lambda/\lambda_m = O(1)$ are Eckhaus stable, and thus at least in this limit λ_m does not seem to be the stability boundary.

Chen and Davis (2001) performed a stability analysis on the steady growth of corrugated eutectic fronts using the method of matched asymptotics. They showed that the Jackson and Hunt scaling, $V\lambda^2 = $ const., is the asymptotic limit of large M_α and M_β. When M_α, M_β decrease, the speed-spacing relationship departs from this power law, and approaches an exponential equation

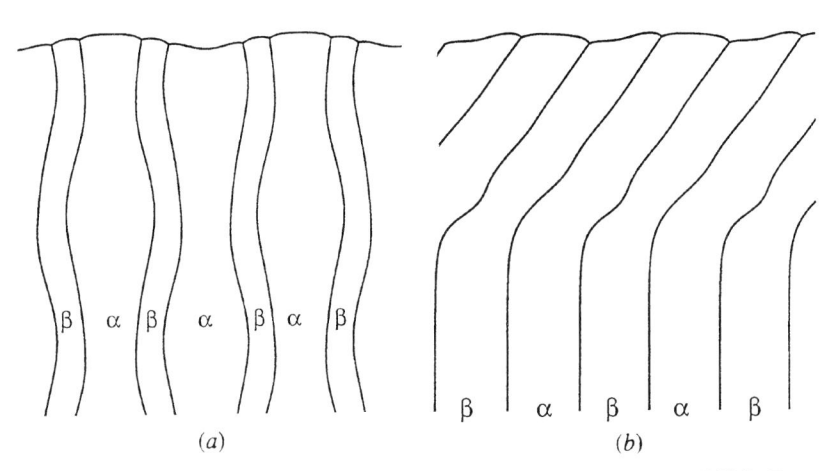

Figure 8.9. A picture of the 2λ oscillatory mode of Datye and Langer (1981). From Karma and Sarkissian (1996). (Reprinted with permission of Metall. mater. Trans.)

$V \exp(K_1\sqrt{G_T}\lambda) = K_2 G_T$ as $M_\alpha, M_\beta \to 0$, where K_1, K_2 are constants depending on the material properties. The observed lamellar spacing may then be smaller than the values predicted by the Jackson and Hunt theory. In fact, early experiments of Racek, Lesoult and Turpin (1974) have shown that the observed spacing follows a correlation, $V\lambda^a$ = const. for many metallic systems, but with a speed-dependent exponent a whose value increases from 2 (the Jackson and Hunt's limit) as V decreases (e.g. $a \approx 3$ Cd-Sn system as $V < 1\,\mu$m), consistent with the results of Chen and Davis.

Karma and Sarkissian (1996) examined the two-dimensional, one-sided FTA model for $\lambda \geq \lambda_m$ in the quasi-static limit, $P_\lambda \to 0$, and for a symmetric phase diagram. They used the boundary-integral method for x-periodic solutions. For CBr_4–C_2Cl_6 they obtained a state diagram, as shown in Figure 8.10, C^∞ versus λ/λ_m; recall that $C^\infty = 0$ corresponds to the far-field concentration

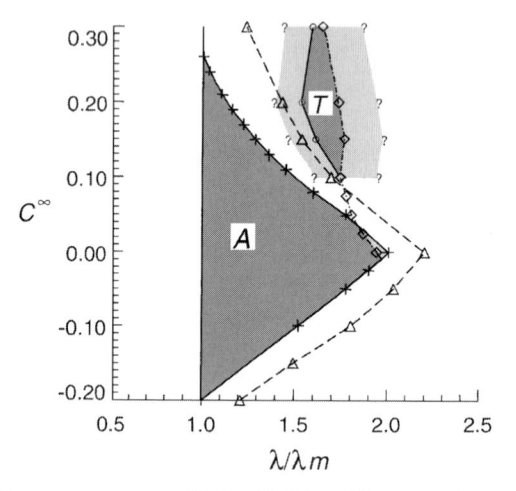

Figure 8.10. Microstructure map of CBr_4–C_2Cl_6, C^∞ versus λ/λ_m, $V = 11.95\,\mu$m/s and $G_T = 187.9$ K/cm. The triangles and dashed line denote the steady tilt bifurcation boundary dividing axisymmetric (to the left) and tilted steady states (to the right), which are only stable within the disjoint dark shaded domains marked repectively by A and T. The A domain is bounded by the primary 2λ-oscillatory (crosses and solid line) and 1λ-oscillatory (diamonds and dashed-dotted line) instabilities. The T domain is bounded by secondary 2λ-oscillatory (open circles and solid line) and 1λ-oscillatory (diamonds and dashed-dotted line) instabilities. Symmetric 1λ and 2λ limit cycles are found, respectively, in the light shaded region at the tip of the A domain and in a narrow (unshaded) strip along the crosses. Tilted 1λ and 2λ limit cycles are found in the light shaded regions to the right and to the left, respectively, of the T domain. From Karma and Sarkissian (1996). (Reprinted with permission of Metall. mater. Trans.)

8.3 Instabilities

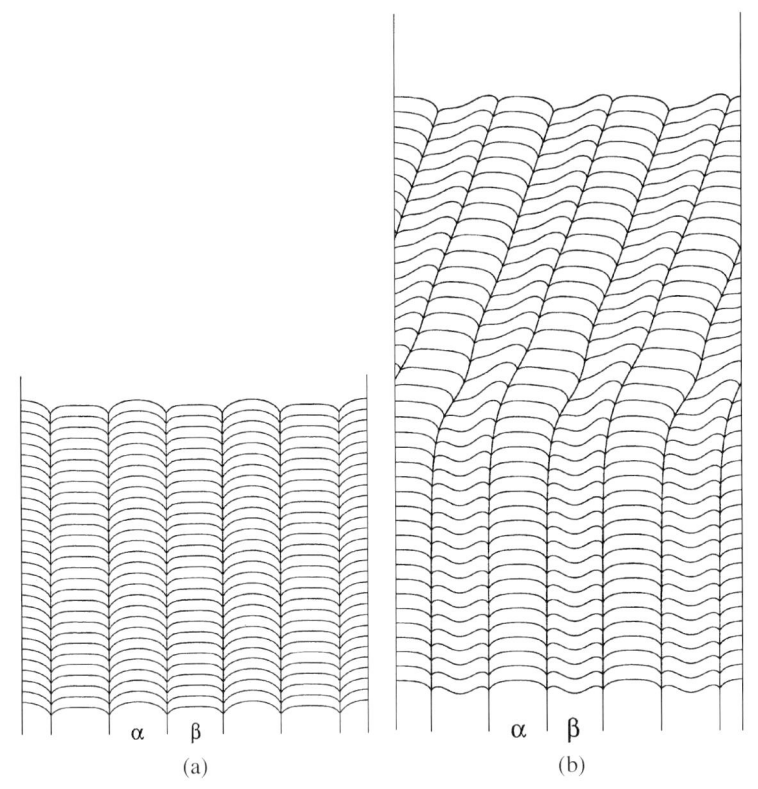

Figure 8.11. A space–time plot illustrating (a) stable, symmetric steady-state growth in CBr$_4$–C$_2$Cℓ_6 with $C^\infty = 0.2$, and $\lambda/\lambda_m = 1.09$, and (b) a dynamical transient from unstable symmetric to stable tilted steady-state growth in CBr$_4$–C$_2$Cℓ_6 with $C^\infty = 0.2$, and $\lambda/\lambda_m = 1.475$, from Karma and Sarkissian (1996). (Reprinted with permission of Metall. mater. Trans.)

being at $C = C_E$. In the region A the untilted modes are stable; in region T there are stable tilted modes, and the caption describes the stability boundaries of these in terms of 1λ and 2λ modes. Figures 8.11 show the $\phi = 0$ mode and the development of the $\phi \neq 0$ mode. Figure 8.12 shows the similar behavior in the experiments of Favre and Ginibre.

For a model metal system Karma and Sarkissian (1996) found (for a symmetric phase diagram) symmetry across $C^\infty = 0$; see Figure 8.13. Again, region A denotes stable, steady eutectics for $\phi = 0$.

The results of unsteady computations for CBr$_4$–C$_2$Cℓ_6, shown in Figure 8.14, predict the two modes for (a) 1λ and (b) 2λ. These should be compared with Figures 8.15, as observed by Favre and Ginibre.

Figure 8.12. Observed transition to a stable, tilted steady-state morphology in thin-sample directional solidification of CBr_4–$C_2C\ell_6$ for $C^\infty \approx 0.21 \pm 0.02$ and $G_T \approx 80$ K/cm. It results from a velocity jump from 0.85 to 3.3 μm/s applied at the time indicated by the arrow, which corresponds to a jump in λ/λ_m from approximately unity to 1.9 ± 0.1 (courtesy of G. Faivre and M. Ginigre). From to Karma and Sarkissian (1996). (Reprinted with permission of Metall. mater. Trans.)

8.4 Remarks

The description of near-eutectic growth is complicated by the presence in steady growth of the scalloped front and the presence of contact lines. As a result, much of the theoretical understanding emanates from numerical simulations in two dimensions in the FTA.

8.4 Remarks

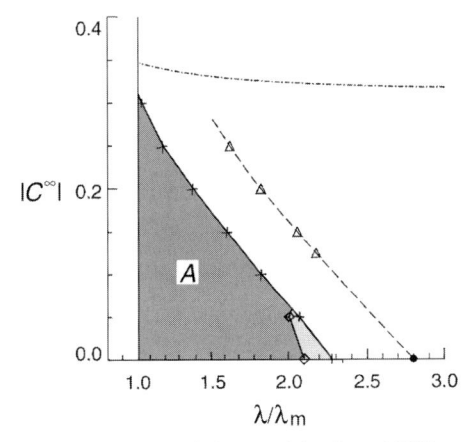

Figure 8.13. Microstructure map of the model alloy, $|C^\infty|$ versus λ/λ_m, where hypoeutectic ($\Delta < 0$) and hypereutectic ($C^\infty > 0$) compositions are equivalent. The symbols and definitions are as in Figure 8.10. The upper dashed-dotted line corresponds to the results of the Datye–Langer stability analysis. From Karma and Sarkissian (1996). (Reprinted with permission of Metall. mater. Trans.)

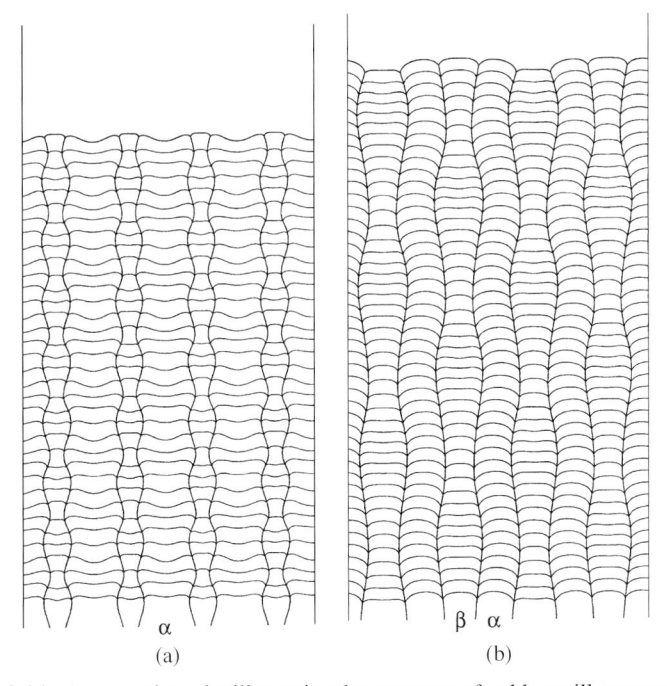

Figure 8.14. A space–time plot illustrating the two types of stable oscillatory morphologies predicted in CBr_4–$C_2C\ell_6$: (a) symmetric 1λ oscillations for $C^\infty = 0$, $\lambda/\lambda_m = 1.952$ and, (b) symmetric 2λ oscillations for $\Delta = 0.20$, $\lambda/\lambda_m = 1.12$. From Karma and Sarkissian (1996). (Reprinted with permission of Metall. mater. Trans.)

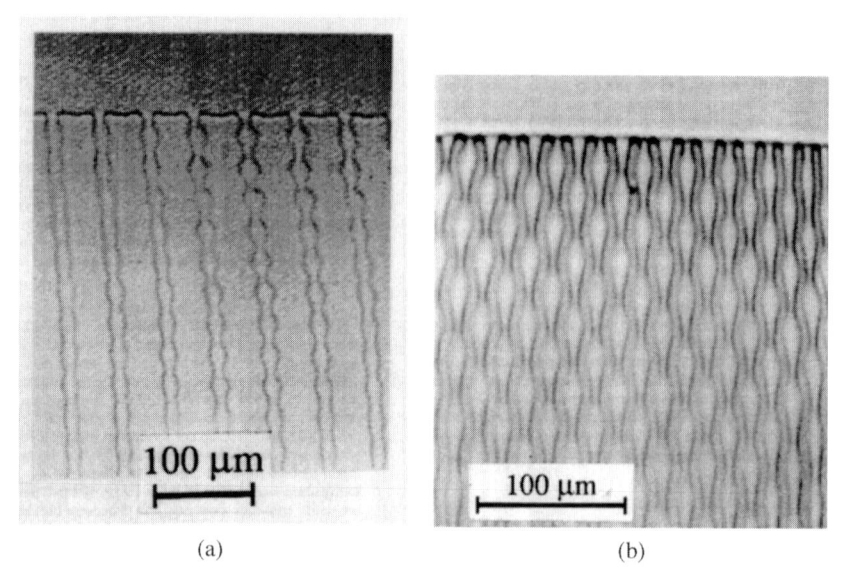

Figure 8.15. Observed symmetric oscillatory morphologies in CBr_4–$C_4C\ell_6$ observed by G. Faivre and M. Ginibre for $G_T \approx 80$ K/cm: (a) 1λ oscillations for $V = 0.51$ μm/s, $\Delta \approx 0 \pm 0.02$, and $\lambda/\lambda_m \approx 2.3 \pm 0.2$; and (b) 2λ oscillations for $V = 2$ μm/s, $\Delta \approx 0.21 \pm 0.02$, and $\lambda/\lambda_m \approx 1.2 \pm 0.1$. From Karma and Sarkissian (1996). (Reprinted with permission of Metall. mater. Trans.)

The observations and the theory for steady growth show that a unique wavelength λ is not selected, but rather a range of λ's can persist. For $\overline{\Delta T}$ versus λ, and M not small wavelengths smaller than λ_m, the minimum, are Eckhaus unstable, leading to overgrowth of some plates and the increase in the effective λ. When λ is sufficiently greater than λ_m, $\lambda > \lambda_M$, tip-splitting limits the λ, but before λ_M is reached, at $\lambda = \lambda'_M < \lambda_M$, a bifurcation to tilted cells occurs, and it is only in the range $\lambda_m < \lambda < \lambda'_M$ that steady growth should be seen. In addition, other instabilities to 1λ and 2λ modes as well as chaotic states can occur. Only when $P_\lambda \to 0$ does $\lambda^2 V$ seem to be constant.

The preceding behaviors occur not only at $C = C_E$ but in a neighborhood about C_E in the hypo- and hypereutectic ranges. If the off-eutectic concentration $|C - C_E|$ is large enough, then, rather than eutectic-like states, cellular or dendritic microstructures will appear.

As in the case of directional solidification of dilute binary alloys, extreme processing or material conditions can lead to more complex behaviors. For pulling speeds is meters per second, departures from equilibrium can give rise to nonequilibrium microstructures (see, e.g., Gill and Kurz 1993). If the anisotropy of one of the phases is large enough, the eutectic structure can display faceted behavior. If significant convection is present in the melt, the range of preferred λ will be altered.

References

Brattkus, K., Caroli, B., Caroli, C., and Roulet, B. (1990). Lamellar eutectic growth at large thermal gradient: I. Stationary patterns. *J. Phys.* (France) **51**, 1847–1864.

Brattkus, K., and Davis, S. H. (1989). Steady and dynamic lamellar-eutectic growth. Preprint cited as Brattkus in Kurz, W., and Trivedi, R. (1990). Solidification microstructure: Recent developments and future directions. *Acta Metall. Mater.* **38**, 1–17.

Caroli, B., Caroli, C., and Roulet, B. (1990). Lamellar eutectic growth at large thermal gradient, II. Linear stability, *J. Phys.* (France) **51**, 1865–1876.

Chen, Y.-J., and Davis, S. H. (2001). Instability of triple junctions in lamellar eutectic growth, *Acta Mater.* (in press).

Datye, V., and Langer, J. S. (1981). Stability of thin eutectic growth, *Phys. Rev. B* **24**, 4155–4169.

Gill, S. C., and Kurz, W. (1993). Rapidly solidified Al–Cu alloys – I. Experimental determination of the microstructure selection map, *Acta Metall. Mater.* **41**, 3563–3573.

Jackson, K. A., and Hunt, J. D. (1966). Lamellar and rod eutectic growth, *Trans. Metall. Soc. AIME* **236**, 1128–1142.

Jordan, R. M., and Hunt, J. D. (1972). Interface under coolings during the growth of Pb-Sn eutectics, *Met. Trans.* **3**, 1385–1390.

Karma, A., and Sarkissian, A. (1996). Morphological instabilities of lamellar eutectics, *Metall. mater. Trans. A* **27**, 635–656.

Kassner, K., and Misbah, C. (1991a). Growth of lamellar eutectic structures: The axisymmetric state, *Phys. Rev. A* **44**, 6513–6532.

———. (1991b). Spontaneous parity-breaking transition in directional growth of lamellar eutectic structures, *Phys. Rev. A* **44**, 6533–6543.

Kurz, W., and Fisher, D. J. (1985). *Fundamentals of Solidification*, Trans. Tech. Publ., Aedermannsdorf, Switzerland.

Langer, J. S. (1980). Eutectic solidification and marginal stability, *Phys. Rev. Lett.* **44**, 1023–1026.

Racek, R., Lesoult, G., and Turpin, M. (1974). The Cd-Sn eutectic structures at low growth rates, *Cryst. Growth* **22**, 210–218.

Seetharaman, V., and Trivedi, R. (1988). Eutectic growth: Selection of interlamellar spacings, *Metall. mater. Trans. A* **19**, 2955–2964.

Zener (1946). Kinetics of the decomposition of austenite, *Trans. Metall. Soc. AIME* **167**, 550–595.

9

Microscale fluid flow

In all the earlier discussions of morphological instabilities, the driver of the morphological changes was the gradient at the front of a diffusive quantity. For the pure material it was an adverse temperature gradient, and for the binary alloy in directional solidification it was an adverse concentration gradient. It seems clear if fluid flow is present in the liquid (melt) that the heat or solute can be convected and hence that its distribution can be altered. Such alterations would necessarily change the conditions for instability. What makes the situation more subtle is that flow over a disturbed (i.e., corrugated) interface will not only alter "vertical" distributions but will also create means of lateral transport, which can stabilize or destabilize the front. The allowance of fluid flow can create *new* frontal instabilities, flow-induced instabilities, that can preempt the altered morphological instability and dominate the behavior of the front.

It has been known since time immemorial that fluid flow can affect solidification; cases range from the stirring of a partially frozen lake to the agitation of "scotch on the rocks." Crystal growers are keenly aware of the importance of fluid flows. Rosenberger (1979) states that "non-steady convection is now recognized as being largely responsible for inhomogeneity in solids."

When one attempts to grow single crystals, the state of pure diffusion rarely exists. Usually, flow is present in the melt; it may be created by direct forcing or it may be due to the presence of convection. Brown (1988) has given a broad survey of the processing configurations and the types of flows that occur.

There are many examples of forced flows. The crystal may be rotated to erase nonaxially symmetric thermal effects, but it creates a von Kármán swirl flow. The use of microgravity environments for the growth of crystals suppresses major buoyancy effects, but the lurching of the spacecraft creates transient accelerations, called g-jitter, that stir the liquid.

There are three types of convection that can occur in the liquid:

1. If the density of the liquid and solid are different, bulk flows in the liquid are driven normal to the interface; this is volume-change convection present even in the absence of gravity.
2. There is buoyancy-driven convection created by density gradients. In binary liquids, the double-diffusive convection can be steady when Rayleigh numbers exceed critical values or at arbitrarily small Rayleigh numbers owing to the presence of, say, heat losses at the sidewalls of the container. If the Rayleigh numbers are high enough, the steady convection can become unstable and, perhaps, unsteady. This is often the case in metallic or semiconductor materials that have very small Prandtl numbers.
3. When fluid–fluid interfaces are present, as in the case of containerless processing by float zone methods or the reprocessing of surfaces by lasers, steep temperature and concentration gradients on these interfaces can drive steady thermosoluto–capillary convection. If these steady flows become unstable, unsteady states can occur.

All of the aforementioned flows are in a sense accidental or at any rate unintentional. They cannot be prevented or else are present, as in the case of the swirl, because the motion is necessary for other reasons. There can also be intentionally imposed flows. In the 1960s Hurle suggested that crystal growers, rather than bemoaning the presence of convection as a source of crystal nonuniformities, should try to "design" natural convection (or forced flows) that would homogenize the solute boundary layer at the interface. In effect, this would decrease the local gradient $|G_c|$ enough to eliminate the possibility of morphological instabilities. This attractive possibility has motivated a good deal of work on the coupling of flow and morphology.

In this chapter, a general formulation of solidification and convection will be given, special cases of which will be studied in subsequent sections. These include *prototype flows* in which standard hydrodynamic-instability models have one or more rigid boundaries replaced with crystal fronts that can melt or freeze. There is also *directional solidification* in the presence of *imposed flows*, either parallel or non-parallel. There is the interaction of directional solidification and convection. If the solid and liquid have different densities, the interface drives a flow. If one solidifies upward and rejects solute of low density, then the liquid is stably stratified thermally but unstably stratified solutally. Thus, there can be convective as well as morphological instabilities. These can occur simultaneously, or the convective field can already be present when the morphological instability begins. Finally, the convection can be turbulent, and this mesoscale convection can strongly affect the gross morphologies of the front and hence the microstructures; this last feature will be discussed in Chapter 10.

9.1 Formulation

Consider an interface $z = h(x, y, t)$ separating solid, $z < h$, from liquid $z < h$. The concentration C of solute, the temperature T of solid and liquid, the velocity field $\mathbf{v}$, and the pressure p in the liquid are strongly coupled to the interface $z = h(x, y, t)$.

In Figure 9.1, the system is shown with solidification upwards, z increasing, and gravity acting downwards. The liquid is described by the Navier–Stokes and continuity equations coupled through buoyancy to T and C.

$$\rho^\ell(\mathbf{v}_t + \mathbf{v} \cdot \nabla \mathbf{v}) = -\nabla p + \mu \nabla^2 \mathbf{v} - \rho^\ell \mathbf{g} \tag{9.1a}$$

$$\rho^\ell_t + \nabla \cdot (\rho^\ell \mathbf{v}) = 0 \tag{9.1b}$$

$$\rho^\ell c^\ell_p (T_t + \mathbf{v} \cdot \nabla T) = k^\ell \nabla^2 T \tag{9.1c}$$

$$C_t + \mathbf{v} \cdot \nabla C = D^\ell \nabla^2 C \tag{9.1d}$$

plus the (linearized) equation of state

$$\rho^\ell = \rho^\ell_0 [1 - \alpha(T - T_0) - \beta(C - C_0)], \tag{9.1e}$$

where T_0 and C_0 are reference values, and $\mathbf{v} = (u, v, w)$ in the Cartesian system (x, y, z).

When the thermal and solutal stratifications are small enough that across the liquid

$$\Delta \rho^\ell / \rho^\ell_0 \ll 1, \tag{9.2}$$

then system (9.1) can be simplified to the Boussinesq equations by assuming inequality (9.2) but taking $g \Delta \rho^\ell / \rho_0$ as appreciable. This leads to taking $\rho^\ell = \rho^\ell_0$

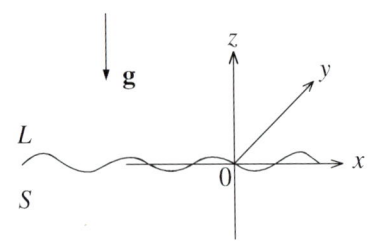

Figure 9.1. A sketch of a disturbed interface between a liquid (L) and a solid (S) subject to a vertical gravity.

9.1 Formulation

everywhere in the system except in the buoyancy term of Eq. (9.1a). If one writes $k_T^\ell/\rho_0^\ell c_p^\ell = \kappa^\ell$ and $\nu = \mu/\rho_0^\ell$, then system (9.3) becomes

$$\mathbf{v}_t + \mathbf{v} \cdot \nabla \mathbf{v} = -\nabla(p/\rho_0^\ell) + \nu\nabla^2\mathbf{v} + [\alpha(T - T_0) + \beta(C - C_0)]\mathbf{g} \quad (9.3a)$$

$$\nabla \cdot \mathbf{v} = 0 \quad (9.3b)$$

$$T_t + \mathbf{v} \cdot \nabla \cdot T = \kappa^\ell \nabla^2 T \quad (9.3c)$$

$$C_t + \mathbf{v} \cdot \nabla C = D^\ell \nabla^2 C, \quad (9.3d)$$

where the constant portion of ρ^ℓ in $\rho^\ell \mathbf{g}$ has been included in p.

This Boussinesq approximation so simplifies the fluid mechanics that it is used by many even when inequality (9.2) is violated. In such cases system (9.3) should be regarded as a model system.

In the solid,

$$T_t^s = \kappa^s \nabla^2 T^s \quad (9.4a)$$

$$C_t^s = D^s \nabla^2 C^s. \quad (9.4b)$$

On the interface $z = h(x, y, t)$, there is the Gibbs–Thomson equation (in the absence of kinetic undercooling, say)

$$T^s = T^\ell = mC + T_m\left[1 + 2H\frac{\gamma}{L_v}\right] \quad (9.5a)$$

and the flux conditions

$$(1-k)CV_n = \left[D^s\nabla C^s - D^\ell \nabla C\right] \cdot \mathbf{n} \quad (9.5b)$$

$$L_v V_n = (k_T^s \nabla T^s - k_T^\ell \nabla T) \cdot \mathbf{n} \quad (9.5c)$$

where the unit normal $\mathbf{n}$ points out of the solid, and V_n is the normal speed of the interface.

Finally, on $z = h$, for each unit tangent $\mathbf{t}^i$, $i = 1, 2$,

$$\mathbf{v} \cdot \mathbf{t}^i = 0, \quad (9.5d)$$

because the front is a no-slip surface and

$$(\rho^\ell - \rho^s)h_t = \rho^\ell(w - uh_x - vh_y), \quad (9.5e)$$

a statement of mass conservation that allows for changes in density upon phase transformation.

In the following, only the one-sided FTA will be discussed so that the temperature T is everywhere given by

$$T = T_0 + G_T z, \quad (9.6)$$

$D^s/D^\ell \to 0$, and for each phase $D/\kappa \to 0$. In this case, Eqs. (9.4a,b) and (9.5c) become redundant.

The remaining system can be scaled as follows: length d, speed V_s, time d/V_s, pressure $\mu V_s/d$, temperature $G_T d$, and concentration is written as $C_\infty[1 + (k-1)C]$; d and V_s will be defined later when specific systems are studied. Then, system (9.3) has the form

$$\text{Re}(\mathbf{v}_t + \mathbf{v} \cdot \nabla \mathbf{v}) = -\nabla p + \nabla^2 \mathbf{v} + R^C C \mathbf{e}_3 \tag{9.7a}$$

$$\nabla \cdot \mathbf{v} = 0 \tag{9.7b}$$

$$\text{ScRe}(C_t + \mathbf{v} \cdot \nabla C) = \nabla^2 C, \tag{9.7c}$$

where $\mathbf{e}_3 = (0, 0, 1)$, the reference concentration C_0 is taken to be C_∞, the thermal buoyancy, proportional to z, is included in p and

$$\text{Re} = \frac{V_s d}{\nu} \qquad \text{Reynolds number}$$

$$\text{Sc} = \frac{\kappa^\ell}{D^\ell} \qquad \text{Schmidt number} \tag{9.8}$$

$$R^C = \frac{g\beta C_\infty (1-k)d^2}{\kappa^\ell \nu V_s} \qquad \text{solutal Rayleigh number.}$$

Equations (9.5a,b) become

$$M^{-1}h = C + 2H\Gamma_1 \tag{9.9a}$$

and

$$[1 + (k-1)C]V_n = \frac{D^\ell}{V_s d} \nabla C \cdot \mathbf{n} \tag{9.9b}$$

where

$$M = \frac{m(k-1)C_\infty}{G_T d}, \quad \Gamma_1 = \frac{T_m \gamma}{L_v d} \frac{1}{m(k-1)C_\infty}. \tag{9.9c}$$

Equation (9.5d) remains unaltered

$$\mathbf{v} \cdot \mathbf{t}^i = 0, \ i = 1, 2 \tag{9.9d}$$

and equation (9.5e) becomes

$$(1-\rho)h_t = w - uh_x - vh_y, \tag{9.9e}$$

where

$$\rho = \frac{\rho^s}{\rho^\ell}. \tag{9.10}$$

Finally, both C and $\mathbf{v}$ need to be prescribed at $z \to \infty$.

9.2 Prototype Flows

9.2.1 Free Convection

Consider free convection in a vertical annular gap filled with SCN, as shown in Figure 9.2. The gap is formed from two coaxial vertical cylinders of radii R_1 and R_2, $R_1/R_2 < 1$. The inner cylinder is a wire heated to constant temperature T_w, whereas the outer metal cylinder is kept, at temperature T_m, the melting point of SCN.

The z-axis acts vertically upward, and the cylindrical coordinates (r, θ, z) have corresponding velocity components (u, v, w); here $C \equiv 0$ because the material is pure and $\rho^s = \rho^\ell$. When the lengths of the cylinders are much larger than the gap width $R_2 - R_1$, one can find a parallel-flow solution $W = \bar{W}(r)$ and $T = \bar{T}(r)$, as shown in Figure 9.2 from Choi and Korpela (1980); the interface between solid and liquid is given by $r = R^I$ where R^I depends on T_w; call $R_1/R^I = \eta < 1$. The Grashof number Gr and Prandtl numbers are

$$Gr = \alpha g (T_w - T_m)(R^I - R_1)^3/\nu^2, \quad Pr = \nu/\kappa^\ell. \tag{9.11}$$

Here the front is *stationary* in the basic state and $\kappa^\ell/(R_2 - R^I)$ is used as the scale for velocity, length $\sim R_2 - R^I$, time $\sim (R_2 - R^I)^2/\kappa^\ell$ and temperature $\sim T_w - T_m$. A linear stability analysis can test the preceding basic state for instability using normal modes of the form $\exp[\sigma t + in\theta + iaz]$.

In the absence of flow and temperature gradients, the cylindrical interface can become unstable when the length ℓ of the cylinders exceeds $\ell_c = 2\pi R^I$, as

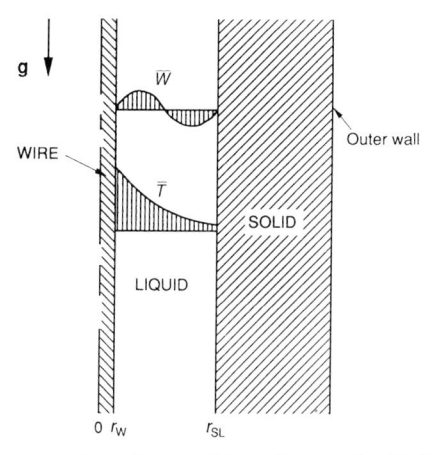

Figure 9.2. A half cross section of concentric cylinders of radii R_1 and R_2, $R_1 < R_2$. The inner cylinder is a heated wire, the outer cylinder is cooled, and $r = r_{SL}$ indicates the basic state cylindrical interface with profiles of $\bar{W}(r)$ and $\bar{T}(r)$. From Davis (1990).

shown by Rayleigh (1879). This capillary instability is axisymmetric, that is, $n = 0$, and leads to the breakup of the cylinder into "droplets." In the presence of the radial temperature gradient as prescribed, bulges of solid are melted back owing to the Gibbs–Thomson effect, and thus the capillary instability is delayed or entirely suppressed. Coriell and McFadden (1992, private communication) found that capillary instabilities do not occur in the range of the experiments described below.

In the absence of interfacial deformation, the parallel flow $\bar{W}(r)$ has a point of inflection, where $d/dr \left[r^{-1} dW/dr \right] = 0$ at a point $r = r_s$, which triggers an inviscid instability that is present even at finite Reynolds numbers. Furthermore, the temperature distribution $\bar{T}(r)$ implies the presence of horizontal density gradients and so is susceptible to buoyancy-driven instabilities. Choi and Korpela (1980) and Shaaban and Özisik (1982) examined the case for $n = 0$, whereas McFadden et al. (1984) examined the spiral modes $n = \pm 1$.

For small Prandtl numbers, convection is weak, and the dominant instability comes from the shear mode in the form of upward propagating waves (opposite in sense to the flow near the interface). For small gaps, $n = 0$ dominates, whereas for larger gaps $n = \pm 1$ does. The buoyancy-driven mode has $Gr_c \to \infty$ as $\Pr \to \infty$ and, at least for small gaps, has $Gr_c \sim \Pr^{-1/2}$ as $\Pr \to 0$.

Now allow the interface to deform. McFadden et al. (1984) neglected surface energy because the disturbance wavelengths are much larger than those, for example, in morphological instability. They found that the ability of the interface to deform by melting or freezing drastically promotes the growth of the shear modes. For example, for small gap, $\eta = 0.02$, $\Pr = 22.8$, the thermal mode has $Gr_c = 2150$, whereas the shear mode has $Gr_c = 176$ for $n = \pm 1$ (compared with 2152 without deformation). The wave propagates upward with phase speed $\sigma_I a = c \approx 7.24 \times 10^{-3}$ in units of $\nu/(R^I - R_1)$, which is consistent with the idea that melting and freezing are sluggish processes. The physical mechanism for this destabilization has not yet been explained.

Experimental observations on SCN by Glicksman and Mickalonis (1982) and Fang et al. (1985) and the preceding predictions are in close accord. Figure 9.3 shows photographs of the spiral modes.

9.2.2 Bénard Convection

In the preceding discussion, the phase transformation at the solid–liquid interface drastically promotes instability and the preferred pattern of instability is determined from linear stability theory. In the Bénard convection problem to be discussed now, linear theory fails to predict the pattern, and the freezing–melting of the interface alters the pattern predicted by a weakly nonlinear theory.

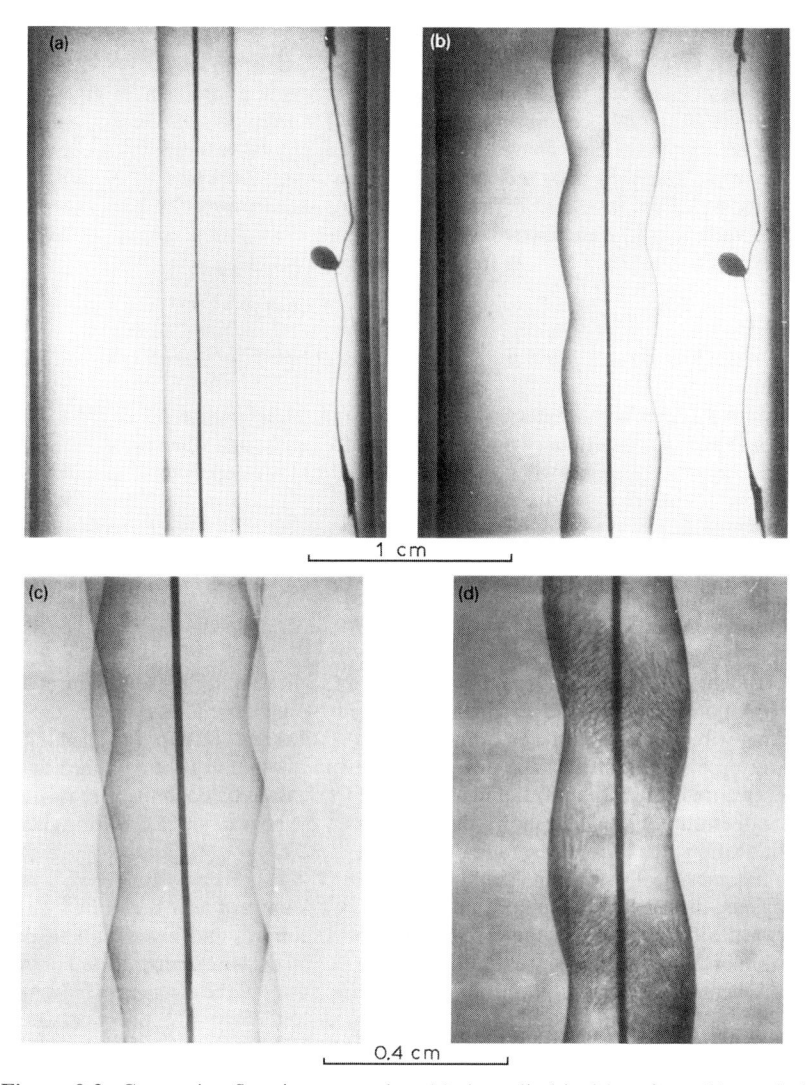

Figure 9.3. Convective flow in an annulus, (a) the cylindrical interface, (b) a spiral interface, (c) multiple exposures of a spiral interface, and (d) a dendritic "belt" formed by abrupt undercooling. From Fang et al. (1985).

Consider two parallel, horizontal, perfectly conducting plates whose gap is filled with a pure liquid. The temperatures T_T and T_B are the fixed values on the top and bottom, respectively. The temperatures are adjusted to span the melting temperature of the liquid so that $T_T < T_m < T_B$, and hence a strip of solidified material occupies the upper portion of the layer, as shown in Figure 9.4(a). The thicknesses h^s and h^ℓ of the two layers are determined by the boundary

9. Microscale fluid flow

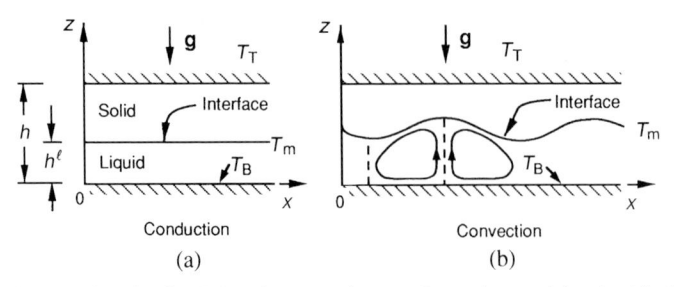

Figure 9.4. A sketch of a Bénard convection configuration at (a) subcritical and at (b) supercritical Rayleigh numbers. From Davis, Müller, and Dietsche (1984).

temperatures and heat conduction. Thus, there is a new parameter A in the problem such that

$$\frac{h^s}{h^\ell} = \frac{k_T^s}{k_T^\ell} \frac{T_m - T_T}{T_B - T_m} \equiv A. \tag{9.12}$$

When $A = 0$, only liquid is present in the gap. The basic state here is composed of a stationary planar interface and piecewise linear temperature profiles due to heat conduction in the gap.

One defines scales for length $\sim h^\ell$, speed $\sim \kappa^\ell/h^\ell$, time $\sim (h^\ell)^2/\kappa^\ell$, pressure $\sim \mu\kappa^\ell/(h^\ell)^2$ and for temperature $\Delta T \sim T_B - T^1$; again $C \equiv 0$. The instability problem involves the second principal parameter, the thermal Rayleigh number,

$$R^T = \frac{\alpha g (T_B - T_m)(h^\ell)^3}{\kappa^\ell \nu} \tag{9.13}$$

where $R^T = Gr\,\mathrm{Pr}^{-1}$.

The linear stability problem was solved by Davis, Müller, and Dietsche (1984), who found that R_c and a_c decrease with A, that is, the thicker the frozen layer the more unstable is the liquid to steady, cellular convection, and the wider are the convection cells; $R_c^T(0) \approx 1708$ and $R_c^T(\infty) \approx 1493$ owing to the interface's becoming a poorer heat conductor as A increases, whereas in the same range of A, a_c decreases from 3.12 to 2.82. The linear theory is silent on the pattern selected.

Figure 9.4(b) is a sketch of the postconvective state. Rising currents of warm liquid melt the solid and hence cause interfacial deflection. This deflection is crucial in the determination of the selected pattern.

Davis, Müller, and Dietsche then examined the weakly nonlinear theory for the competition between hexagons and rolls. They obtained bifurcation equations in the steady state equivalent to system (4.15). When $A = 0$, the quadratic coefficients $b_0 = 0$, and thus the standard result for Bènard convection is recovered; only rolls are stable for $R^T > R_c^T$, and no subcritical convection is present (Schlüter, Lortz, and Busse 1965). As A is increased, $|b_0|$ increases. This situation is analogous to the work of Davis and Segel (1968), who studied Bènard convection with an upper free surface. The surface deflections give rise to non-Boussinesq effects reflected in $b_0 \neq 0$. Now hexagons, hexagons and rolls, and rolls are successively stable; see Section 4.1.3. This sequence of transitions is seen in the experiment of Davis, Müller, and Dietsche (1984), as shown in Figure 9.5. For technical reasons, only qualitative comparisons between theory and experiment are possible. These comparisons show that the thicker the solid, the larger the interface deflection and the more likely one is to see hexagonal states.

There is one prediction of the theory that was not verifiable in the experiment. As R^T is increased, the onset of hexagons from pure conduction occurs via a subcritical bifurcation and hence a jump transition. When this occurs, the rapid rise in vertical heat transport will increase the *mean* height of the interface. Further, R^T must be decreased below R_c^T before there is a jump transition back to pure conduction. Dietsche and Müller (1985) performed another set of experiments on a layer, one of whose horizontal dimensions was comparable to its gap width. Hexagons are not regular, but Dietsche and Müller observed the hysteresis predicted and the jump-like nature of the onset of polygonal convection. At higher R^T, bimodal patterns are seen.

If the two plates of the original configuration are poor heat conductors, the pattern selection is different. Hadji, Schell, and Riahi (1990) predicted square patterns for small A and squares and rolls at higher A.

Many other systems involving hydrodynamic instabilities with freezing–melting boundaries exist. Zimmermann, Müller, and Davis (1992) included Soret effects in the liquid in the Bénard cell. McFadden et al. (1989) have examined Couette flow in which the free convection configuration discussed earlier has the inner cylinder rotating. Other examples can be found in the conference proceedings of Davis et al. (1992).

9.3 Directional Solidification and Volume-Change Convection

The interface at $z = h(x, y, t)$ is the site of a phase transformation in which the liquid of density ρ^ℓ changes to solid of density ρ^s. Morphological changes are driven by solute rejection there and, in addition, if $\rho^\ell \neq \rho^s$, the interface drives a convective flow. For example, a dilute lead–tin alloy shrinks upon solidification, $\rho^s/\rho^\ell \approx 1.05$, and a weak flow from "infinity" is generated to conserve mass. On the other hand, silicon expands upon solidification, $\rho^s/\rho^\ell \approx 0.91$, and the generated flow is from the interface to "infinity."

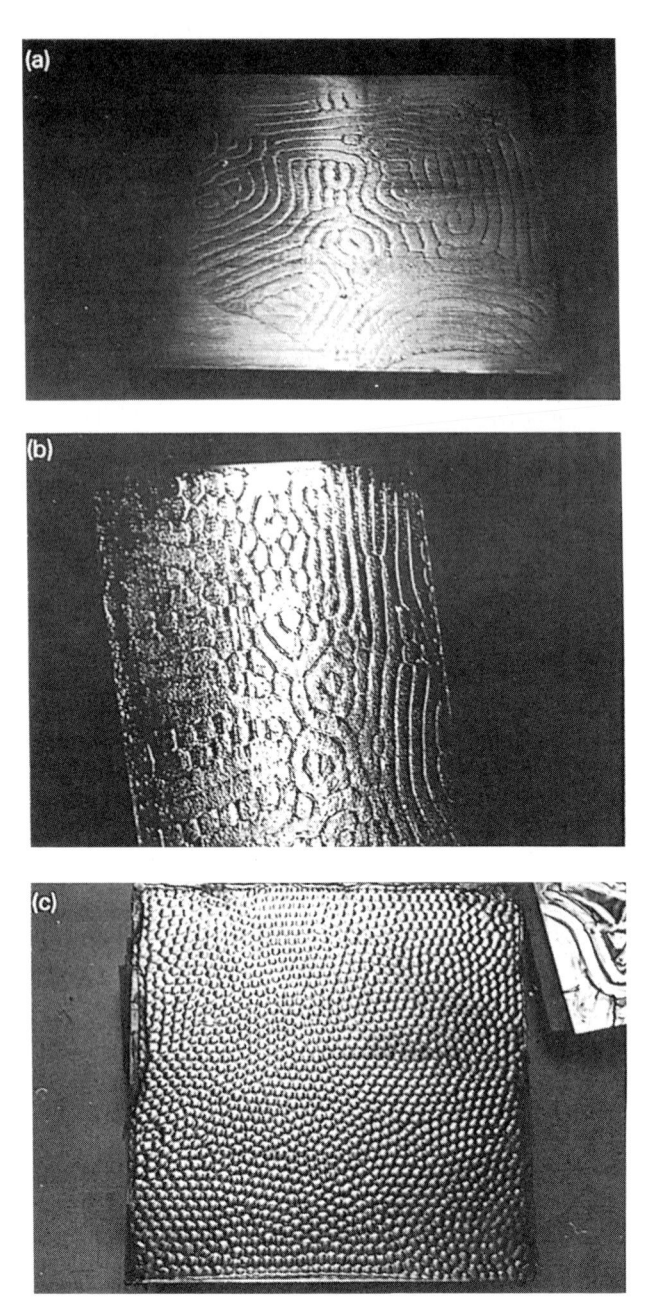

Figure 9.5. The morphology of the interface when the solid layer is (a) thin, (b) moderate, and (c) thick. From Davis, Müller, and Dietsche (1984).

9.3 Directional solidification and volume-change convection

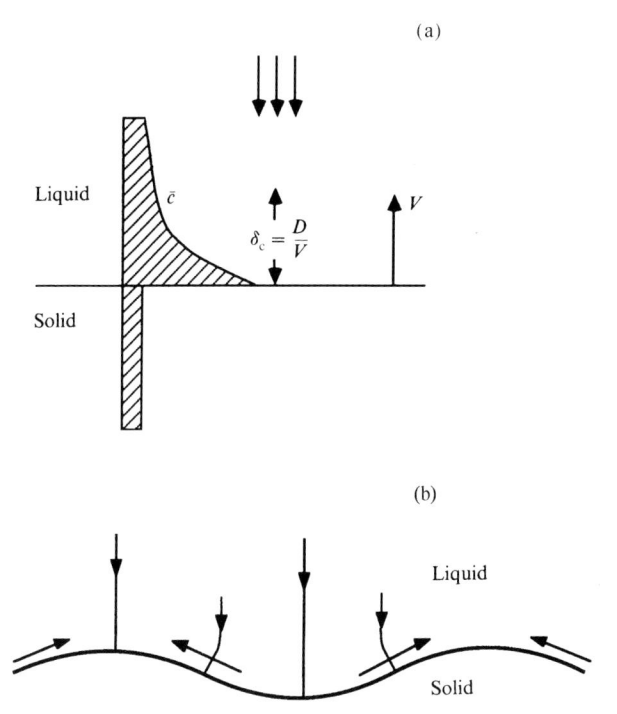

Figure 9.6. Transport mechanisms in volume-change convection for $\rho^s > \rho^\ell$. (a) Boundary-layer alteration. (b) Lateral transport. From Davis (1990).

When $\rho^\ell = \rho^s$, the interface is impermeable to flow, but when $\rho^\ell \neq \rho^s$, it acts as a porous surface that produces or consumes liquid at the rate required by the volume changes; see Eq. (9.9e). In the basic state there is a vertical velocity component W such that $\bar{W} = (1 - \rho)V$ in excess of the translational speed of the front. The basic state thus consists of $\bar{h} = 0$, $\bar{C} = C_\infty \left[1 + (1-k)e^{-z/\delta_c}\right]$ and $\bar{\mathbf{v}} = (0, 0, \bar{W})$.

The linear theory for morphological instability with volume change convection was given by Caroli et al. (1985b). The following explanation is a variant of theirs. Consider the case of shrinkage $\rho^s > \rho^\ell$, as shown in Figure 9.6(a). Two competing effects are present.

On the one hand, the flow from infinity will cause the concentration boundary layer of thickness $\delta_c = D/V$ to be compressed so that the local gradient $|G_c|$ is increased. As discussed in Chapter 3, this will enhance the morphological instability through what can be called *boundary-layer alteration*. Thus, shrinkage is destabilizing, whereas expansion is stabilizing.

On the other hand, consider the result of an initial corrugation of the interface, as shown in Figure 9.6(b). Because the interface is a no-slip surface,

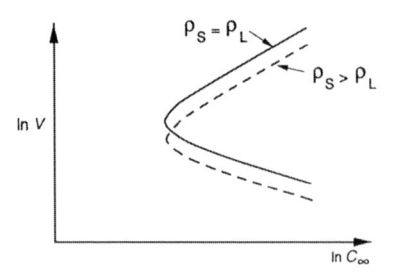

Figure 9.7. Sketch of V versus C_∞ for fixed G_T with moderate k for $\rho^s = \rho^\ell$ and $\rho^s > \rho^\ell$. From Davis (1993).

all streamlines cross the interface normally. At the crest or trough, the streamlines are vertical, but elsewhere they are curved. The curvature is accompanied by transverse velocities that, for $\rho^s > \rho^\ell$, carry the solute from the troughs to the crests, as shown, which homogenizes the solute and decreases $|G_c|$. (Recall that the "pure" morphological instability leads to excess solute in the troughs.) Thus, for the case of shrinkage, the induced *lateral transport* of solute is stabilizing.

Caroli et al. (1985b) found that low solidification speeds V, which correspond to "thick" concentration boundary layers, promote destabilization by shrinkage because the boundary-layer alteration is more effective than is lateral transport. At high speeds V, the opposite is the case. Figure 9.7 is a sketch of this result.

Brattkus (1988) took another point of view by comparing different materials. Materials with small segregation coefficient k reject nearly all their solute. Because $|G_c|$ is monotonically decreasing with k, systems with small k are destabilized when compared with the constant-density case; the reverse is true in materials with moderate k. This switchover of effect with material properties is important to recognize because, for convenience, many experiments substitute transparent organics for metallics; such a switch may also reverse the influence of volume-change convection.

Wheeler (1991) has examined the coupling of morphological instability and volume-change convection by deriving a strongly nonlinear evolution equation for the interface.

Clearly, volume–change convection has small effects if $|\rho^\ell - \rho^s|/\rho^\ell$ is small. It should be of importance only when other modes of convection are absent or else of very small magnitude. Such would be the case in a microgravity environment in space. Volume–change convection may also be important in the deep roots between cells, where dimensions are small and diffusion is very slow; see the mushy-zone analogy of this in Chapter 10.

9.4 Directional Solidification and Buoyancy-Driven Convection

Consider the directional solidification of a binary liquid in which the front moves vertically upward. The rejected solute profile determines not only the possibility of morphological instability but also the possibility of buoyancy-driven convective instability. If the solute is of *low density* compared with the alloy, the steady basic state consists of a "heated-from-above" temperature field and an unstably stratified concentration field. Thus, there can be a double-diffusive Bénard instability on a semi-infinite domain containing an exponentially decaying concentration profile, and an increasing temperature distribution. Take $V_s = V$ and $d = \delta_c$ for the scales.

In Section 9.2, the FTA is used, and thus the thermal buoyancy is linear in z and is represented by a conservative body force; therefore, it is included in the pressure gradient. If the FTA were abandoned, there would be two Rayleigh numbers, the solutal one, (9.8) and a thermal one $R^T = g\alpha G_T \delta_c^4 / \kappa^\ell \nu$. Under these more general conditions, one can ask when the thermal buoyancy is negligible, that is, when is $R^T/R^C \ll 1$. If one uses the definition (9.9c) for M, then

$$\frac{R^T}{R^C} = \frac{|\alpha|}{|\beta|} \frac{mD}{\kappa^\ell} M^{-1}. \tag{9.14}$$

In Eq. (9.14), the ratio of expansion coefficients is near unity. Figure 9.8 shows a typical morphological stability curve (labeled MI) for fixed G_T, ln V versus ln C_∞. The upper branch asymptotically approaches to slope unity, which represents the absolute stability boundary. The lower branch asymptotically tends to the constitutional undercooling limit, $M \to 1$, which has a slope of -1. Plotted here as well is a typical convective instability curve (labeled CI). The line $R^T/R^C = $ const. is shown using Eq. (9.14); typically $D/\kappa^\ell \ll 1$, and the estimate, $M \approx 1$. The line has a slope of -1 and is *always* well below the lower

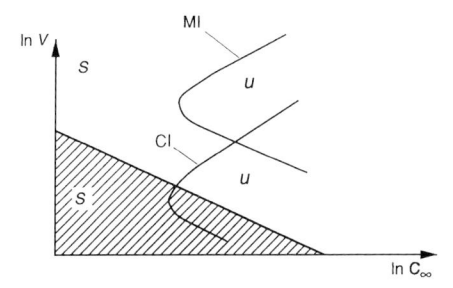

Figure 9.8. Sketch of V versus C_∞ for fixed G_T showing the morphological and convective neutral curves. In the shaded region, thermal buoyancy is important. From Davis (1993).

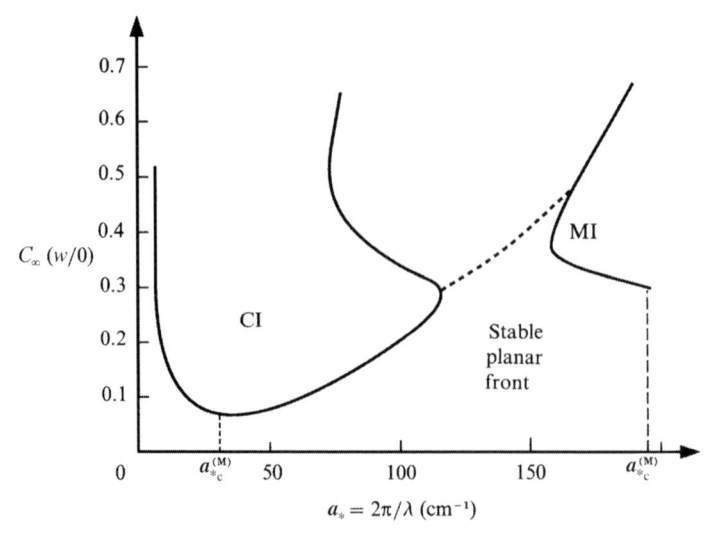

Figure 9.9. Netural curves for Pb–Sn with $V = 30$ μm/s and $G_T = 200$ K/cm for upward solidification with the rejection of low-density solute. The convective (CI) and morphological instabilities (MI) have preferred wave numbers $a_{*_c}^{(C)}$ and $a_{*_c}^{(M)}$, respectively. Solid (dashed) curves denote steady (time-periodic) instability. From Coriell et al. (1980). (Reprinted with permission of North-Holland.)

branch of the morphological instability. *Below* this line, thermal buoyancy is important, whereas above this the convection is driven solely by solutal buoyancy, justifying the use of the FTA.

The linear stability theory for soluto–convective instability with $\rho^\ell = \rho^s$ has been examined in great detail (Coriell et al. 1980; Hurle, Jakeman, and Wheeler 1982,1983; Hennenberg et al. 1987). Such theories give results typified by Figure 9.9, which shows, for fixed G_T and V, the mean solute concentration C_∞ versus (dimensional) wave number $a_* = 2\pi/\lambda$. There are two coexisting neutral curves, one for the "pure" soluto–convective mode and one for the "pure" morphological. Typically, the critical wave numbers $a_{*_c}^{(C)}$ and $a_{*_c}^{(M)}$ for the two are widely separated, and the convective mode is of much longer scale. In the figure, $a_{*_c}^{(M)}/a_{*_c}^{(C)} \approx 6.5$, and thus there is a large-scale convective flow with a small-scale morphology. As V is increased, the length scale δ_c in the solutal Rayleigh number decreases, the convective curve rises, and the morphological curve falls.

At a specific value of V, the critical C_∞ of both modes coincides, giving the possibility of a coupled instability. The bifurcation theory (Jenkins 1985a, b; Caroli et al. 1985a), however, shows that the wavelength disparity mentioned above leads to only a very weak coupling between the instabilities.

9.4 Direction solidification and buoyancy-driven convection

This conclusion, however, is based upon the examination of only a small set of material systems, mainly lead–tin and SCN–ethanol (Schaefer and Coriell 1982). Clearly, if the linear theory critical value of C_∞ is common, and if $a_c^{(M)} = a_c^{(C)}$, then the two-dimensional problem has a codimension-two bifurcation structure, and there is the possibility that strong interactions of the two modes would be possible; secondary oscillations may also occur. More general interactions could give rise to oscillations; for example, if $a_c^{(M)} = 2a_c^{(C)}$, then strong interactions would also be possible. Riley and Davis (1989a) have undertaken a systematic study of such possibilities for both these cases over wide classes of solutes and solvents and their material properties. They found that such interactions do occur but only at temperature gradients and speeds too small to be of physical interest.

Generally, each instability stabilizes the other mode, albeit slightly (Caroli et al. 1985a; Riley and Davis 1989b,1991). Consider the pure morphological instability, as in Figure 9.10(a). Here the corrugated interface, a cell, has concentration excesses (deficiencies) at troughs (crests) relative to the basic state. A slight buoyancy causes the lighter fluid in troughs to flow uphill towards crests, as shown in Figure 9.10(b). The $+$ and $-$ indicate changes in C from Figure 9.10(a). These changes alter the solidification rates such that troughs (crests)

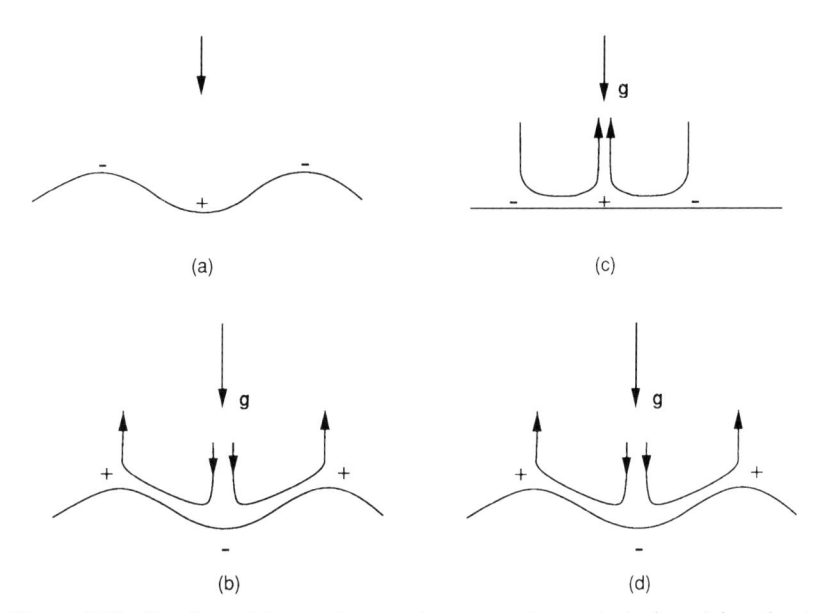

Figure 9.10. Sketches of flow patterns and concentration perturbations (given by $+$ and $-$) for (a) pure morphological instability, (b) morphological instability plus small buoyancy, (c) pure convective instability, and (d) convective instability with some morphological changes. From Davis (1993).

grow faster (slower), leading to a delay in morphological instability. Figure 9.10(c) shows the result of pure convective instability. Below an upwelling (downwelling), there is a concentration excess (deficiency). When morphological changes are allowed, the concentration excess (deficiency) creates a trough (crest); buoyancy and the curved interface create perturbations of C and velocity relative to those in Figure 9.10(c), as shown in Figure 9.10(d). These counteract the original convective instability and lead to a stabilization.

The soluto–convective mode can be sketched for fixed G_T on a V versus C_∞ plot; this is shown in Figure 9.8. The full curve is nonmonotonic with the nose and lower branch lying in the shaded region, where thermal buoyancy is important.

The results shown in Figure 9.9 indicate that, in addition to the weakly coupled convective and morphological instabilities predicted, there is a new mode, denoted by the dashed curve, which represents a time-periodic instability generated by the coupling of the two pure modes (Coriell et al. 1980). This periodic branch occurs near the point in Figure 9.8 where the two modes cross. For the lead–tin system, this time-periodic instability is not the first one to appear under ordinary operating conditions, though it can be under extreme conditions of low G_T. Coriell (private communication) found that it is the first to appear for $G_T = 0.1$ K/cm and $V = 10\,\mu$m/s. Schaefer and Coriell (1982) showed for SCN–ethanol that the mode is present and can, under ordinary conditions, be preferred theoretically, though it was not observed in their experiment. Jenkins (1990) found that the oscillatory mode is preferred for systems with sufficiently high surface energy. His weakly nonlinear theory showed that this mode can be subcritical or supercritical and can correspond to either traveling or standing waves. These theoretical results are very interesting but ones that caution the crystal grower. The "attempt" to impress convective flow on a growing crystal may give rise to unwanted oscillations rather than the homogenization sought. The physical mechanism responsible for this oscillatory "mixed" mode has not been explained though a partial explanation in terms of lateral transport is given by Davis (1990).

For most common materials, then, $a_c^{(M)} \gg a_c^{(C)}$. However, there is the possibility that limiting cases exist that display a stronger coupling. Sivashinsky (1983) showed for the pure morphological problem that, for $k \to 0$

$$a_c^{(M)} \sim k^{1/4}. \tag{9.15a}$$

Riley and Davis (1989b) showed for the pure convective problem that, for $k \to 0$,

$$a_c^{(C)} \sim k^{1/4}. \tag{9.15b}$$

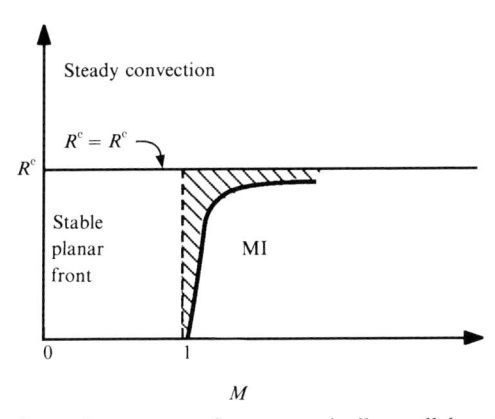

9.4 Direction solidification and buoyancy-driven convection

Figure 9.11. The linear-theory curves for asymptotically small-k materials during upward solidification with the rejection of low-density solute. The hatched area corresponds to the delay of morphological instability by buoyancy. From Young and Davis (1986). (Reprinted with permission of the American Physical Society.)

as well. Young and Davis (1986) examined the small-k limit for the coupled convective–morphological problem and found the effect shown in Figure 9.11. Above the line, Rayleigh number $R^c = R^c = 2(1 + Sc^{-1})$, there is convective instability, as shown by Hurle et al. (1982,1983). However, there is another curve $R^c = R(M)$, to the right of which lies the region of morphological instability (MI) as modified by buoyancy. Thus, for $R^c < R^c$, the morphological instability is *delayed* by buoyancy owing to effects of lateral transport. When the basic state is perturbed by an interface corrugation, the rejected solute, if it is of low density, moves, via a baroclinic motion, from troughs to crests, lowering the local $|G_c|$; an effective segregation coefficient is produced that has the value $k[1 - (R/R_c)]^{-1}$, which is larger than k and gives an effectively smaller $|G_c|$. This delay of interfacial instability is small but is one of the few found to date. However, see the discussion following in the next paragraph of thermally induced convective instability.

The convective instability can be delayed as well by the imposition of solid body rotation on the system. Öztekin and Pearlstein (1992) showed that the convective curve of Figure 9.8 shifts to the right as the rotation rate is increased, whereas the morphological curve is unaffected.

There can be a strong coupling in non-FTA between *thermal convection* and morphological instability (Coriell and McFadden 1989). This occurs at wavelengths much larger than δ_c, and thus the estimate of Eq. (9.14) is invalid. When solutal buoyancy is absent ($R^C = 0$ or $\beta = 0$), Coriell and McFadden found a greatly promoted morphological instability driven by thermal buoyancy. In this case, the conductivity ratio $k_T = k_T^s/k_T^\ell$ is a crucial parameter. Moreover,

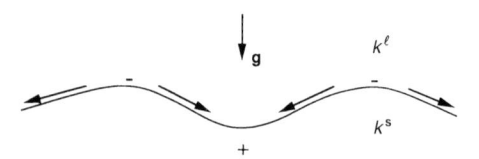

Figure 9.12. Sketches of flow patterns and concentration perturbations (given by + and −) for effects of thermal buoyancy. From Davis (1993).

when the rejected fluid is heavy (light) and solutal buoyancy is present, the destabilization is retarded (promoted).

Figure 9.12 show a corrugated interface at a fixed time and the concentration perturbations given by morphological instability. Say that, typical of metals, $k_T > 1$. This will result in a convective flow driven laterally, as shown, from crest to trough. (The solid acts as a cold, conductive body in a warm fluid.) The flow drives lateral transport that augments the concentration at the troughs, depletes it at the crests, and thus promotes morphological instability. Coriell and McFadden found for $R^C = 0$ that M_c can decrease by orders of magnitude. When the rejected solute is dense, the excess in the troughs creates a stably stratified fluid that retards the thermally driven flow and, hence, the modifications of M_c. When the rejected solute is less dense than the melt, it should augment the lateral flow, which is consistent with the mechanism suggested by Young and Davis (1986) at small k and in the FTA.

When $k_T < 1$, as in semiconductors, the flow of Figure 9.12 is reversed, and morphological instability is retarded. In addition Coriell and McFadden (1989) found that a new time-periodic mode is now present and that this mode grows at $M < M_c$ as well. The inability of researchers to find beneficial couplings between morphological instability and buoyancy-driven convection has led some to examine simpler interactions in the hope that detailed understandings of coupling mechanisms will lead to effective control of morphological instability. Thus, rather than attempting to couple complex convective flows to the front, researchers have examined prescribed flows of different types. Apart from representing the "poor man's convection," these flows can reflect actual process conditions such as the effect of the rotation of the crystal.

9.5 Directional Solidification and Forced Flows

The study of forced flows over solidifying interfaces aims at understanding how the solute redistribution by the flow alters morphological instabilities or creates new instabilities. Further, forced flows serve as a surrogate, allowing one to isolate certain effects of convection in the melt and to focus on one-way couplings.

The first studies of forced flows were those of Delves (1968,1971), who imposed a Blasius boundary layer on the interface, but, by examining wavelengths that were short compared with the spreading length of the boundary layer, he really focused on locally parallel flow. Arguments of local parallelism led Coriell et al. (1984) to impose plane Couette flow upon the interface.

If one pictures an constant imposed flow across an interface at "infinity" and allows the solidification to proceed, the forced flow has the asymptotic suction profile, a boundary-layer flow on the scale δ_v, the viscous boundary layer thickness,

$$\delta_v = \nu/V. \tag{9.16}$$

The velocity component along the interface has the form of the asymptotic suction profile, $u \propto 1 - e^{-z/\delta}$, which couples with morphological instability (Forth and Wheeler 1989). Given that the concentration and thermal boundary layers scale on δ_c and δ_T, three lengths are involved in such problems.

In the case of an organic alloy,

$$\delta_c \ll \delta_T \ll \delta_v \tag{9.17a}$$

because $\delta_v/\delta_T = \text{Pr}$, and Pr is large. However, for a small-Prandtl-number metallic alloy,

$$\delta_c \ll \delta_v \ll \delta_T. \tag{9.17b}$$

These inequalities are relevant when one considers disturbances of various wavelengths λ. The "normal" situation described has $\lambda < \delta_c$ and thus the thermal field and the velocity field can be represented in locally linear form by

$$T = T_0 + G_T z \tag{9.18a}$$

and

$$u \sim \frac{V}{\nu} z, \tag{9.18b}$$

the latter being the plane Couette flow considered by Coriell et al. (1984). However, when λ becomes large enough, these localizations are no longer valid because the disturbances are affected by the curvatures of the profiles. The distance L^+ from the interface to the upper heat source can then be a relevant length scale as well.

Coriell et al. (1984) examined two-dimensional disturbances (periodic in the flow direction) and found that the flow stabilizes the interface because the presence of the flow increases M_c. However, if one considers longitudinal-roll disturbances (periodic cross stream), the flow decouples from the problem and M_c is unchanged. Thus, under "normal" circumstances, the flow leaves M_c unchanged but selects the cellular state (longitudinal rolls) that appears.

Forth and Wheeler (1992) have examined the weakly nonlinear development of two-dimensional waves and found that the flow promotes supercritical bifurcation and narrows the bandwidth of two-dimensional structures allowable in the nonlinear regime.

Forth and Wheeler (1989) have looked at a wider range of conditions and found for long waves that the flow destabilizes the interface for two-dimensional disturbances that propagate against the flow. However, Forth and Wheeler did not determine when this mode is preferred.

Hobbs and Metzener (1991) have examined the effect of the asymptotic suction profile on the interface near the absolute stability boundary $V \simeq V_A$. Here the wavelengths are long compared with δ_c, and the two-dimensional waves lower M_c and, hence, are preferred. Hobbs and Metzener established, via linear stability theory, conditions that determine neutral stability of the interface. Figure 9.13 shows that the neutral curve on the lower branch and beyond the nose corresponds to the flow-free values, but that a substantial destabilization occurs on the upper branch.

Hobbs and Metzener (1992) have extended their work into the nonlinear regime by examining parallel flow over an interface moving at $V \simeq V_A$. Here

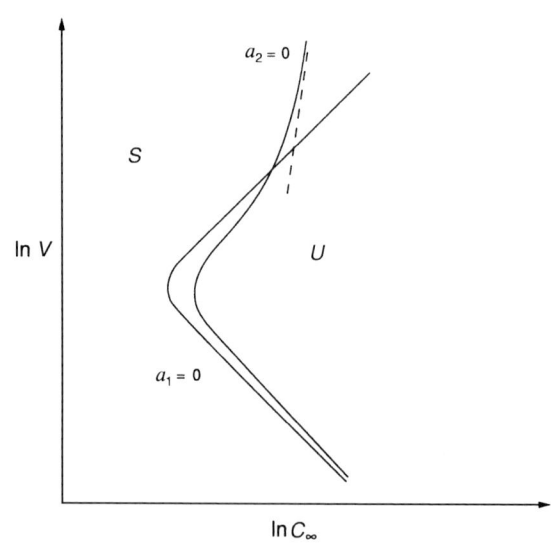

Figure 9.13. Sketch of V versus C_∞ for fixed G_T for imposed parallel shear flow. $S \equiv$ stable, $U \equiv$ unstable, and $(\mathbf{a}_1, \mathbf{a}_2)$ is the disturbance wave vector. From Hobbs and Metzener (1981). (Reprinted with permission of North-Holland.)

the long-wave structure of the solution induces a flow correction to the equation of Brattkus and Davis (1988a). The resulting interface equation contains an extra linear term that accounts for the destabilization, but, otherwise, the equation is left unchanged. When Brattkus and Davis specialized their equation to the weakly nonlinear regime, they obtained a modified Newell–Whitehead–Segel equation. Further, when this equation is examined for phase evolution, a modified Kuramoto–Sivashinsky equation is obtained, showing various sequences of "normal," as well as chaotic, solutions.

Most imposed flows are not parallel. These range from the von Kármán swirl flow generated by the rotation of the crystal to the locally hyperbolic flows present when cellular convection exists at the interface.

Brattkus and Davis (1988b,c) studied flows with hyperbolic streamlines directed upon a solidifying interface, namely, a von Kármán swirl flow and stagnation point flows. The simplest of these is two-dimensional stagnation point flow. Figure 9.14 shows the Hiemenz flow, which as $z \to \infty$ has the form

$$u \sim (K\nu)^{1/2} x F'(z),$$
$$w \sim -(K\nu)^{1/2} F(z), \qquad (9.19)$$

where F is a function that is obtained numerically and K measures the strength of the flow. The linear stability problem is made tractable by assuming that the viscous boundary-layer thickness δ_v is much larger than the concentration boundary-layer thickness δ_c and that the Schmidt number Sc is very large. Explicitly, it was assumed that

$$\delta_c/\delta_v \equiv \left(\frac{D}{V}\right)\left(\frac{\nu}{K}\right)^{-1/2} = O(Sc^{-1/3}) \quad \text{as} \quad Sc \to \infty. \qquad (9.20)$$

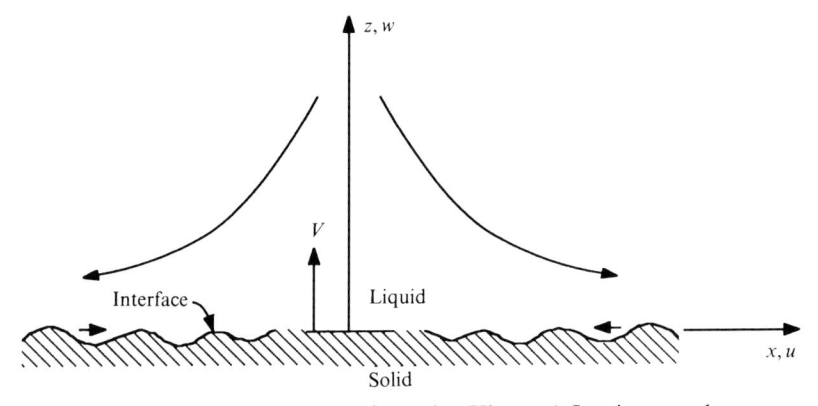

Figure 9.14. Two-dimensional stagnation-point (Hiemenz) flow impressed upon a solidifying interface. The waves on the interface propagate towards the stagnation point. From Davis (1990).

Note that the δ_v here, is distinct from that in Eq. (9.16). Given the smallness of δ_c, the interface senses only the local forms of the imposed flow, and the flow senses a flat interface; thus,

$$u \sim \beta x \zeta, \quad w \sim -\frac{1}{2}\beta \zeta^2, \tag{9.21}$$

where $\zeta = z/\delta_c$, and β measures $F''(0)$, the local shear. Finally, Brattkus and Davis considered waves long compared with δ_c though smaller than δ_v. They solved the modified diffusion problem

$$C_{\zeta\zeta} + \left(1 + \frac{1}{2}\beta\zeta^2\right)C_\zeta - \beta x \zeta C_x = C_\tau, \quad 0 \ll \zeta < \infty,$$

$$C_\zeta + \left(1 - M^{-1}\right)^{-1} C_\tau + \left[k(1 - M^{-1})^{-1} + (1-k)\right]C = 0, \quad \zeta = 0,$$

$$C(\infty) = 0,$$
(9.22)

where τ is a scaled (slow) time. Note that the longwave approximation leads to the neglect of the lateral diffusion term C_{xx}. This neglect is justified only far away from the stagnation point at $x = 0$ but makes the solution of the linear-stability problem (9.22) tractable.

The nonparallel flow gives rise to the term xC_x, which is scale invariant and thus survives the longwave approximation. The system (9.22) can be solved by employing quasi-normal modes as follows:

$$C(x, \zeta, \tau) = e^{\sigma\tau + ia \ln x}\hat{C}(\zeta), \tag{9.23}$$

which converts system (9.22) into a constant-coefficient eigenvalue problem for growth rate $\sigma = \sigma(k; M; a)$. The system is solved numerically, and Figure 9.15 shows the neutral curves.

The flow produces a longwave instability that creates waves that travel inward toward the stagnation point, as shown in Figure 9.14. The flow is locally periodic in x but by the structure of the normal modes is not globally so. The instability exists for long waves in a region where the Mullins and Sekerka condition gives only stability. Thus, it is called *flow-induced morphological instability*. The largest growth rate Reσ occurs for $a \to \infty$, where the longwave theory is invalid and surface energy should help stabilize the interface. Thus, "longish" waves would be preferred, and these would grow for M just above unity, that is, for any degree of constitutional undercooling; thus, it is a morphological instability. The neutral stability curve, sketched for all wave numbers, would be as shown in Figure 9.16. When the wave numbers are large, the disturbances see only the local velocities, and the flow appears to be locally parallel. The dashed curve of Figure 9.16 shows the analog of the results of Coriell et al. (1984)

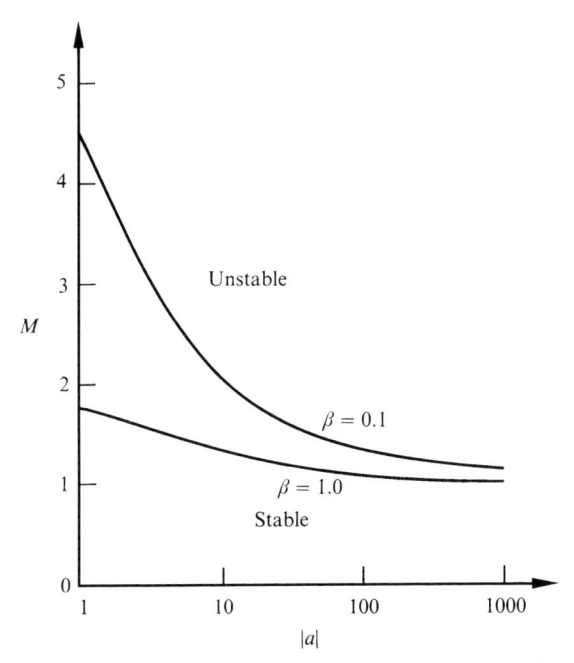

9.5 Directional solidification and forced flows   297

Figure 9.15. Neutral curve, M versus $|a|$, for long two-dimensional disturbances on a Hiemenz flow against a solidifying interface for $k = 0.3$. Here β is a nondimensional measure of the shear stress exerted by the Hiemenz flow in the interface. From Davis (1990).

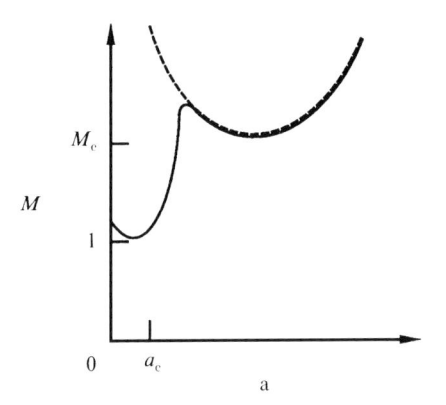

Figure 9.16. Conjectured neutral curve, M versus $|a|$, shown as the solid curve for general two-dimensional disturbances on a Hiemenz flow against a solidifying interface. For small a, nonparallel effects dominate; for other a the curve coincides with the locally parallel theory of Coriell et al. (1984), as shown by the dashed curve. (Reprinted with permission of North-Holland.)

appropriate to locally parallel flows. When the wave numbers are small, the disturbances see the curvature of the streamlines and, hence, the nonparallel flow effects, as shown in the solid curve of Figure 9.16. The connection between the small-$|a|$ loop and the ordinary morphological instability loop is conjectured.

The destabilization by nonparallel flow depends on both velocity components. The component normal to the mean position of the interface is directed inward. Its presence causes boundary-layer alteration. The concentration boundary layer is compressed, steepening the local gradient $|G_c|$. The lateral component of velocity (linear in x) varies with distance from the stagnation point and produces horizontal concentration gradients that drive the traveling cells propagating into the oncoming flow.

The destabilization of long waves in the x-direction may be negated by "end" effects that disallow the "fitting" of such waves in the system. In this case, the results of McFadden, Coriell, and Alexander (1988) would be regained. One could then allow disturbances of the form

$$C(y, z, t) = e^{\sigma t + iby}\hat{C}(z) \tag{9.24}$$

for cross-stream periodic waves that are x-independent ($a = 0$). The full stagnation-point flow linear stability problem has been examined in this case by McFadden et al. (1988). They found in this case that the flow would slightly delay morphological instability.

The effects of unsteadiness in the melt flow have been investigated by Merchant and Davis (1989). They considered plane stagnation-point flow against the interface but allowed the flow at infinity to be time periodic, where the strength K of the flow in Eq. (9.21) is replaced by a time-periodic function as follows:

$$K \to K\Theta(\omega t) = K[1 + \delta \cos \omega t]. \tag{9.25}$$

Merchant and Davis again considered longwave, two-dimensional disturbances and found that system (9.22) is replaced by the following:

$$C_{\zeta\zeta} + \left[1 + \frac{1}{2}\beta\Theta^{3/2}(\tau)\zeta^2\right]C_\zeta - \beta\Theta^{3/2}(\tau)x\zeta C_x = \Omega C_t,$$

$$C_\zeta + \Omega(1 - M^{-1})^{-1}C_t + [k(1 - M^{-1})^{-1} + (1 - k)]C = 0, \quad \zeta = 0,$$

$$C(\infty) = 0, \tag{9.26}$$

where $\Omega = \omega D/V^2$ is the scaled forcing frequency. They found that modulation at low frequency stabilizes the interface against flow-induced morphological instabilities, whereas high frequency promotes the instabilities. The response

9.5 Directional solidification and forced flows

of the system to instability is quite complex, for a disturbance is composed of two independent frequencies: the imposed frequency and the traveling-wave frequency modified by the modulation.

In an effort to construct a flow that would delay morphological instability Schulze and Davis (1994) examined directional solidification when the crystal undergoes lateral harmonic oscillations, say along the x-axis with speed $u_w = U_0 \cos \omega t$. Using scales for length D/V, speed U_0, time ω^{-1}, pressure $\rho U_0 V$ and the usual concentration and temperature units, they obtained the governing equations

$$\Omega \mathbf{v}_t + \epsilon \mathbf{v} \cdot \nabla \mathbf{v} - \mathbf{v}_z = -\nabla p + Sc \nabla^2 \mathbf{v} \quad (9.27\text{a})$$

$$\nabla \cdot \mathbf{v} = 0 \quad (9.27\text{b})$$

$$\Omega C_t + \epsilon \mathbf{v} \cdot \nabla C - C_z = \nabla^2 C \quad (9.27\text{c})$$

and

$$T = z. \quad (9.27\text{d})$$

Here,

$$\Omega = \frac{\omega D}{V^2} \quad \text{and} \quad \epsilon = \frac{U_0}{V} \quad (9.28)$$

As $z \to \infty$

$$\mathbf{v} \to 0, \ C \to 1, \quad (9.29)$$

and on $z = h(x, t)$

$$u = \cos t, \quad (9.30\text{a})$$

$$w = 0 \quad (9.30\text{b})$$

$$M^{-1} h = C + 2\Gamma h \quad (9.30\text{c})$$

$$(1 + \Omega h_t)[1 + (k - 1)C] = C_z - C_x h_x. \quad (9.30\text{d})$$

Schulze and Davis first used perturbation theory in ϵ,

$$M^{-1} = M_c^{-1} + \epsilon^2 M_2 + \cdots \quad (9.31\text{a})$$

$$\sigma = \epsilon \sigma_1 / \Omega + \cdots \quad (9.31\text{b})$$

and employed Floquet theory where σ is the Floquet exponent, which on the neutral curve has Re $\sigma = 0$.

At $O(\epsilon^2)$ they found that M_2 is determined; Figure 9.19 shows the result for domains of stabilization (S) and destabilization (D) for lead–tin. When ϵ is

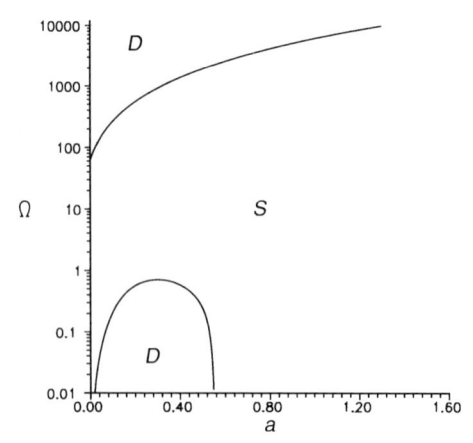

Figure 9.17. Direction solidification in compressed stokes layers (CSL). Regions of the $a - \Omega$ plane where the flow stabilizes (S) or destabilizes (D) the interface relative to the case without flow; $Sc = 8.10$ and $k = 0.3$; results are independent of Γ. From Schulze and Davis (1995). (Reprinted with permission of North-Holland.)

small, there is a window of stabilization, here from about $\Omega = 1$ to $\Omega = 50$ (0.50 – 25.0 cm/s for $V = 100\,\mu$m/s); the morphological instability is delayed.

Schulze and Davis (1994) argued that the stabilization is due to the lateral transport of solute by the flow. In the absence of flow, one can write $C(x, h', t) \sim \bar{C}(0) + h'(x, t)\bar{C}_z(0) + C'(x, 0, t)$. Here $\bar{C}(0)$ is part of the basic state, and the term $h'\bar{C}_t$ represents the destabilization of the interface in which bulges in solid encounter smaller $\bar{C}$ because $\bar{C}_z$ is negative. The term $C'(x, 0, t)$ is stabilizing because C' raises the solute concentration at frontal peaks, resulting in higher melting points there.

When the oscillatory flow is present and ϵ is small, the concentration equation has the form $LC_2 = \bar{\mathbf{v}} \cdot \nabla C' + w'\bar{C}_z$ and only time-independent forcings enter the stability relation. The advection flux $\bar{\mathbf{v}} \cdot \nabla C'$ for lateral transport reduces to $\bar{u}C'_x$. For the case shown in Figure 9.17, the solute concentration is greater at frontal peaks and so is stabilizing. The vertical component w' of disturbance velocity is directed downward on peaks and upward from valleys, and thus above the peak the liquid is diluted and the front is destabilized. In general these two fluxes can compete or cooperate. When the stability window is created, the stabilization by lateral transport dominates.

That this stabilization is due to lateral transport can be demonstrated directly. Let an overbar denote an average of over one period in time and one period in x. Under neutral stability conditions, the solute concentration equation has the form

$$\bar{C}_{zz} + \bar{C}_z = \overline{C'w'_z}$$

9.5 Directional solidification and forced flows

If one substitutes the calculated C' and w subject to $\bar{C} = 0$ at $z = 0$ and ∞, one finds that the local gradient at the front $\bar{C}_z|_0$ actually steepens because of the flow – even when stabilization is obtained.

The modulation of the crystal, as described, creates a window of stability for small ϵ and two-dimensional disturbances. However, if three-dimensional disturbances were allowed, as before, longitudinal rolls would be unaffected by the flow and hence preferred.

In the three-dimensional case, in order to obtain stabilization one must "confuse" the system so that longitudinal rolls cannot be defined. This can be done by following a suggestion of Kelly and Hu (1993), who considered nonplanar modulations in Bénard convection and succeeded in delaying the onset of convection. One thus imposes velocities on the crystal, relative to the liquid at infinity, in two orthogonal components, namely,

$$u_W = U_0 \cos \omega t, \quad v_W = \beta U_0 \cos(\omega t - \delta). \tag{9.32}$$

Schulze and Davis (1994) examined the linearized system for small ϵ and found that the modification M_2 to the critical morphological number has the form

$$M_2 = \left[(\cos \theta + \beta \cos \delta \sin \theta)^2 + \beta^2 \sin^2 \theta \sin^2 \delta\right] M_2^{2D}, \tag{9.33}$$

where M_2^{2D} is that obtained earlier (see Figure 9.17), and $\theta = \tan^{-1}(a_2/a_1)$ is the angle of the disturbance wave vector with the x-axis. This 2D→3D relation has the same form as that found by Kelly and Hu in the Bénard problem. The range of stabilization shown in Figure 9.17 for the two-dimensional case now applies to the three-dimensional case as long as δ is not an integral multiple of π. The maximal stabilization occurs when $\delta = \pi/2$, and if β is unity, the crystal is *translating* on circular orbits and no preferred direction is distinguished for the cellular morphology.

Schulze and Davis (1995) then examined the linearized stability for arbitrary ϵ by representing the eigenfunctions as $(C', w') = \sum_{m=-\infty}^{\infty} [C_m(z), w_m(z)] e^{imt}$ times a Floquet multiplier.

Figure 9.18 shows M^{-1} versus ϵ for $a = a_c$ (no flow) in the case of lead–tin. In the window of stability, as ϵ increases, M_c^{-1} decreases through zero, which corresponds to complete stabilization. Note that if ϵ is increased even further, the tendency to stabilization can be reversed. Figure 9.19 shows the stability boundary in the M^{-1} versus Γ plane for fixed k and Ω. Notice that the intercept at (0,1) never changes but that there is a substantial stabilization near the absolute stability limit. Figure 9.20 shows the same information in physical coordinates. The stabilization is substantial near the nose of the curve and near the absolute stability boundary. Table 9.1 shows the values calculated for top and bottom

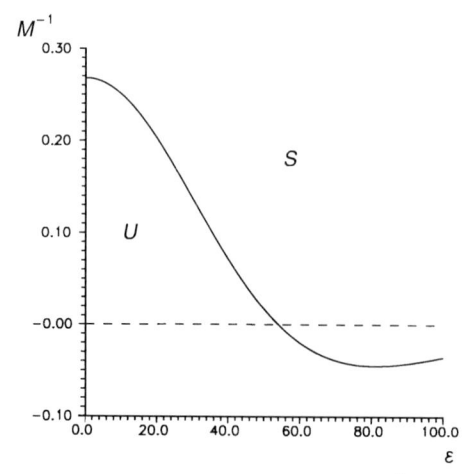

Figure 9.18. Directional solidification into CSL. Plot of M^{-1} versus ϵ with $a = a_c$, the critical value for the no-flow case; $k = 0.3$, $S = 81.0$, and $\Omega = 10.0$. This plot shows that the stabilizing trend eventually reverses as ϵ is increased. From Schulze and Davis (1995). (Reprinted with permission of North-Holland.)

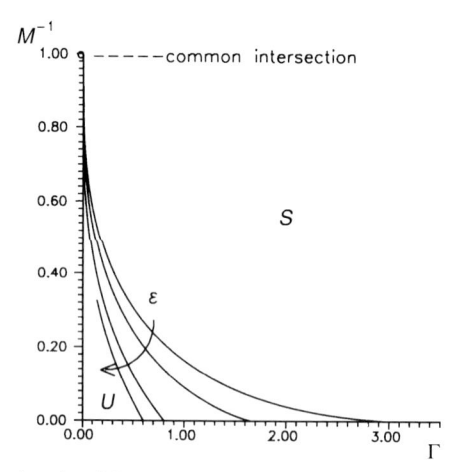

Figure 9.19. Directional solidification into CSL. Plot of the critical value of M^{-1} as a function of Γ for $k = 0.3$, $S = 81.0$, $\Omega = 10.0$, and $\epsilon = (0, 20, 40, 60)$. The arrow indicates the direction in which ϵ increases. The interface is linearly stable (S) when the inverse morphological number is above the neutral curve. All of the curves pass through the point $(\Gamma = 0, M^{-1} = 1)$; however, the calculations for the $\epsilon = 60$ curve were terminated before reaching that point. Notice that the range of parameter values for which the interface is stable increases with ϵ. From Schulze and Davis (1995). (Reprinted with permission of North-Holland.)

9.5 Directional solidification and forced flows

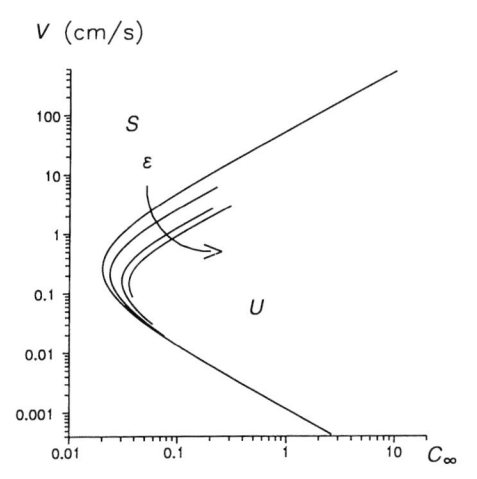

Figure 9.20. Directional solidification into CSL. Plot of the neutral curve in dimensional form: V versus C_∞ for $k = 0.3$, $S = 81.0$, $\Omega = 10.0$, and $\epsilon = (0, 20, 40, 60)$. The arrow indicates the direction in which ϵ increases. The temperature gradient G_T is 200 K/cm. All of the curves extend infinitely along tangents to the portions shown. The interface is linearly stable (S) when the far-field concentration is to the left of the neutral curve. Notice that the stability of the interface increases with ϵ. From Schulze and Davis (1995). (Reprinted with permission of North-Holland.)

of the stability window for $\epsilon \to 0$ and the values ϵ^* and Ω^* above which the instability is suppressed for all M for Pb–Sn, Si–Sn, and SCN–acetone. Note that $\epsilon^* = \epsilon^*(k, Sc, \Gamma, \Omega)$ and that the corresponding Ω^* is not necessarily the optimal frequency for stabilization.

Given that the modulational stabilization is greatest near the absolute stability limit (but without effects of disequilibrium), Schulze and Davis (1996) rederived the Brattkus–Davis evolution equation (see Section 4.2.3) and performed a bifurcation analysis. They found that the presence of the flow promotes

Table 9.1. *Window of stabilization, $\Omega_{min} < \Omega < \Omega_{max}$, and the approximate value of the modulation amplitude ϵ^* at the frequency Ω^* above which the instability is suppressed for all morphological numbers. From Schulze and Davis (1995).*

Material	Ω_{min}	Ω_{max}	Ω^*	ϵ^*
Pb–Sn	0.7	66	10	60
Si–Sn	0	2000	5	>500
SCN–Ace	1.5	2500	5	500

Reprinted with permission of North-Holland.

subcritical instability. Hence, a portion of the predicted stabilization may be lost at finite amplitude.

These analyses in some sense validate the comment made by Delves (1971), who considered steady flow over a crystal that "it might be possible to ensure complete stability of the interface by rapid and continuous changes in the flow pattern."

9.6 Directional Solidification with Imposed Cellular Convection

The interaction of simultaneous instabilities discussed in Section 9.4, valid near the codimension-two point, the intersection point of the two neutral curves of Figure 9.8, yielded only small mutual stabilizations of the morphological and convective instabilities. This smallness is due to the disparity of the critical wave numbers; see Figure 9.9. The common situation in crystal growth involves systems, shown in Figure 9.8, in which one operates at values of C_∞ well to the right of the intersection of the two neutral curves. In such cases, as V is increased from zero, there is first a convective instability whose convective amplitudes increase to well-developed cellular convection. It is in this convective region that the neutral curve for morphological instability lies. Thus, the normal situation has the onset of morphological instability occurring in the presence of a pre-existing long-scale convective field.

One can "manufacture" a convective field from the Boussinesq equations with the solutal buoyancy removed and replaced by a "fictitious" body force that generates the convection. Given this convection, one can then determine the interfacial response by analyzing the convection–diffusion equation for C. Suppose that $\rho^s = \rho^\ell$ and that the one-sided model using the FTA is used, and consider the two-dimensional problem. A solution of the Navier–Stokes equations for a particular choice of body force $2\pi/\hat{\alpha}$-periodic in x is

$$\bar{u} = -U_0(1 - e^{-z/s})\sin \hat{\alpha} x \qquad (9.34a)$$

$$\bar{w} = \hat{\alpha} U_0 \left[z - s(1 - e^{-z/s}) \right] \cos \hat{\alpha} x, \qquad (9.34b)$$

where V and δ_c are chosen as speed and length scales and the parameter s measures the decay length normal to the interface – in effect the viscous boundary-layer thickness in units of δ_c (Schulze and Davis 1994). It is convenient to map the interface to a plane by writing $\zeta = e^{H-z}$. Then, the streamfunction for Eqs. (9.34) is

$$\bar{\psi}(x, \zeta) = \left[-s + \ln \zeta - s\zeta^{1/s} \right] U_0 \sin \hat{\alpha} x. \qquad (9.34c)$$

9.6 Directional solidification with imposed cellular convection

Write asymptotic expansions for $\bar{C}$ and $\bar{H}$ valid for $\hat{\alpha}\hat{\delta}$ small

$$\bar{C}(\alpha x, \zeta) \sim \zeta + \hat{\alpha}\hat{\delta}\bar{C}_1(\zeta) \cos \hat{\alpha} x$$

$$\bar{H}(\alpha x) \sim 0 + \hat{\alpha}\hat{\delta}\bar{H}_1 \cos \hat{\alpha} x, \tag{9.35}$$

where

$$\hat{\delta} = \frac{U_0}{1+s}. \tag{9.36}$$

When these are substituted into the governing system, one finds that

$$\bar{C}_1 = \zeta \left[\frac{1}{k} + (s^2 - 1) \ln \zeta + \frac{1}{2}(s+1) \ln^2 \zeta + s^3(1 - \zeta^{1/s}) \right]$$

$$\bar{H}_1 = -\frac{M}{k}, \tag{9.37}$$

which describe the solute and interface as forced by the flow.

To describe the morphological instability, as altered by the presence of the flow, write

$$M^{-1} = M_c^{-1} - \mu_1 \epsilon^2 m, \quad \hat{\delta} = \mu_2 \epsilon^2 \delta$$

$$T = \mu_3 \epsilon^2 t, \quad X = \epsilon \beta_c x, \quad \hat{\alpha}/\beta_c = \epsilon \alpha \tag{9.38}$$

so that δ represents the convective amplitude and α represents the unit-order scaled $\hat{\alpha}/\beta_c$. The parameter ϵ measures the ratio of the wave numbers and $\epsilon \to 0$ governs the system of long-scale convection forcing small-scale morphology. Here the X represents a slow spatial scale, and T represents the slow time. When $m > 0$, then the flow delays the morphological instability. The μ_i are unit order real and positive coefficients to be determined later.

Now write

$$\mathbf{v} = \bar{\mathbf{v}} + \left\{ \epsilon \mathbf{v}_1(T, X_i, \zeta) + \epsilon^2 \mathbf{v}_2 + \epsilon^3 \mathbf{v}_3 + \cdots \right\} e^{i\beta_c \cdot (x, y)} \tag{9.39}$$

where $\mathbf{v} = (C, H, v)$; $\mathbf{v}_0$ is taken to be $O(\epsilon^2)$, namely $\mathbf{U}_0 = \epsilon^2 \mathbf{v}_0$. At $O(\epsilon)$ one obtains the no-flow linear problem, symbolically $L\mathbf{v}_1 = 0$, which gives (β_c, M_c) and $\mathbf{v}_1$. At $O(\epsilon^2)$, $L\mathbf{v}_2 = F_2(\mathbf{v}_1)$ can be solved directly. At $O(\epsilon^2)$, $L\mathbf{v}_3 = F_3(\mathbf{v}_1, \mathbf{v}_2)$ has solutions only if an orthogonality condition holds, namely (with the subscript dropped),

$$H_T = [m + i\delta \sin \alpha x] H + H_{XX} - b_1 H |H|^2$$

$$+ O(\epsilon \alpha \delta H \cos \alpha x, \; \epsilon \delta H_X \sin \alpha X) = 0. \tag{9.40}$$

Here b_1 is the Landau coefficient for the case with no flow, as first computed by Wollkind and Segel (1970). Notice that when $\delta = 0$ (no flow) that $m = 0$

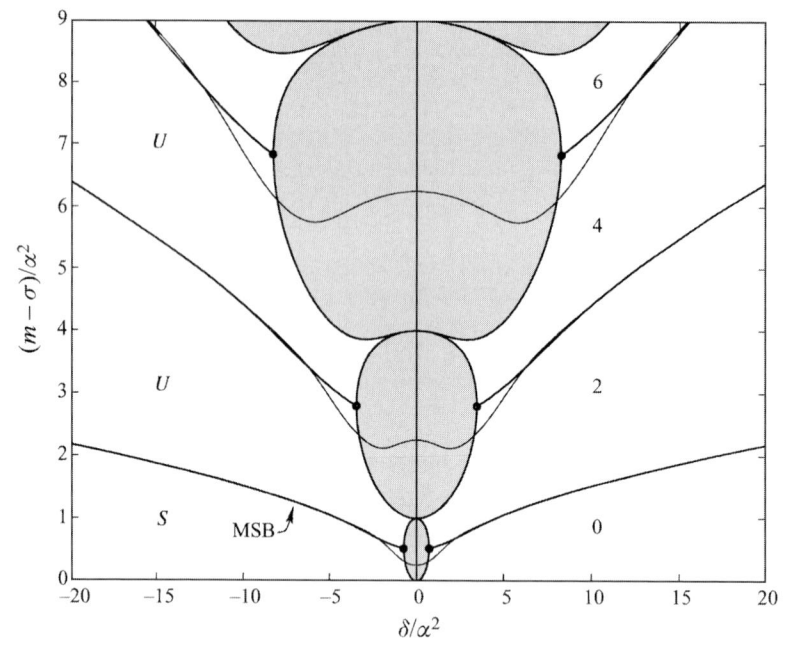

Figure 9.21. Characteristic curves for the two-dimensional-cell equation. Thick lines are period-$(2\pi/a)$ (harmonic) solutions, and thin lines are period-$(4\pi/a)$ (subharmonic) solutions. Solutions in the shaded area have spatial structures incommensurable with the flow period. Solid circles represent doubly degenerate points; S denotes stable, and U denotes unstable regions; MSB denotes the marginal stability boundary. From Chen and Davis (1999).

(Mullins and Sekerka) and hence H is constant. When $\delta \neq 0$, then the spatial structure is governed by a Mathieu-like differential operator.

In the linearized theory $H = \hat{H} e^{(\sigma + i\omega)t}$, where σ is real, and hence

$$\hat{H}_{XX} + [(m - \sigma) - i\omega + i\delta \sin \alpha X] \hat{H} = 0. \tag{9.41}$$

Bühler and Davis (1998) and Chen and Davis (1999) solved Eq. (9.41) using Floquet theory for systems with complex coefficients. The characteristic curves are shown in Figure 9.21 for $2\pi/\alpha$-periodic solutions. Inside the boundaries, the solutions (shown as shaded) are aperiodic, and $H(x)$ is incommensurable with the flow. The harmonic solutions (thick lines) correspond to stationary modes ($\omega = 0$) when δ/α^2 is small but change to time-dependent modes at the doubly degenerate points. These oscillatory modes are left-traveling cells localized near $\alpha X = (2n + 1/2)\pi$, or right-traveling cells localized near $\alpha X = (2n - 1/2)\pi$. The superposition of these gives solutions localized between surface stagnation points, as shown in Figure 9.22.

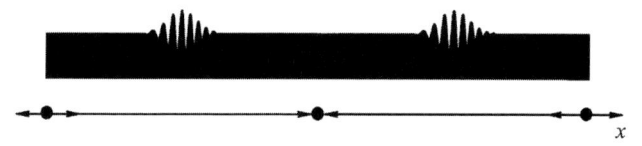

9.6 Directional solidification with imposed cellular convection

Figure 9.22. Interfacial disturbance of the two-dimensional-cell morphology (side view) predicted by the linear stability analysis: (a) pure, (b) incommensurable, and (c) localized morphologies. Solid circles are stagnation points, and arrows indicate the flow direction. From Chen and Davis (1999).

Consider now neutral curves, $\sigma = 0$. In the absence of flow, there is a continuous spectrum along the axis $m \geq 0$ with its minimum, $m = 0$, corresponding to the Mullins and Sekerka value. When flow is present, parametric resonance breaks the spectrum, as shown in Figure 9.21, for increasing values of δ. The bands of the spectrum contract into lines or very thin layers as δ increases, each of which has localized spatial structure and a quantized morphological number given by $m^{(\nu)}$, the sites of Hopf bifurcations; $m = m^{(1)} \equiv m_c$ is the critical value and, as shown near the origin, the flow delays the morphological instability in two dimensions. A WKB analysis for $\delta/\alpha^2 \ll 1$ gives

$$m_c \sim \frac{1}{2}\alpha\delta^{\frac{1}{2}}, \quad \omega_c \sim \delta - \frac{1}{2}\alpha\delta^{\frac{1}{2}}$$

$$H \sim \exp\left\{\frac{1}{4}(-1 \pm i)\left(\frac{\delta}{\alpha^2}\right)^{\frac{1}{2}}(\alpha X - \alpha X_\pm)^2 \pm i\omega T\right\}, \quad (9.42)$$

where $\alpha X_\pm = \left(2n \pm \frac{1}{2}\right)\pi$. The requirement for representing the basic state, $\hat{\alpha}\hat{\delta} \ll 1$, sets an upper bound on the validity of the asymptotics. The localization occurs where the X-component of the flow speed is maximum, and the results should apply in a Hele–Shaw geometry in which three-dimensional effects are minimized.

The correct linear combination of the right-traveling and left-traveling modes can be obtained by a weakly nonlinear theory. Chen and Davis (1999) wrote $m = m_c + \lambda$, $\omega = \omega_L + \varpi(\lambda)$ and found that

$$\lambda A_1 - bA_1[\kappa_1|A_1|^2 + \kappa_2|A_2|^2] = 0$$
$$\lambda A_2 - bA_2[\kappa_1|A_2|^2 + \kappa_2|A_1|^2] = 0. \quad (9.43)$$

Because the envelopes of the eigenfunctions Φ_1 and Φ_2 are separated in space, κ_2 is exponentially small whereas κ_1 is of unit order and positive. Straightforward analysis shows that the mixed mode $|A_1| = |A_2| \neq 0$ is stable when the no-flow morphology bifurcates supercritically, $b_1 > 0$, in which case the pure modes are unstable. When $b_1 < 0$, the stability characteristics are reversed. Thus, Figure 9.24, showing both localizations, depicts the stable, steady state when $b_1 < 0$.

When the flow remains two-dimensional, but three-dimensional morphological disturbances are allowed, localization still occurs, but the structure is severely altered.

Consider plane disturbances inclined by an angle θ to the x-axis. Thus, $\theta = 0$ corresponds to two-dimensional cells, and $\theta = \pi/2$ corresponds to longitudinal rolls. The generalized version of Eq. (9.53) then becomes

$$H_T = [m + i\delta \cos\theta \sin\alpha X]H + H_{XX}\cos^2\theta - b_1 H |H|^2$$
$$+ O(\epsilon\alpha\delta H \cos\alpha X, \epsilon\delta H_X \cos\alpha X). \quad (9.44)$$

As θ increases from zero, two changes occur. First, the tangential component of velocity is reduced by $\cos\theta$, as shown, so that as $\theta \to \pi/2$, this component vanishes. Before this happens, it becomes comparable to the normal component, which is included in the error term. Thus, the system has to be rescaled, and when this is done, it turns out that the front becomes increasingly unstable as $\theta \to \pi/2$; the longitudinal roll is preferred in the three-dimensional case (with two-dimensional convection). Second, as $\theta \to \pi/2$ the coefficients of H_{XX} in Eq. (9.62) vanish; hence, one has to proceed to higher order to obtain the governing system.

The rescaling and higher-order analysis (Chen and Davis 1999) can be done when $U_0 = O(\epsilon)$, $\delta^* = \epsilon\delta$. The result of the multiple-scale analysis is

$$H_T = [m + \chi\alpha\delta^* \cos\alpha X]H + \delta^* H_X \sin\alpha - bH |H|^2 - \frac{1}{4}\epsilon^2 H_{XXXX}, \quad (9.45)$$

which contains a singular perturbation in ϵ. Here the coefficient $\chi(k, \Gamma, s)$ is a parameter that characterizes the strength of the normal component of convection relative to the tangential component. It is a monotone function of s, $s > 1$; as the viscous boundary layer is made thicker, s increases and hence χ increases.

In the linear theory,

$$-\frac{1}{4}\epsilon^2 \hat{H}_{XXXX} + \delta^* \hat{H}_X \sin\alpha X + [(m - \sigma) - i\omega + \chi\alpha\delta^* \cos\alpha X]\hat{H} = 0. \quad (9.46)$$

In Figure 9.23 the characteristic curves are given for this case; all the longitudinal rolls are stationary, $\omega = 0$, with incommensurable bands shrinking like $\delta^*/\alpha^3\epsilon^2 \gg 1$.

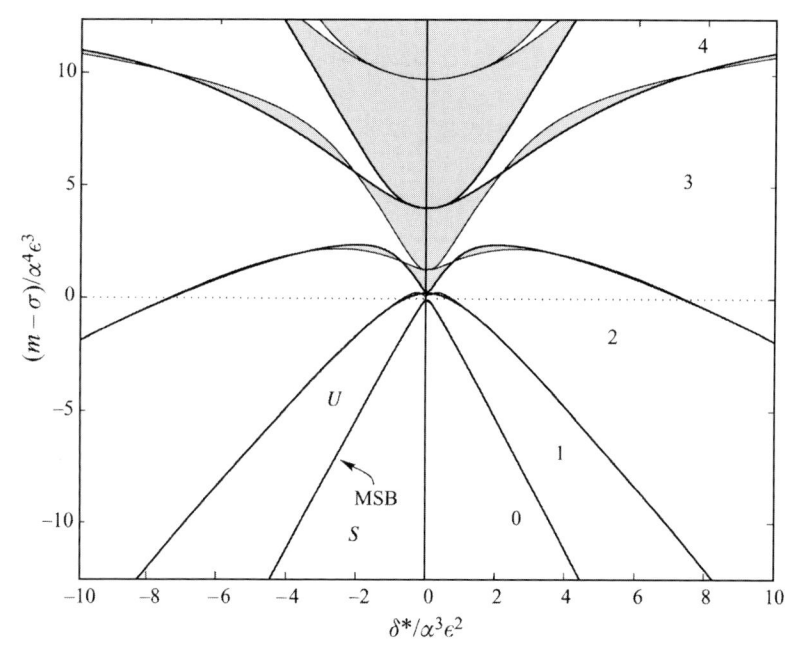

9.6 Directional solidification with imposed cellular convection

Figure 9.23. Characteristic curves for the longitudinal-roll equation when $\chi = 4.0$; otherwise see the caption to Figure 9.21. From Chen and Davis (1999).

Chen and Davis (1999) used methods of matched asymptotic expansions to analyze Eq. (9.46). They found that

$$m_c \sim (1 - \chi)\alpha\delta^*, \quad \omega_c = 0$$

$$\hat{H} \sim \int_0^\infty \cos \xi t e^{-\frac{1}{4}t^4} dt \tag{9.47}$$

$$\delta^*/\alpha^3\epsilon^2 \gg 1,$$

where the envelope function is characterized by

$$\xi = \left(\frac{4\delta^*}{\alpha^3\epsilon^2}\right)^{\frac{1}{4}} (\alpha X - 2\pi n). \tag{9.48}$$

Note that Eq. (9.48) is an eigensolution for $\nu = 1$ of the "comparison" equation

$$H_{\xi\xi\xi\xi} - \xi H_\xi - \nu H = 0, \quad \nu = 1, 2, 3, \tag{9.49}$$

which describes the local behavior of the system when $\cos \alpha \, X \to 1$, $\sin \alpha \, X \to \alpha X$, and $m/\alpha\delta^* + \chi = \nu$. The method of steepest descent applied to

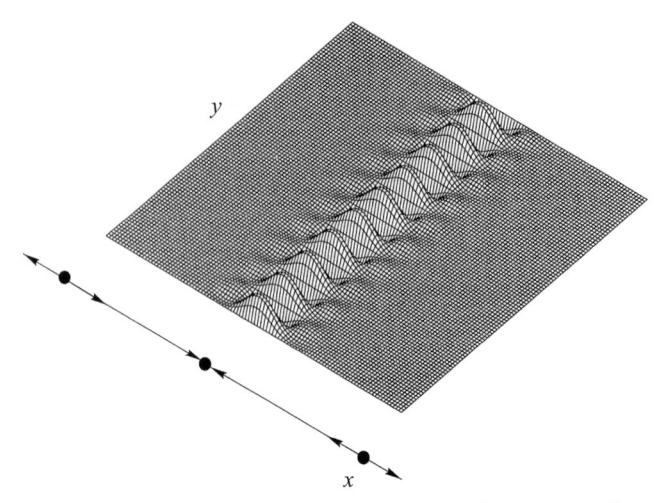

Figure 9.24. Interfacial disturbance of the longitudinal-cell morphology in perspective as predicted by the linear stability theory. Solid circles are stagnation points, and arrows indicate the flow direction. From Chen and Davis (1999).

Eq. (9.49) gives

$$H \sim \sqrt{\frac{2\pi}{3}} \xi^{-\frac{1}{3}} \exp\left(-\frac{3}{8}\xi^{\frac{4}{3}}\right) \cos\left(\frac{3\sqrt{3}}{8}\xi^{\frac{4}{3}} - \frac{\pi}{6}\right) \quad (9.50)$$

valid for $1 \ll \xi \ll [\delta^*/\alpha^3\epsilon^2]^{1/4}$ or, equivalently, $[\delta^*/\alpha^3\epsilon^2]^{-1/4} \ll \alpha X \ll 1$.

Figure 9.24 shows that localizations are now at the inward stagnation points and contain standing longitudinal rolls. Figure 9.25 shows a perspective of the predicted localized morphologies in two and three dimensions.

Weakly nonlinear theory now involves the growth of a longitudinal roll and gives the steady bifurcation equation

$$\lambda A - b_1 |A|^2 A = 0. \quad (9.51)$$

Again, the no-flow theory determines b_1, and the bifurcation is subcritical (supercritical) when $b_1 < 0$ ($b_1 > 0$).

When the **flow** is three-dimensional, the analogous morphological instability becomes a Floquet problem but with multiple degrees of freedom in space (Chen and Davis 2000). In this case, the flow field near the crystal plane is characterized by inward–outward stagnation points or lines, which consequently result in local focus- or ridge-like morphologies, respectively. The induced localization has a spatial structure well-correlated with that of the flow profile. Near the onset of morphological instability, for example, an inward hexagonal

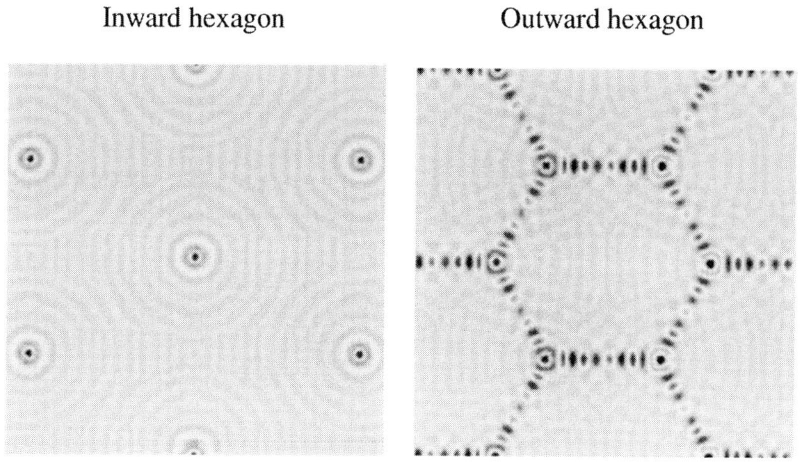

9.7 Flows over ivantsov needles

Inward hexagon   Outward hexagon

Figure 9.25. Surface deflections for directional solidification with an imposed hexagonal convective field with the surface flow (a) inward and (b) outward. From Chen and Davis (2000).

flow produces localized foci at the center of each hexagonal flow cell (Figure 9.25(a)), whereas an outward flow produces localized ridges at the rims (boundaries) of each hexagon (Figure 9.25(b)). Compare these with Figure 4.11. Those stagnation regions are the positions where the interfacial flow converges and inhomogeneities accumulate. The crystal so produced then possesses a type of "composite" structure induced by the cellular convective field. Its structure can also be constructed by a WKB analysis similar to those performed in the two-dimensional system.

The local flow patterns in the vicinity of stagnation regions can easily be analyzed and classified, and the insights from analyses on the cellular flow fields presents a way to understand the flow-induced morphological structures in complex flow configurations.

9.7 Flows over Ivantsov Needles

In the past few sections flows over arrays of cells were discussed. When the morphological number is varied so that dendrites are present, one can anticipate at least two important studies. One can examine how an imposed fluid flow or natural convection affects a single needle crystal. One can also examine flow or convection in arrays of dendrites, a so-called mushy zone. This latter topic will be discussed in Chapter 10.

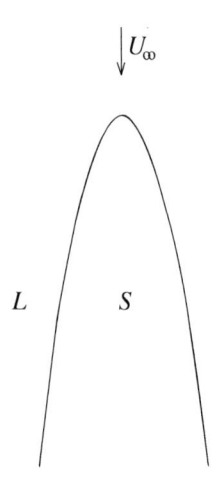

Figure 9.26. An isolated needle in an oncoming uniform flow.

Consider the system shown in Figure 9.26 in which an Ivantsov needle crystal is subjected to an oncoming flow that is uniform at large distances. In a coordinate system moving with the constant-speed needle the energy equation in the liquid has the form

$$T_t + \mathbf{v} \cdot \nabla T - V T_z = \kappa \nabla^2 T. \tag{9.52}$$

When **v** represents Oseen or potential flow, it is still possible to obtain similarity solutions for parabolic or paraboloidal needles (Ananth and Gill 1988a,b; Saville and Beaghton 1988; Ben Amar, Bouissou, and Pelce 1988). In particular, McFadden and Coriell (1986) found solutions for the case of convection driven by density changes, which is a potential flow; see Section 9.3. They found

$$S^{-1} \sim -\rho Pe\,[\ell n\,Pe + \gamma], \quad Pe \ll 1 \tag{9.53a}$$

and

$$S^{-1} \sim 1 - \frac{1}{\rho^2 Pe}, \quad Pe \gg 1, \tag{9.53b}$$

where γ is Euler's constant, $\gamma \approx 0.5772$ and $\rho^s \neq \rho^\ell$. Roughly speaking, ρ modifies the Peclet number, and thus shrinkage, $\rho > 1$, decreases Pe for a given S^{-1}.

Ananth and Gill solved the Oseen-flow problem and used a thermal-convection analogy by writing $Re = \sqrt{Gr}$, $Gr = \beta g(\Delta T)\rho^3/\nu^2$, taking a fixed value of $\sigma^* = 0.075$, and comparing to the data of Huang and Glicksman

9.7 Flows over ivantsov needles

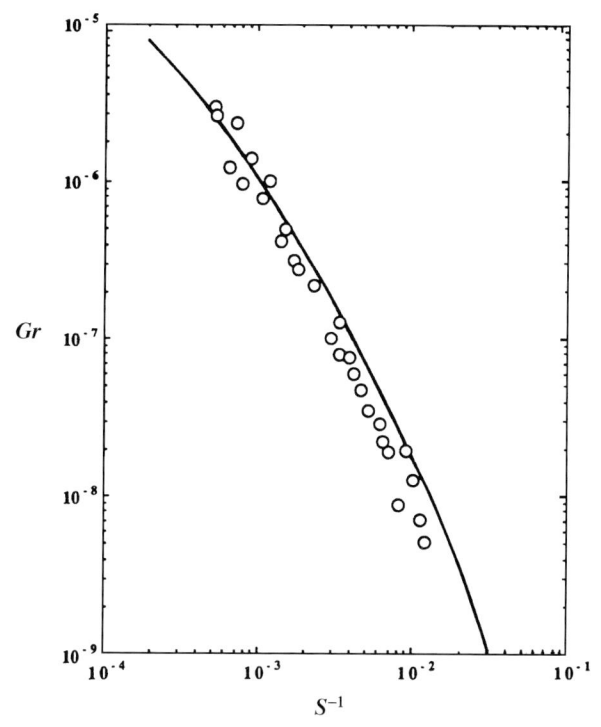

Figure 9.27. Use of thermal convection analogy, $Gr = Re_m^2$, with selection parameter, $\sigma^* = 0.075$ to estimate level of natural convection. Comparison of Grashof numbers observed experimentally with those predicted. From Gill (1993). (Reprinted with permission of the American Society of Engineering Education.)

(1981) on SCN. Figure 9.27 shows that the trends of the theory and experiment are the same. The question is, What σ^*, if any, should be used?

Experiments by Gill et al. (1992) on SCN forced past a needle show that σ^* increases by 50% with U_∞ in the range U_∞/V from 3 to 300 for Sc^{-1} in the range 0.230 to 1.000. The trend that σ^* increases with U_∞ is opposite to that predicted by Bouissou and Pelcé (1989) using MST and thus suggests that MST may be inapplicable. On the other hand, experiments by Bouissou, Perrin, and Tabeling (1989) on alloys of PVA and ethanol gave opposite dependence with U_∞. No clear explanation of this disparity now exists.

When the flows are not of the special forms mentioned, no similarity solution is possible. Xu (1994) considered a flow U_∞ of any magnitude but only in the limit $Pr \to \infty$, appropriate to the transparent plastic crystals used in experiment. He used matched asymptotic expansions and found that the modifications

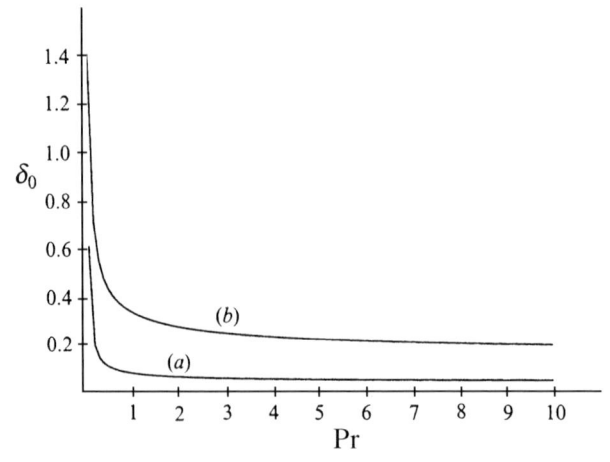

Figure 9.28. The variation of the parameter δ_0 with Pr for (a) $U_\infty = 0.1$, $Pe_0 = 0.2$; (b) $U_\infty = 0.5$, $Pe_0 = 0.1$. From Xu (1994).

depend on the parameter δ_0

$$\delta_0 = \frac{2U_\infty}{\ell n \left[\dfrac{\text{Pr}}{(1+U_\infty)Pe_0}\right] + 1 - \gamma}, \tag{9.54}$$

where Pe_0 is the Peclet number of the Ivantsov solution and δ_0 is a function of the undercooling, the far-field flow speed, and Pr. Figure 9.28 shows how δ_0 varies with Pr. In particular as Pr $\to \infty$, the departure of the solution from similarity is $O(1/\text{Pr}\,\ell n\,\text{Pr})$. Figure 9.28 shows the final growth characteristics, Pe versus S^{-1}. For a given undercooling, as the flow increases in magnitude, the Peclet number increases.

When the needle grows downward into a buoyancy field, the convection and growth are tightly coupled. Huang and Glicksman (1981) observed for SCN that when S^{-1} is small, convective effects are important, and that dendrites growing downward grow faster, with finer tips, than those growing upward. Ananth and Gill (1988b) took the needle to be a paraboloid and applied a coordinate expansion formally valid with a fraction of a tip radius from the tip. They included the nonlinear effects of buoyancy and derived a pair of coupled nonlinear ordinary differential equations for the temperature and stream function in this region. They solved these equations numerically for various choices of the parameters, the Peclet number and the Grashof number, for the value Pr = 23.1, and determined the Stefan number. To compare the results with the experiments of Huang and Glicksman (1981), for each Stefan number, they chose a Grashof number in order to match the experimental Peclet number.

9.7 Flows over ivantsov needles 315

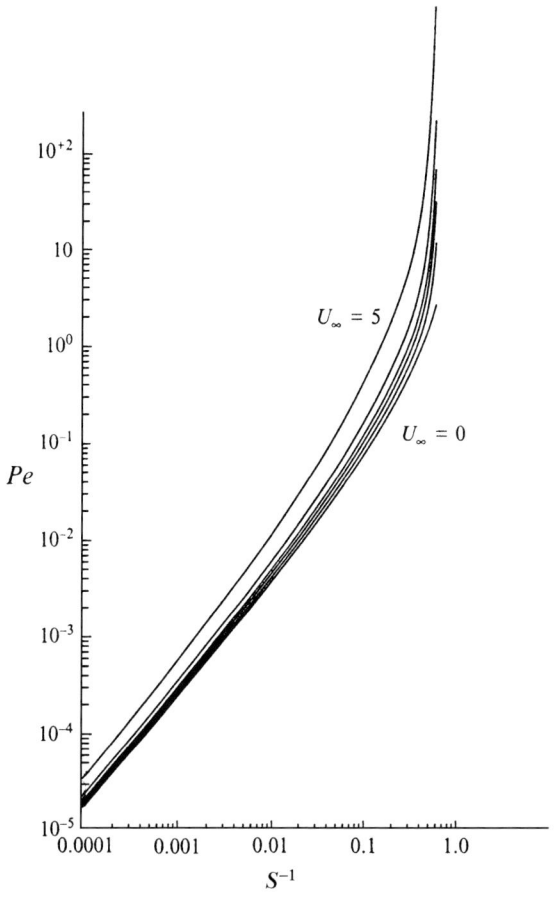

Figure 9.29. The variation of the growth parameter Pe with S^{-1} for $Pr = 13.5$ and various flow parameters $U_\infty = (0, 0.4, 0.8, 1.2, 2.0, 5.0)$. From Xu (1994).

From this, they determined the tip radius and the growth speed; these results compare well with the experiments over the whole range of the data.

Canright and Davis (1991) sought perturbation solutions valid all along the dendrite but for small amplitude convection seeking changes in tip radius and speed due to the presence of flow. Their model presumes that $\mathbf{v} \to \mathbf{0}$ far from the needle, which implies that the needle is growing in a large, permeable box.

The natural parameter that emerges in this analysis is G,

$$G = \frac{\alpha g \Delta T \delta_T^3}{\kappa^2} = Gr \, Pr^{-2}. \tag{9.55}$$

There is a $O(G^{-1})$ near-tip region, a free-convection boundary layer along the

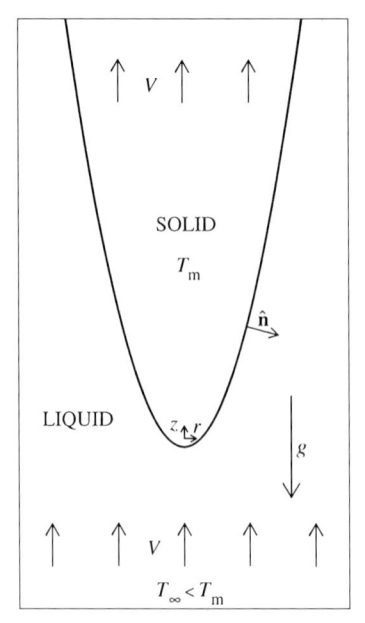

Figure 9.30. The flow relative to the growing needle surface moves steadily up and through the interface as it become solid. Gravity is directed downward, and far from the dendrite the liquid is undercooled and in uniform motion. From Canright and Davis (1991). (Reprinted with permission of North-Holland.)

needle, and a potential flow at large distances. Their results are only locally valid because the last two regions are not examined.

They found that the results are strongly dependent on Prandtl number. Figure 9.31 shows Pe versus S^{-1} and the signs of the Peclet number perturbation $Pe^{(1)}$ from $Pe \sim Pe^{(1)} + GPe^{(2)}$. When Pr is large, convection enhances the tip growth (Huang and Glicksman 1981), but when Pr is small, as it would be for metals or semiconductors, buoyant convection *diminishes* growth at the tip. If one uses the *same* σ^* condition as for the isothermal case, the tip is larger than in the convection-free case. This result suggests that growth information learned on transparent alloys must be used with caution on systems using metals or semiconductors.

In the preceding the convection is imposed antiparallel to gravity with the result that both V and ρ depend on Gr and Pr. The experiments of Huang and Glicksman (1981) show what happens when SCN dendrites are grown at various angles to gravity; see Figure 9.32. Both V and ρ would now depend on Gr, Pr, and the angle between the finger axis and **g**.

To date no quantitative predictions are available for the cases shown in Figure 9.32(a),(b), and (d). However, insights are available. Murakami et al.

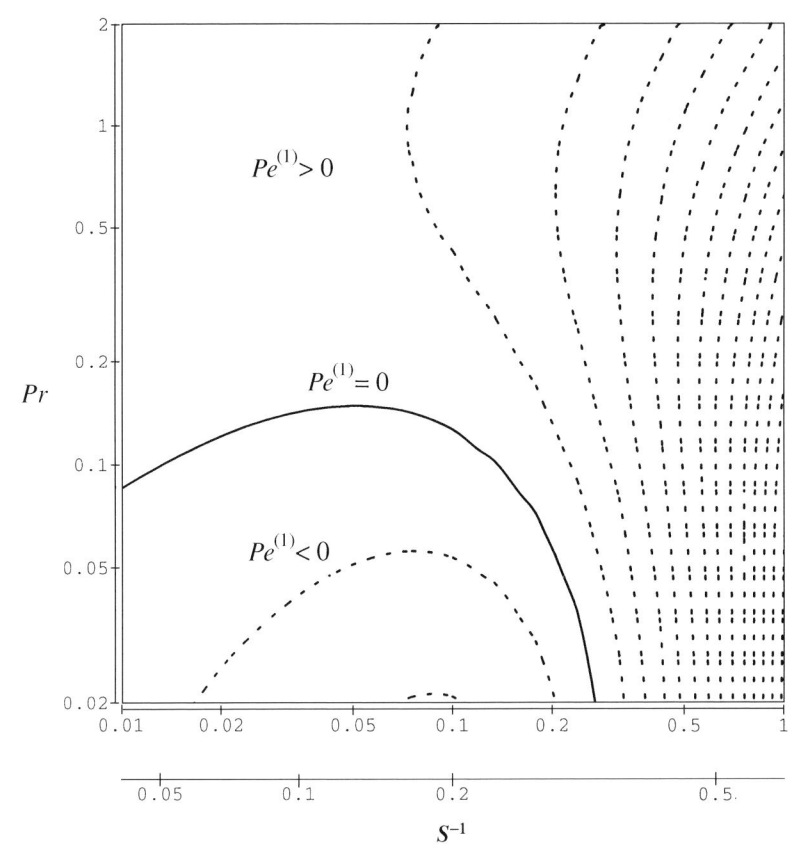

Figure 9.31. In the S^{-1} and Pr parameter plane the curve $Pe^{(1)} = 0$ separates $O(G)$, the region where buoyancy enhances tip growth $(Pe^{(1)} > 0)$, from the other region where buoyancy reduces tip growth $(Pe^{(1)} < 0)$. Also shown are other contours (dashed) of constant $Pe^{(1)}$ at equal intervals of 0.005. Note that the scale for S^{-1} shown for comparison is not a regular logarithmic scale. From Canright and Davis (1991). (Reprinted with permission of North-Holland.)

(1983), Murakami, Aihara, and Okamoto (1984), and Fredrikkson et al. (1986) observed that a forced shear flow across an array of needle crystals causes the needles to tilt into the flow. Dantzig and Chao (1990) argued on theoretical grounds that a single needle in a cross flow will grow into the flow, as shown in Figure 9.33. Because the dendrite diameter is quite small, the local Reynolds number for the flow is often very small; therefore, the Stokes flow past an axisymmetric needle will be symmetric front to back. However, because D/κ is often 10^{-3}, the Peclét number Pe can be substantial. Thus, the concentration field is not symmetric front to back. Rather, the isopycnals are

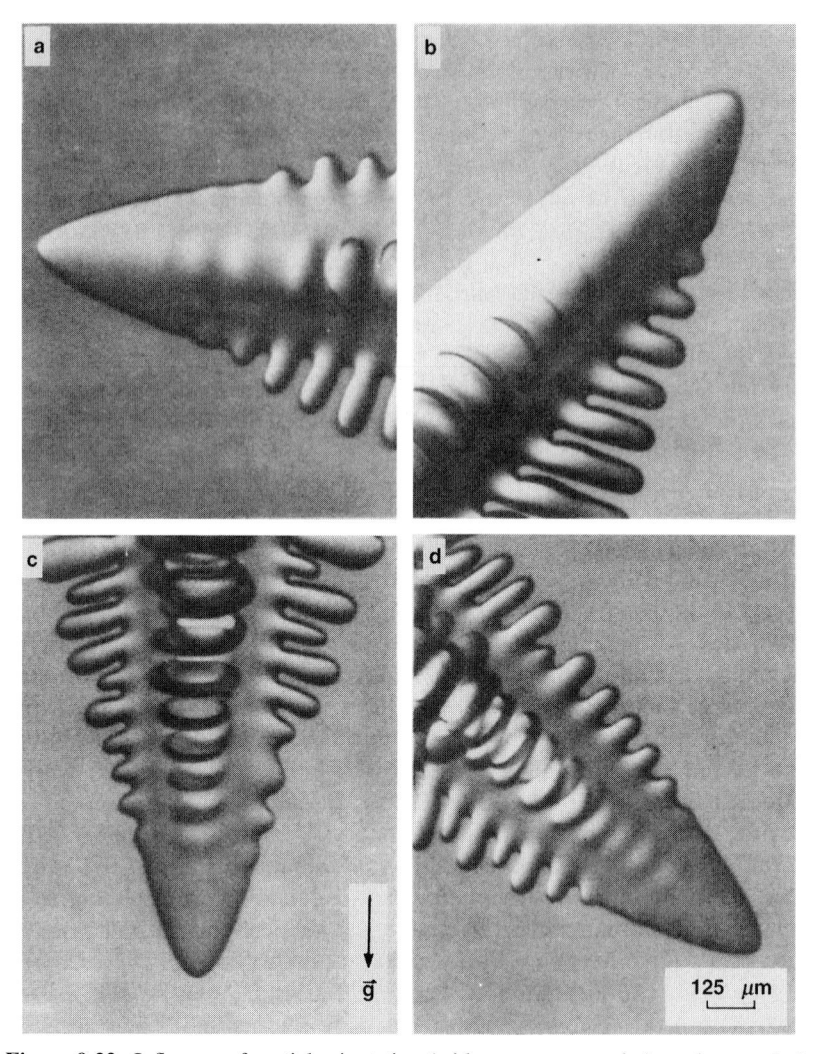

Figure 9.32. Influence of spatial orientation (with respect to gravity) on the morphology of dendrites growing into undercooled succinonitrile as determined by Glicksman and Huang (1982). From Glicksman, Coriell, and McFadden (1986). (Reprinted with permission of Annual Reviews Inc.)

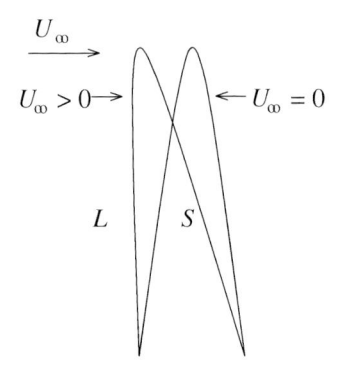

Figure 9.33. A needle crystal grows into a cross flow and tilts toward the flow.

more tightly bunched in front than in back. In front the concentration gradient is steeper than in the back; hence, the needle will grow towards the flow.

Dantzig and Chao (1990) revealed through numerical simulations that two- and three-dimensional cases are quantatively different because, in the former, heat, solute, or both convect over the top of the needle "strip," whereas there is also convection laterally in the latter case. Tönhardt and Amberg (1998) have used phase-field methods (see Chapter 11) to study the two-dimensional case of shear past a solid, growing nucleus.

9.8 Remarks

A large variety of situations have been examined in which fluid flows play major roles in the solidification process. These flows alter stability criteria for existing mechanisms and create new, coupled phenomena. The responses to flow provide a template for the interpretation of microstructures and their antecedents. They may, as well, provide the means of microstructure design if one could create a flow on demand of a given pattern and intensity.

In part of the discussion, velocity fields were prescribed, and the mass or heat transfer problem was solved without accounting for the effect of the evolving morphology on the fluid flow. This one-way decoupling is attractive for giving relatively simple models that one can analyze. However, these analyses constitute merely approximate systems, and the predictions that come from them are suggestive only. One must now confront the fully coupled problems and determine the concomitant system responses.

References

Ananth, R., and Gill, W. N. (1988a). The effect of convection on axisymmetric parabolic dendrites, *Chem. Eng. Comm.* **68**, 1–14.

Ananth, R., and Gill, W. N. (1988b). Dendritic growth with thermal convection, *J. Crystal Growth* **91**, 587–598.

Ben Amar, M., Bouissou, P. H., and Pelce, P. (1988). An exact solution for the shape of a crystal growing in a forced flow, *J. Crystal Growth* **92**, 97–100.

Bouissou, P., and Pelcé, P. (1989). Effect of a forced flow on dendritic growth *Phys. Rev. A* **40**, 6673–6680.

Bouissou, P., Perrin, B., and Tabeling, P. (1989). Influence of an external flow on dendritic crystal growth, *Phys. Rev. A* **40**, 509–512.

Brattkus, K. (1988). Directional solidification of dilute binary alloy, Ph.D. Dissertation, Northwestern University, Evanston, IL.

Brattkus, K., and Davis, S. H. (1988a). Cellular growth near absolute stability, *Phys. Rev. B* **38**, 11452–11460.

——— (1988b). Flow-induced morphological instabilities: The rotating disc. *J. Crystal Growth* **87**, 385–396.

——— (1988c). Flow-induced morphological instabilities: Stagnation-point flows, *J. Crystal Growth* **89**, 423–427.

Brown, R. A. (1988). Theory of transport processes in single crystal growth from the melt, *AIChE J.* **34**, 881–911.

Bühler, L., and Davis, S. H. (1998). Flow-induced changes of morphological instability in directional solidification: Localized morphologies, *J. Crystal Growth* **186**, 629–647.

Canright, D., and Davis, S. H. (1991). Buoyancy effects of a growing, isolated dendrite, *J. Crystal Growth* **114**, 153–185.

Caroli B., Caroli, C., Misbah C., and Roulet, B. (1985a). Solutal convection and morphological instability in directional solidification of binary alloys, *J. Phys.* (Paris) **46**, 401–413.

——— (1985b). Solutal convection and morphological instability in directional solidification of binary alloys. II. Effect of the density difference between the phases, *J. Phys.* (Paris) **46**, 1657–1665.

Chen, Y.-J., and Davis, S. H. (1999). Directional solidification of a binary alloy into a cellular convective flow: Localized morphologies, *J. Fluid Mech.* **395**, 253–270.

——— (2000). Flow-induced patterns in directional solidification: Localized morphologies in three-dimensional flows *J. Fluid Mech.* **421**, 369–380.

Choi, I. G., and Korpela, A. A. (1980). Stability of the conduction regime of natural convection in a tall vertical annulus, *J. Fluid Mech.* **99**, 725–738.

Coriell, S. R., Cordes, M. R., Boettinger, W. S., and Sekerka, R. F. (1980). Convective and interfacial instability during unidirectional solidification in a binary alloy, *J. Crystal Growth* **49**, 13–28.

Coriell, S. R., and McFadden, G. B. (1989). Buoyancy effects on morphological instability during directional solidification, *J. Crystal Growth* **94**, 513–521.

Coriell, S. R., McFadden, G. B., Boisvert, R. F., and Sekerka, R. F. (1984). Effect of a forced Couette flow on coupled convective and morphological instabilities during unidirectional solidification, *J. Crystal Growth* **69**, 15–22.

Dantzig, J. A., and Chao, L. S. (1990). The effect of shear flows on solidification microstructure, in C. F. Chen, editor, *Proc. 10th U.S. Nat. Cong. of Appl. Mech.*, pages 249–255. ASME Press, N.Y.

Davis, S. H. (1990). Hydrodynamic interactions in directional solidification, *J. Fluid Mech.* **212**, 241–262.

——— (1993). Effects of flow on morphological instability, in *Handbook of Crystal Growth*, Vol. 1b, pp. 861–897, North-Holland, Amsterdam.

Davis, S. H., Huppert, H. E., Müller, U., and Worster, M. G. (1992). Interactive dynamics on convection and solidification, NATO ASI Series E, Vol. 219, Kluwer Publishers, Dordrecht.

Davis, S. H., Müller, U., and Dietsche, C. (1984). Pattern selection in single-component systems coupling Bénard convection and solidification, *J. Fluid Mech.* **144**, 133–151.

Davis, S. H., and Segel, L. A. (1968). Effects of surface curvature and property variation on cellular convection. *Phys. Fluids* **11**, 470–476.

Delves, R. T. (1968). Theory of stability of a solid liquid interface during growth from stirred melts, *J. Crystal Growth* **3–4**, 562–568.

——— (1971). Theory of the stability of a solid liquid interface during growth from stirred melts II, *J. Crystal Growth* **8**, 13–25.

Dietsche, C., and Müller, U. (1985). Influence of Bénard convection on solid–liquid interfaces, *J. Fluid Mech.* **161**, 249–268.

Fang, Q. T., Glicksman, M. E., Coriell, S. R., McFadden, G. B., and Boisvert, R. F. (1985). Convective influence on the stability of a cylindrical solid–liquid interface, *J. Fluid Mech.* **151**, 121–140.

Forth, S. A., and Wheeler, A. A. (1989). Hydrodynamic and morphological stability of the unidirectional solidification of a freezing binary alloy: A simple model, *J. Fluid Mech.* **202**, 339–366.

——— (1992). Coupled convective and morphological instability in a simple model of the solidification of a binary alloy, including a shear flow, *J. Fluid Mech.* **236**, 61–94.

Fredriksson, H., El Mahallawy, N., Taha, M., Liu, X., and Wänglöw, G. (1986). *Scand. J. Metall.* **15**, 127–137.

Gill, W. N. (1993). Interactive dynamics of convection and crystal growth, *Chem. Eng. Educ.* **27**, 198–205.

Gill, W. N., Lee, Y. W., Koo, K. K., and Ananth, R. (1992). Interaction of thermal and forced convection with the growth of dendritic crystals, in S. H. Davis, H. E. Huppert, U. Müller, and M. G. Worster, editors, *Interactive Dynamics of Convection and Solidification*, Kluwer Academic Publishers, Dordrecht.

Glicksman, M. E., and Mickalonis, J. I. (1982). Convective coupling of a solid–liquid interface in an internally heated vertical cylinder, in T. Negat Veziroglon, editor, *Thermal Sciences*, **16**, 505–515, Hemisphere Publications, Washington, DC.

Glicksman, M. E., Coriell, S. R. and McFadden, G. B. (1986). Interaction of flows with the crystal-melt interface, *Ann. Rev. Fluid Mech.* **18**, 307–335.

Hadji, L., Schell, M., and Riahi, D. N. (1990). Interfacial pattern formation in the presence of solidification and thermal convection, *Phys. Rev. A* **41**, 863–873.

Hennenberg, M., Rouzaud, A., Favier, J. J., and Camel, D. (1987). Morphological and thermosolutal instabilities inside a deformable boundary layer during directional solidification. I. Theoretical methods, *J. Phys.* (Paris) **48**, 173–183.

Hobbs, A. K., and Metzener, P. (1991). Long-wave instabilities in directional solidification with remote flow, *J. Crystal Growth* **112**, 539–553.

——— (1992). Directional solidification: Interface dynamics and weak remote flow, *J. Crystal Growth* **118**, 319–332.

Huang, S. C., and Glicksman, M. E. (1981). Fundamentals of dendritic solidification-I. Steady–state growth, *Acta Metall. Mater.* **29**, 701–717.

Hurle, D. T. J., Jakeman E., and Wheeler, A. A. (1982). Effect of solutal convection on the morphological stability of a binary alloy *J. Crystal Growth* **58**, 163–179.

——— (1983). Hydrodynamic stability of the melt during solidification of a binary alloy *Phys. Fluids* **26**, 624–626.

Jenkins, D. R. (1985a). Nonlinear interaction of morphological convective instabilities during solidification of a dilute binary alloy, *IMA J. Appl. Math.* **35**, 145–147.

———. (1985b). Nonlinear analysis of convective and morphological instability during solidification of a dilute binary alloy, *Physiochem. Hydrodyn.* **6**, 521–537.

———. (1990). Oscillatory instability in a model of direction of solidification, *J. Crystal Growth* **102**, 481–490.

Kelly, R. E., and Hu, H.-C. (1993). The onset of Rayleigh–Bìnard convection in non-planar oscillatory flows, *J. Fluid Mech.* **249**, 373–390.

McFadden, G. B., and Coriell, S. R. (1986). The effect of fluid flow due to the crystal-melt density change on the growth of a parabolic isothermal dendrite, *J. Crystal Growth* **74**, 507–512.

McFadden, G. B., Coriell, S. R., and Alexander, J. I. D. (1988). 'Hydrodynamic and free-boundary instabilities during crystal growth – The effect of a plane stagnation flow, *Comm. Pure Appl. Math.* **41**, 683–706.

McFadden, G. B., Coriell, S. R., Boisvert, R. F., Glicksman, M. E., and Fang, Q. T. (1984). Morphological stability in the presence of fluid flow in the melt, *Metall. Trans.* **A15**, 2117–2124.

McFadden, G. B., Coriell, S. R., Boisvert, R. F., Glicksman, M. E., and Selleck, M. E. (1989). *PhysicoChem. Hydrodyn.* **11**, 387.

Merchant, G. J., and Davis, S. H. (1989). Flow-indiced morphological instabilities due to temporally-modulated stagnation-point flow, *J. Crystal Growth* **96**, 737–746.

Murakami, K., Fujiyama, T., Koike, A., and Okamoto, T. (1983). Influence of melt flow on the growth directionals of columnar grains and columnar dendrites, *Acta Metall. Mater.* **31**, 1425–1432.

Murakami, K., Aihara, H., and Okamoto, T. (1984). Growth direction of columnar crystals solidified in a flowing melt, *Acta Metall.* **32**, 933–939.

Özetkin, A., and Pearlstein, A. J. (1992). Coriolis effects on the stability of plane-front solidification of dilue Pb–Sn binary alloys, *Metall. Trans.* **B23**, 73–80.

Lord Rayleigh (1879). On the instability of jets, *Proc. Lond. Math. Soc.* **10**, 4–13.

Riley, D. S., and Davis, S. H. (1989a). Applied Mathematics Technical Report No. 8838 (Northwestern University).

———. (1989b). Hydrodynamic stability of the melt during the solidification of a binary alloy with small segregation coefficient, *Physica D.* **39**, 231–238.

———. (1991). Long-wave interactions in morphological and convective, instabilities *IMA J. Appl. Math.* **45**, 267–285.

Rosenberger, F. (1979). *Fundamentals of Crystal Growth I*, Springer-Verlag, New York.

Saville, D. A., and Beaghton, P. J. (1988). Growth of needle-shaped crystals in the presence of convection, *Phys. Rev. A* **37**, 3423–3430.

Schaefer, R. J., and Coriell, S. R. (1982). Convective and interfacial instabilities during solidification of succinonitrile containing ethanol, in G. E. Rindone, editor, *Materials Processing in the Reduced Gravity Environment of Space*, pages 479–489, Materials Research Society Symposium Proceedings, Vol. 9, North-Holland Press, New York.

Schlüter, A., Lortz, and Busse, F. H. (1965). On the stability of steady finite amplitude convection, *J. Fluid Mech.* **23**, 129–144.

Schulze, T. P., and Davis, S. H. (1994). The influence of oscillating and steady shears on interfacial instability during directional solidification, *J. Crystal Growth* **143**, 317–333.

———. (1995). Shear stabilization of morphological instability during directional solidification, *J. Crystal Growth* **149**, 253–265.

——— (1996). Shear stabiligation of a solidifying front: Weakly nonlinear analysis in the long-wave limit *Phys. Fluids* **9**, 2319–2336.

Shaaban, A. H., and Özisik, M. N. (1982). Effect of curvature on the thermal stability of a fluid between two long vertical coaxial cylinders, in U. Grigull, E. Hahne, K. Stephan and J. Straub, editor, *Heat Trans. for* 1982, Munich, Vol. 2, 281–286, Hemisphere Publishers, Washington, DC.

Sivashinsky G. I. (1983). On cellular instability in the solidification of a dilute binary alloy, *Physica D*. **8**, 243–248.

Tönhardt, R., and Amberg, G. (1998). Phase-field simulation of dendritic growth in a shear flow, *J. Crystal Growth* **194**, 406–425.

Wheeler, A. A. (1991). A strongly nonlinear analysis of the morphological instability of a freezing binary alloy–solutal convection, density changes, and nonequilibrium effects, *IMA J. Appl. Math.* **47**, 173–192.

Wollkind, D. J., and Segel, L. A. (1970). A nonlinear stability analysis of the freezing of a dilute binary alloy, *Philos. Trans. Roy. Soc. London A* **268**, 351–380.

Young, G. W., and Davis, S. H. (1986). Directional solidification with buoyancy in systems with small segregation coefficient, *Phys. Rev. B* **34**, 3388–3396.

Xu, J.-J. (1994). Dendritic growth from a melt in an external flow: Uniformly valid asymptotic solution for the steady state, *J. Fluid Mech.* **263**, 227–243.

Zimmermann, G., Müller, U., and Davis, S. H. (1992). Bénard convection in binary mixtures with Soret effects and solidification, *J. Fluid Mech.* **238**, 657–682 and Corrigendum **254**, 720.

10

Mesoscale fluid flow

In Chapter 9 the discussion of the interaction of fluid flows with solidification fronts focused on how individual cells or dendrites are affected by the motion of the melt. The discussion was confined to the onset, or near onset, of morphological instability and the influence of laminar flows because the systems had small scale and the rate of solidification could be readily controlled. There are many situations in industrial or natural situations in which the systems have large scale and the freezing rates are externally provided.

When the freezing rate is not carefully controlled near $M = M_c$, the typical morphology present is dendritic (or eutectic) and strongly nonlinear in the parameter space $V - C_\infty$ of Figure 3.6. The region within which there are both dendrites and interstitial liquid is called a *mushy zone*. Such zones should not be described pointwise in the same sense that one would not want to describe flow pointwise in a porous rock. Instead, the zone is treated as a porous region that is reactive in the sense that the matrix melts and the liquid freezes and whose properties are described in terms of quantities averaged over many dendrite spacings. There still is a purely liquid region and a purely solid region, but now they are separated by an intermediate layer, the mushy zone.

When the length scale of the system is large and the fluid is subjected to gravity, the fluid in the fully liquid system will undergo buoyancy-driven convection. The large scale may imply that the Rayleigh number is large enough that the convection is unsteady, laminar, or turbulent.

If one simultaneously has strongly supercritical morphological instability, turbulent flow in the liquid, and finally (weaker) convective flow in the mush, the dynamics is so complex that direct large-scale numerical simulations give too much data and not enough information. Alternatively, one can attempt to understand dominant balances by drastically simplifying the systems and learning the underlying physics step-by-step.

In this chapter, models systems will be discussed, and the predictions from these will be compared with the observations from table-top experiments. In contrast to the experiments in previous chapters, which use dilute organic alloys, here the typical experiment uses aqueous solutions near their eutectic concentrations. Although most of the experiments have addressed the cooling from one boundary of a bulk solution, and the concomitant time-varying front position, the theories often regard directional solidification. One must then compare carefully different types of systems.

10.1 Formulation

The earlier formulations on solidification of a melt are generally applicable here with two alterations: one regarding the phase diagram and the other involving the transport mechanisms.

Figure 10.1 shows an idealized phase diagram appropriate to aqueous solutions. When a hypoeutectic aqueous solution is frozen, the ice that results is nearly pure water. This is reflected in the phase diagram in which the solidus is represented a vertical line, and hence the segregation coefficient $k = 0$. Also indicated on the diagram are lines of constant liquid density ρ^ℓ, which are nearly vertical, reflecting the greater sensitivity of the density to changes in solute concentration than to changes in temperature. An implication of this property (Huppert 1993) involves the cooling of a hypoeutectic ($C < C_E$) solution. As the temperature is lowered, the density of the residual fluid decreases,

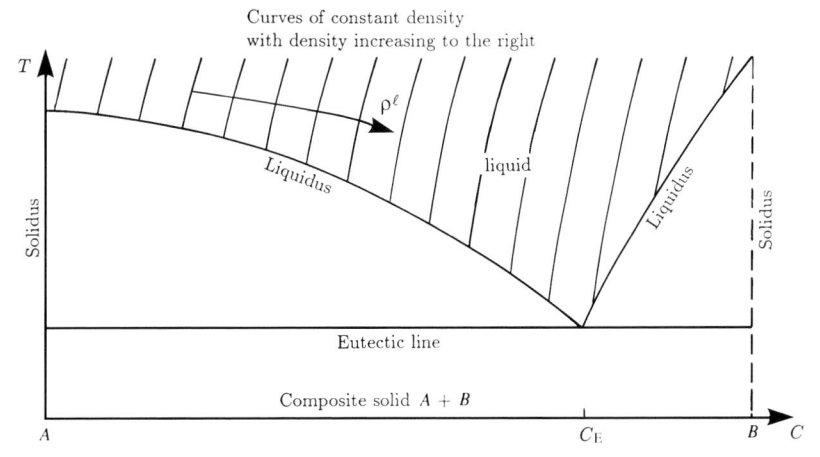

Figure 10.1. A typical phase diagram for aqueous solutions with the second component having concentration C. The solidi are vertical so that the $k = 0$. Superposed on the liquid region are curves of constant liquid density. From Huppert (1993). (Reprinted with permission of Elsevier Science.)

contrary to the "normal" behavior. This is because the liquid becomes depleted of the heavy component (B) and the density is more sensitive to concentration decreases than to temperature decreases. Similarly, the cooling of a hypereutectic, $(C > C_E)$ solution results in successively decreasing densities of the rejected material.

When vigorous convection is present and only large-scale properties of the interface are sought, it is sufficient to represent the thermal transport (beyond conduction) by a simplified functional. In this case the interfacial thermal balance might be written as

$$\left[c_p^\ell(\bar{T} - T_m) + L_v\right] h_t = -k^s T_n^s - F_T, \tag{10.1}$$

where F_T is a correlation for convective heat transport (Turner 1979) of the form

$$F_T = \hat{c} k^\ell \left(\frac{\alpha g}{\kappa^\ell \nu}\right)^{\frac{1}{3}} (\bar{T} - T_m)^{\frac{4}{3}}, \tag{10.2}$$

$\bar{T}$ is the mean temperature of the liquid, and $\hat{c}$ is an empirical constant, $\hat{c} \approx 0.14$.

In addition, it turns out that effects of attachment kinetics, important in melts only at high pulling speeds, enter the description of aqueous solutions at much lower speeds, and thus at the interface

$$V_n = \mu(T^\ell - T^I) \tag{10.3}$$

where, for example in aqueous isopropanol, $\mu \approx 2.2 \times 10^{-4}$ cm s^{-1} $(C^0)^{-1}$ (Kerr et al. 1990).

10.2 Planar Solidification between Horizontal Planes

Consider two horizontal plates that enclose an aqueous solution. Huppert and Worster (1985) classified the various possibilities as functions of composition and cooling position into six cases; Table 10.1 from Huppert (1993) summarizes these.

When a eutectic solution is cooled from below, Case A, Figure 10.2(a), one has a classical, one-component, Stefan problem, as discussed in Section 2.1. Solutions exist for all values of the Stefan number S, the planar front $h(t)$ moves with speed proportional to $\sqrt{t}$, and the configuration is thermally and morphologically stable.

When a eutectic solution is cooled from above, Case B, one has a eutectic solid created above liquid, precisely the problem studied by Davis, Müller, and Dietsche (1984) and discussed in Section 9.2.2; no solute is rejected. Because the liquid is heated from below, the system is susceptible to Bénard instabilities

Table 10.1. *A summary of the six different regimes that occur when a two-component solution is cooled at a horizontal boundary. The three different compositional conditions are tabulated in the first column, and the effect this has on the fluid released by the resultant solidification is tabulated in the last column. The two different thermal conditions are tabulated in the middle two columns, and the effect this has on the stability of the resulting thermal field in the melt is tabulated at the bottom. The interior 3×2 array summarizes sequentially the major effect in each case and indicates under which classification of the test is treated.*

Compositional constraint	Cooling from		Effect of composition
	Below	Above	
$C < C_E$	C. Stagnant melt and mushy layer	F. Convection in melt driven by both thermal and compositional effects, $T < T_E$	Relatively heavy fluid released
$C = C_E$	A. Classical Stefan problem	B. Thermally driven convection in liquid possible	Fluid density depends only on temperature
$C > C_E$	E. Compositional convection in melt, $T < T_E$	D. Thermally driven convection in melt beneath a stagnant mushy layer	Relatively light fluid released
Resulting thermal profile in melt	Stable	Unstable	

327

for Rayleigh number $R^T \approx R_c^T$. Turner, Huppert, and Sparks (1986) and Huppert and Worster (1991) studied situations in which $R^T \gg R_c^T$ and the lower boundary is changed from a perfect conductor to a perfect insulator (Figure 10.2(b)). This change results in a system that completely solidifies. By using thermal condition (10.1), Huppert and Worster (1991) identified three distinct asymptotic ranges for the solid thickness h/d versus time $\tau = \kappa^\ell t/d^2$, where d is the spacing between the plates. The changes occur at $\tau \sim [\hat{c} R_0^{1/3}]^{-2}$ and $\tau \sim [\hat{c} R_0^{1/3}]^{-6/7}$, where R_0 is the Rayleigh number at $t = 0$. There is no morphological instability in this case.

When a hypoeutectic aqueous solution is cooled from below, Case C, Figure 10.2(c), the thermal field is stably stratified, and the rejection creates a negative solute gradient at the front. Thus, the front is morphologically unstable. Because $C < C_E$, the solutal field is also stably stratified. When $M \gg M_c$, the front becomes dendritic, and a mushy layer is formed, as shown in Figure 10.2(d). The interest, then, is the prediction of the thickness $h(t)$ of the mush.

When a hypereutectic solution is cooled from above, Case D, Figure 10.2(e), the picture is superficially the same as that of Case B, Figure 10.2(b), where ice is above liquid, but now there is solute rejection, a morphological instability, and a mushy zone above the interface. When $R^T \gg R_c^T$, the interface between mush and liquid is subjected to augmented heat transfer like that represented by condition (10.2).

When a hypereutectic solution is cooled from below, $T_B < T_E$, Case E, Figure 10.2(f), the thermal field is stably stratified, but there is solutal convection. Woods and Huppert (1989) estimated by analogy with the thermal case that the convective compositional flux F_c has the form

$$F_c = \hat{c}_c (g\beta D^2/\nu)^{\frac{1}{3}} (\Delta C)^{\frac{4}{3}} \qquad (10.4)$$

where ΔC is the change in composition across the boundary layer above the mush-liquid interface and $\hat{c}_c$ is an empirical constant. Both the thickness of the mush and the microstructure are affected by the convection.

When a hypoeutectic solution is cooled from above, $T_B < T_E$, Case F, Figure 10.2(g), both the compositional and thermal fields are unstably stratified, and strong double-diffusion convection occurs under a mushy layer formed at the top. This convection is dominated by the solutal buoyancy, because, if the thermal-equilibrium assumption is valid, the buoyancy $g\Delta\rho$ has the form (Worster 1992b)

$$g\Delta\rho = -g[\alpha \Delta T + \beta \Delta c] = -g(\alpha m + \beta)\Delta c, \qquad (10.5)$$

where $m < 0$. Generally speaking $|\beta| \gg |m\alpha|$, and thus thermal buoyancy is

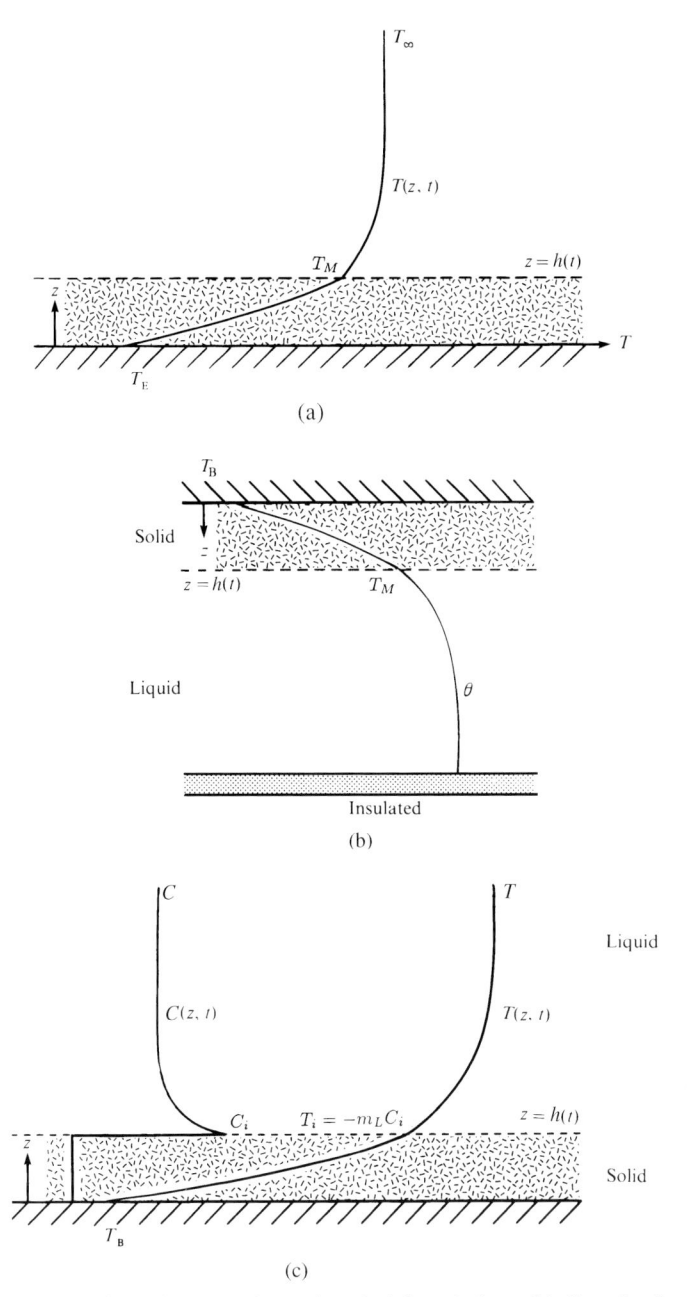

Figure 10.2. (a) Case A, $C = C_E$ and cooled from below. (b) Case B, $C = C_E$ and cooled from above with (1) the lower boundary at fixed temperature, and (2) the lower boundary allowing zero heat flux. (c) Case C, $C < C_E$ and cooled from below. (d) Case D, $C < C_E$ and cooled from below with $M \gg M_c$. (e) Case E, $C > C_E$ and cooled from above. (f) Case F, $C > C_E$ and cooled from below. (g) Case G, $C < C_E$ and cooled from above. From Huppert (1993). (Reprinted with permission of Elsevier Science.) (*Continued*)

Bulk Models of Solidification

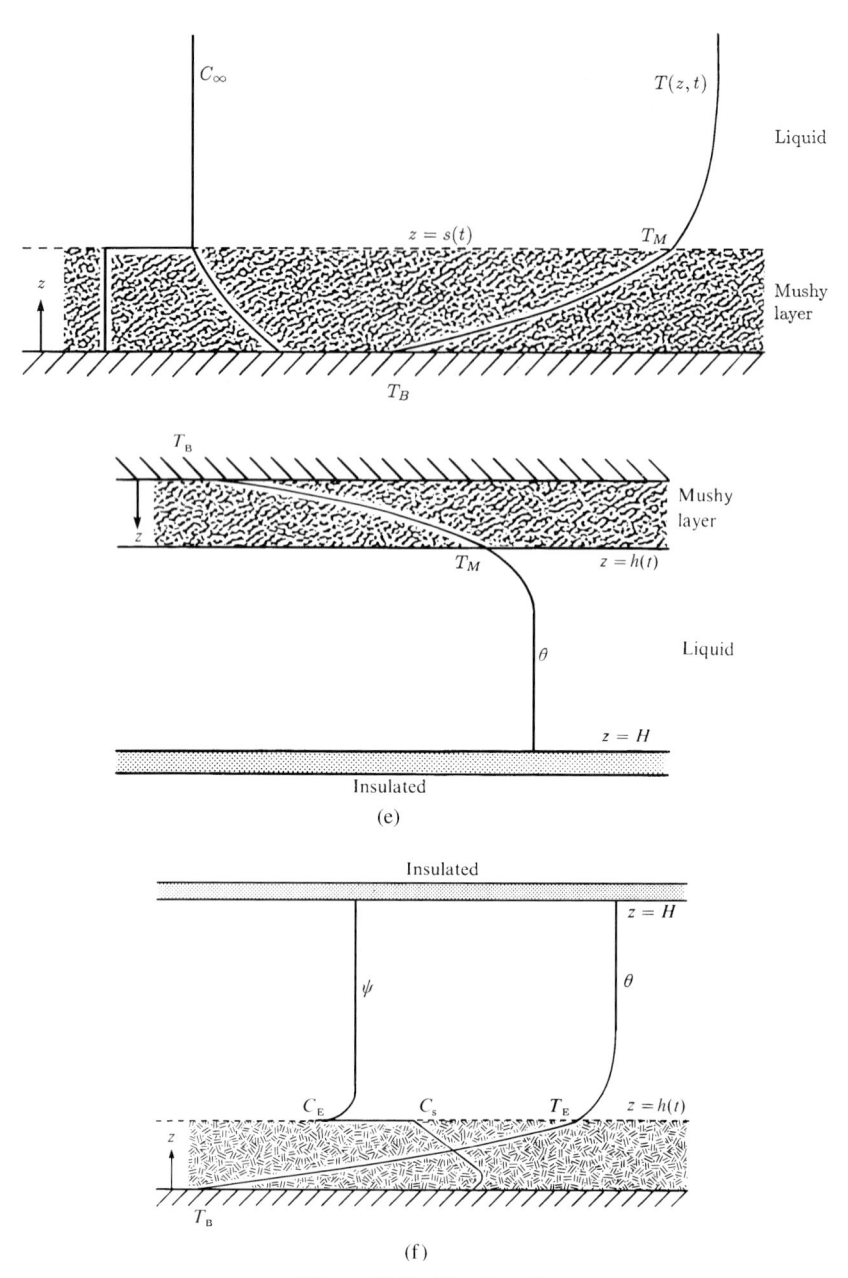

Figure 10.2. (*Continued*)

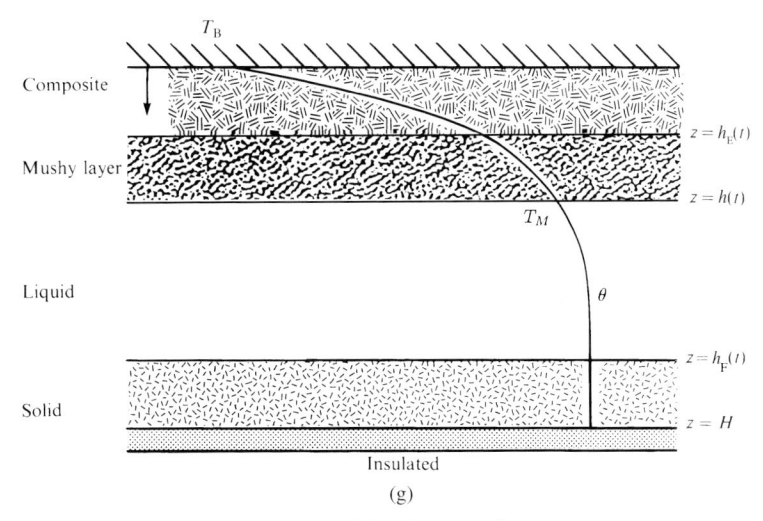

Figure 10.2. (*Continued*)

negligible. This is the mesoscale version of the microscale criterion (9.14) for the neglect of thermal buoyancy.

10.3 Mushy-Zone Models

The hypoeutectic example shown in Figure 10.2(c) involves a planar solid–liquid front susceptible to morphological instability. When $M \gg M_c$, one can assume a planar mush–liquid front and calculate the solution of the appropriate Stefan problem to find the position $h(t)$. Figure 10.3 shows this curve with superposed data for aqueous NaNO$_3$ with $T_B = -17°C$ and $T_\infty = 15°C$. Thus, despite the simplest geometry, the simplest forcings, and the simplest question (where is the average interface?), ignoring the mush gives dreadful results.

Even in the absence of flow, a mush is an entity involving three-dimensional, unsteady diffusion of heat and solute, phase transformation, and complex thermodynamics. One can approach such a problem directly by formulating what is in effect a generalized mixture theory involving general interactions and the concomitant requirement to pose many constitutive relations. This has been done quite nicely by Hills, Loper, and Roberts (1983), who have followed this initial attempt by more refined models and analyses. The strengths of such models are their generality and the ability to eliminate thermodynamically unallowable relations systematically. The weakness is their generality, making

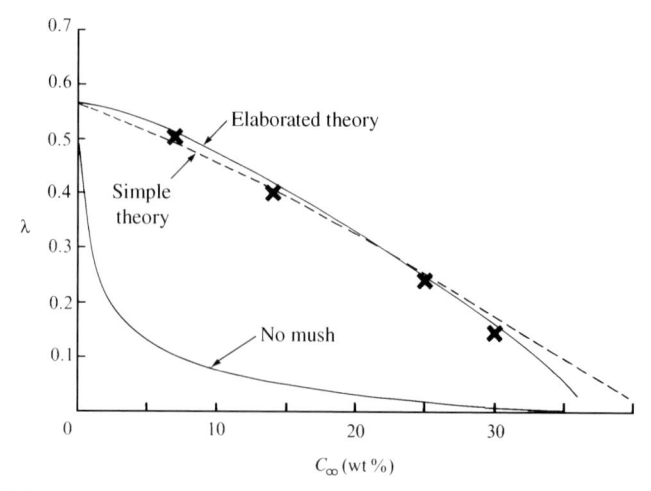

Figure 10.3. The nondimensional coefficients λ for the interface speed as a function of C_∞ for $NH_4C\ell$ solutions, the "simple theory" of Huppert and Worster (1985), and the elaborated theory of Worster (1990). From Huppert (1993). (Reprinted with permission of Elsevier Science.)

the solution of concrete problems quite formidable and dependent on the many functional assumptions required.

An alternative approach is to begin with the simplest of models and make these increasingly general as the need to explain new physical effects arises. It is this second path that will be followed here.

A simple mushy-zone model is that of Huppert and Worster (1985) in which the solid fraction is uniform and the individual dendrites have generators oriented normal to the mean of the interface at $z = h(t)$. In terms of the situation in Figure 10.2(c), there is the usual thermal diffusion in the liquid

$$T_t^\ell = \kappa^\ell T_{zz}^\ell, \quad z > h(t) \tag{10.6a}$$

and an average diffusion in the mush

$$T_t^m = \kappa^m T_{zz}^m, \quad 0 < z < h(t). \tag{10.6b}$$

The heat balance across the interface (between liquid and mush) is

$$L_v \phi \frac{dh}{dt} = k_T^m T_z^m - k_T^\ell T_z^\ell, \tag{10.6c}$$

where the coefficients in the mush are defined by weighted averages,

$$k_T^m = \phi k^s + (1-\phi)k^\ell = c_p^m \kappa^m \tag{10.6d}$$

10.3 Mushy-zone models

and

$$c_p^m = \phi c_p^s + (1 - \phi)c_p^\ell. \tag{10.6e}$$

Here ϕ is the (constant) volume fraction of solid and c_p is the specific heat per unit volume. Huppert and Worster argued that (10.6d) well describes a random array of dendrites whose phase boundares are sensibly vertical and whose solid fraction is uniform; relation (10.6e) is exact.

The model involves the assumption that "all" the rejected solute produced by growth is trapped within the mush and that the boundary layer above the mush accounts for "zero" rejected solute, which is consistent with $D^\ell \ll \kappa^\ell$. As will be seen, this implies that the overall thickness of the mushy zone scales with thermal – rather than solutal – diffusion rates.

A major assumption in these theories is that the mush is in thermodynamic equilibrium because the concentration and the temperature are always confined to lie on the liquidus, which is consistent with the intimate intermingling of solid and liquid and the mush's being described on scales of many dendrites. Thus,

$$C^m = T^m/m, \quad 0 < z < h(t) \tag{10.6f}$$

and

$$C^\ell = C_\infty, \quad z \geq h(t). \tag{10.6g}$$

Equation (10.6g) is consistent with a thin concentration boundary layer.

Finally, there is a global solute balance,

$$(1 - \phi) \int_0^{h(t)} C^m(z, t) dz = h(t) C_\infty. \tag{10.6h}$$

Huppert and Worster solved the Stefan problem for this system in which $h(t) = 2\lambda^m \sqrt{\kappa^\ell t}$. For the conditions of their experiment, their predictions, shown in Figure 10.3 and labelled "simply" theory, are quite good. The success of this model in predicting the depth of the mushy layer with $h(t) \propto \sqrt{\kappa^\ell t}$ establishes that the depth of the mush scales on thermal properties of the fluid even though it is the solute distribution in the boundary layer that produces the initial morphological instability and hence the dendrites.

If all that needed predicting were $h(t)$, one would be finished. However, the microstructure in the resulting solid is also of interest. In this case, one would need to allow $\phi = \phi(\mathbf{x})$, and if one were to consider the approach to steady state, then one would need $\phi = \phi(\mathbf{x}, t)$. Finally, many features of the microstructure are created by instabilities in the mush, and any consideration of stability theory requires a model in which $\phi = \phi(\mathbf{x}, t)$, as well.

One such elaborated theory was given by Worster (1990). He wrote

$$\rho^m = \phi \rho^s + (1-\phi)\rho^\ell, \tag{10.7a}$$

where now ϕ is the (variable) volume fraction, and if $\mathbf{V}$ is the Darcy velocity field (valid over many dendrite spacings), then mass conservation in the liquid in the mush requires that

$$\rho_t^m + \nabla \cdot (\rho^\ell \mathbf{V}) = 0. \tag{10.7b}$$

If one defines again $\rho = \rho^s/\rho^\ell$, then Eq. (10.7b) implies that

$$\nabla \cdot \mathbf{V} = (1-\rho)\phi_t, \tag{10.7c}$$

and thus expansion or contraction upon phase transformation destroys the solenoidal nature of $\mathbf{V}$, as discussed in Section 9.3.

The local average concentration C^m in the mush is given by

$$\rho^m C^m = \rho^s \int_0^\phi C^s(\phi')d\phi' + \rho^\ell(1-\phi)C, \tag{10.7d}$$

and the balance equation for solute takes the form

$$(\rho^m C^m)_t + \nabla \cdot (\rho^\ell C \mathbf{V}) = \rho^\ell \nabla \cdot (D^m \nabla C), \tag{10.7e}$$

where D^m is the local (average) solute diffusivity in the mush. One can simplify equation (10.7e), using Eq. (10.7c), to yield

$$(1-\phi)C_t + \mathbf{V} \cdot \nabla C = \nabla \cdot (D^m \nabla C) + \rho(1-k)C\phi_t. \tag{10.7f}$$

Notice in Eq. (10.7f) that the last term represents a sink of solute driven by rejection.

The temperature field in the mush has an enthalpy $\hat{H}^m$ such that

$$\rho^m \hat{H}^m = \phi \rho^s \hat{H}^s + (1-\phi)\rho^\ell \hat{H}^\ell, \tag{10.7g}$$

and the heat balance has the form

$$(\rho^m \hat{H}^m)_t + \nabla \cdot (\rho^\ell \hat{H}^\ell \mathbf{V}) = \nabla \cdot (k_T^m \nabla T), \tag{10.7h}$$

where k_T^m is the mean thermal conductivity in the mush. One can simplify equation (10.7h) (Worster 1992a), using Eqs. (10.7b,g), giving

$$c_p^m T_t + c_p^\ell \mathbf{V} \cdot \nabla T = \nabla \cdot (k_T^m \nabla T) + \rho_s L \phi_t \tag{10.7i}$$

where

$$c_p^m = \phi c_p^s + (1-\phi)c_p^\ell \tag{10.7j}$$

and in both the solid and liquid c_p is the density times $d\hat{H}/dT$. The latent heat per unit mass L can be identified with $\hat{H}^\ell - \hat{H}^s$ and is generally a function of the dependent variables.

Equations (10.7c,f,i) constitute a coupled system through the following assumption: if the mush obeys the equilibrium relationship, $T^m = T_L(C)$, always lying on the liquidus, then the two fields T and C are required to vary in tandem and in effect are no longer independent functions.

Finally, the flow field is defined in the porous mush by Darcy's equation

$$\mu^\ell \mathbf{V} = \Pi[(\rho^\ell - \rho_0)\mathbf{g} - \nabla p] \qquad (10.7k)$$

where ρ_0 is a reference density and Π is the permeability of the medium. All the mush properties, D^m, k_T^m, and Π, depend on ϕ, and some assumptions need to be made regarding these variations. Typically, one takes $D^m = (1-\phi)D^\ell$, the one-sided model, and $k_T^m = (1-\phi)k_T^\ell + \phi k_T^s$ and some relation is required for $\Pi = \Pi(\phi)$. The permeability Π has units of length squared, where the length scales with the pore diameter (dendrite spacing) of the mush.

Given these bulk equations, one can use "pillbox" arguments to obtain jump conditions across the liquid–mush interface. If $[J] \equiv J^\ell - J^m$, then

$$[\mathbf{V} \cdot \mathbf{n}] = -(1-\rho)v_n[\phi], \qquad (10.7l)$$

$$\rho(1-k)C[\phi]v_n = [D^m \nabla C \cdot \mathbf{n}] \qquad (10.7m)$$

$$\rho^s L_v[\phi]v_n = [k^m \nabla T \cdot \mathbf{n}], \qquad (10.7n)$$

if it is assumed that T is continuous across the boundary. Worster (1986) suggested that the equilibrium relation $T = T_L(C)$ be used at the interface and internal equilibrium applied, in which case

$$[\nabla T \cdot \mathbf{n}] = m[\nabla C \cdot \mathbf{n}]. \qquad (10.7o)$$

Commonly, Eqs. (10.7m–o) imply that $[\phi] = 0$, but not always. In the limit $D/\kappa \to 0$, Eq. (10.7o) implies that C and T are comparable, and thus Eq. (10.7m) becomes on this limit

$$\rho(1-k)C[\phi]v_n = 0,$$

and thus,

$$[\phi] = 0. \qquad (10.7p)$$

Equation (10.7n) then shows that $\nabla T \cdot \mathbf{n}$ is continuous across the interface.

In the limit $D/\kappa \to 0$ then, the governing system is Eqs. (10.7c,f) with solute diffusivity neglected,

$$(1 - \phi)C_t + \mathbf{V} \cdot \nabla C = \rho(1 - k)C\phi_t, \qquad (10.8)$$

and Eqs. (10.7i, k).

This elaborated model should of course be consistent with the simple experiment of Figure 10.2d. Worster (1986) re-solved the Stefan problem using this new formulation. Figure 10.3 shows the position of the front, which predicts the data for NaNO$_3$ about as well as the simple theory.

10.4 Mushy Zones with Volume-Change Convection

The elaborated system includes the possibility that $\rho \neq 1$. For example if $\rho > 1$, so that material shrinks upon solidification, small changes will occur in the system, including an additional "flow from infinity" of magnitude $|1 - \rho|$. In the absence of mush, such effects are normally quite small – perhaps a few percent change in the stability conditions; see Section 9.3. However, in a mush the changes are potentially more important because shrinkage can alter the permeability of the mush and substantially redirect the interstitial flows.

Chiareli, Huppert, and Worster (1994) recalculated the solution to the Stefan problem with ρ arbitrary, using the similarity variable $\eta = z/2\sqrt{\kappa^\ell t}$ and interface position $h(t) = 2\lambda\sqrt{\kappa^\ell t}$, and applied the results to the measurements of Chen and Chen (1991) and Shirtcliffe, Huppert, and Worster (1991). In the case of the latter data on aqueous NaNO$_3$, Figure 10.4 shows that the accounting for density changes can give remarkable quantitative agreement between theory and experiment for the volume-fraction *distribution* as a function of z and t.

Chiareli and Worster have shown that a mushy layer is susceptible to a flow-focusing instability due to shrinkage. A binary alloy undergoes directional solidification at speed V subject to the FTA and one-sided solute diffusion. The speeds are scaled on V, the lengths on *thermal length* $\delta_T = \kappa^\ell/V$, and time on δ_T/V. The pressure is scaled on $\mu V \delta_T / \Pi^*$, where Π^* characterizes the permeability.

The height $z = 0$ is taken to be at the eutectic temperature T_E and concentration C_E; the temperature is T_∞ and the concentration is C_∞ far from $z = 0$. If the liquidus is taken to be a straight line, then, for this hypereutectic case, the liquidus has the form

$$T = T_E + m(C - C_E), \quad m > 0. \qquad (10.9a)$$

At large z, $T(C_\infty) = T_E + m\Delta C$, where

$$\Delta C = C_\infty - C_E \qquad (10.9b)$$

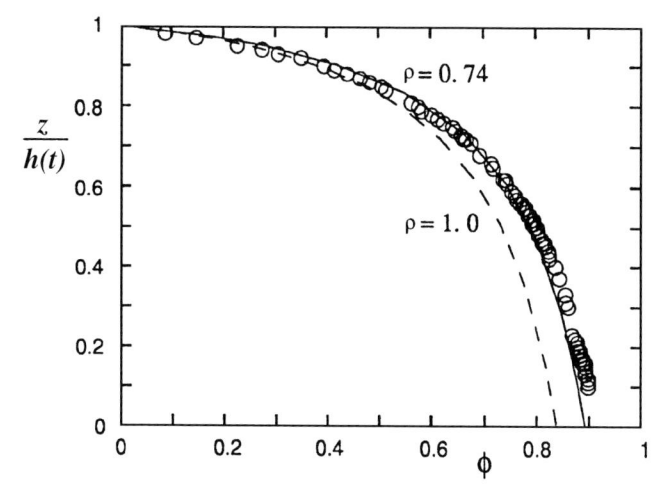

Figure 10.4. The volume fraction of solid ϕ in a mushy layer as a function of the relative depth in the layer, $z/h(t)$, where $h(t)$ is the height of the layer. The circles denote data from experiments of Shirtcliffe, Huppert, and Worster (1991) using aqueous solutions of sodium nitrate as reinterpreted by Chiareli and Worster. The dashed line is the prediction of a model that neglects solidification shrinkage ($\rho = 1$), whereas the solid line takes full account of the interactions between solidification and the velocity field induced by shrinkage ($\rho = 0.74$). From Worster (1992a).

and one can define ΔT to be

$$\Delta T = m\Delta C. \tag{10.9c}$$

One then defines

$$\theta = \frac{T^m - T_E - m\Delta C}{\Delta T} \tag{10.10a}$$

and

$$\Theta = \frac{(C^m - C_\infty)}{\Delta C}. \tag{10.10b}$$

The governing mushy-layer system in a frame moving with speed V then reduces to to the following in the *liquid*:

$$\theta_t - \theta_z + \mathbf{v} \cdot \nabla\theta = \nabla^2\theta \tag{10.11a}$$

$$\Theta_t - \Theta_z + \mathbf{v} \cdot \nabla\Theta = \epsilon\nabla^2\Theta \tag{10.11b}$$

$$Pr^{-1}(\mathbf{v}_t - \mathbf{v}_z + \mathbf{v} \cdot \nabla\mathbf{v}) = \nabla^2\mathbf{v} - \mathcal{H}\nabla p \tag{10.11c}$$

$$\nabla \cdot \mathbf{v} = 0 \tag{10.11d}$$

with the interface conditions on $z = h$,

$$\theta = \Theta \tag{10.11e}$$

$$\nabla \theta \cdot \mathbf{n} = \nabla \Theta \cdot \mathbf{n} \tag{10.11f}$$

$$[\mathbf{V} \cdot \mathbf{n}] = 0 \tag{10.11g}$$

$$\left[\mathbf{V} \cdot \mathbf{t}^i\right] = 0, \quad i = 1, 2 \tag{10.11h}$$

and as $z \to \infty$

$$\Theta \to 0; \quad \theta = \theta_\infty, \quad \mathbf{v} \to (1 - \rho)\mathbf{z}. \tag{10.11i}$$

The nondimensional parameters are the Prandtl number Pr, the permeability ratio

$$\mathcal{H} = \delta_T^2 / \Pi^* \tag{10.12a}$$

the far-field transport θ_∞,

$$\theta_\infty = \frac{T_\infty - T_E}{\Delta T} + 1 \tag{10.12b}$$

and

$$\epsilon = D^\ell / \kappa^\ell. \tag{10.12c}$$

In the mush, the assumption of thermodynamic equilibrium gives in $0 < z < h$,

$$\theta \equiv \Theta \tag{10.13a}$$

$$\theta_t - \theta_z + \mathbf{V} \cdot \nabla \theta = \nabla^2 \theta + S_1 (\phi_t - \phi_z) \tag{10.13b}$$

$$(1 - \phi)(\theta_t - \theta_z) + \mathbf{V} \cdot \nabla \theta = \rho (\theta - \mathcal{C})(\phi_t - \phi_z) \tag{10.13c}$$

$$\nabla \cdot \mathbf{V} = (1 - \rho)(\phi_t - \phi_z) \tag{10.13d}$$

and Darcy's law

$$\mathbf{V} = -\Pi(\phi)\nabla p. \tag{10.13e}$$

Here,

$$S_1 = \frac{\rho^s L}{\rho^\ell c_p \Delta T}, \tag{10.14a}$$

and

$$\mathcal{C} = \frac{C_s - C_E}{\Delta C} - 1, \tag{10.14b}$$

10.4 Mushy zones with volume-change convection

where C_s is the solute concentration in the solid matrix, and $\mathbf{V} = (1-\phi)\mathbf{v}$. Chiareli and Worster chose the empirical law for permeability,

$$\Pi(\phi) = (1-\phi)^q \tag{10.15}$$

where q indicates the sensitivity to changes in permeability.

Finally, at $z = 0$,

$$\theta = -1, \quad \mathbf{V} \cdot \mathbf{z} = (1-\rho)(1-\phi) \tag{10.16}$$

and at $z = h$

$$[\theta] = 0, \quad \phi = 0, \quad [\nabla\theta \cdot \mathbf{n}] = 0, \quad [p] = 0. \tag{10.17}$$

As seen in Section 9.3, the flow produced by density changes is determined by the continuity equation only. The Navier–Stokes equation, not shown, merely determines the pressure field.

A basic state for this system is

$$\bar{\mathbf{v}} = (0,\ 0,\ 1-\rho) \quad \text{in the liquid} \tag{10.18a}$$

$$\bar{\mathbf{V}} = \{0,\ 0,\ (1-\rho)(1-\bar{\phi})\} \quad \text{in the mush.} \tag{10.18b}$$

In the liquid,

$$\bar{\theta} = \theta_\infty + \left(\theta^{\mathrm{I}} - \theta_\infty\right) e^{-\rho(z-\bar{h})} \tag{10.18c}$$

and

$$\bar{\theta} = \theta^{\mathrm{I}} e^{-\rho(z-\bar{h})/\epsilon}. \tag{10.18d}$$

The interfacial temperature is θ^{I},

$$\theta^{\mathrm{I}} = -\frac{\epsilon}{1-\epsilon}\theta_\infty. \tag{10.18e}$$

In the mush,

$$\bar{\phi} = \frac{\bar{\theta}}{C - \bar{\theta}} \tag{10.18f}$$

and so

$$\frac{d\bar{\theta}}{dz} = \rho\theta_\infty - \bar{\theta} - \frac{S_1\bar{\theta}}{C-\bar{\theta}} - (1-\rho)C\ln\left(\frac{C-\bar{\theta}}{C}\right), \tag{10.18g}$$

which determines $\bar{h}$. Finally,

$$\frac{d\bar{p}}{dz} = -(1-\rho)\frac{1-\bar{\phi}}{\Pi(\bar{\phi})}. \tag{10.18h}$$

In the basic state the flow is vertical; in the liquid there are exponential profiles of temperature and concentration, and in the mush the profiles are given by Eqs. (10.18f–h).

Chiareli and Worster (1995) identified a near-eutectic limit that simplifies the dynamics. When $\Delta C \to 0$, then $C_\infty \to C_E$ and one obtains the limiting forms of the Fowler (1985) model generalized to $q \neq 1$.

The basic state suggests considering (for $\Delta C \to 0$)

$$C \to \infty, \quad \theta_\infty \to \infty, \quad C/\theta_\infty = O(1), \quad \hat{q} = q/C \tag{10.19a}$$

and rescaling as follows:

$$\hat{\phi} = C\phi \quad \text{and} \quad \hat{z} = \rho\theta_\infty z. \tag{10.19b}$$

If one considers the linear stability problem in the unscaled variables and uses normal modes $\exp(\sigma t + iax)$, then

$$\hat{a} = \frac{a}{\rho\theta_\infty}, \quad \hat{\sigma} = \frac{\sigma}{\rho\theta_\infty}. \tag{10.19c}$$

The leading-order linear stability problem reduces in the mushy layer to

$$\left(\hat{D}^2 - \hat{a}^2\right)\theta = 0 \tag{10.20a}$$

$$(\hat{\sigma} - \rho\hat{D})\theta + W = \rho\left(\hat{D} - \hat{\sigma}\right)\hat{\phi} \tag{10.20b}$$

$$\left(\hat{D}^2 - \hat{q}\hat{D} - \hat{a}^2\right)W = \hat{a}^2(1-\rho)\hat{p}\hat{\phi}. \tag{10.20c}$$

A special case is considered in which the mush–liquid interface is unperturbed. This decouples the mush from the bulk liquid. The boundary conditions on the mush are then

$$\theta = 0, \quad W = 0 \quad \text{at} \quad \hat{z} = 0. \tag{10.20d}$$

$$\theta = 0, \quad \hat{D}W = 0, \quad \hat{\phi} = 0 \quad \text{at} \quad \hat{z} = 1. \tag{10.20e}$$

In conditions (10.20 d,e), θ, W, and $\hat{\phi}$ represent normal-mode amplitudes, and $\hat{D} = d/d\hat{z}$.

Clearly, Eq. (10.20a) and $\theta = 0$ on $\hat{z} = 0, 1$ imply that $\theta \equiv 0$. The remaining system is solved numerically. The results are shown in Figure 10.5. Only when $\hat{q}$ is sufficiently large is there an instability justifying the rescaling (10.19a). Extreme changes in permeability result in an instability that has the streamlines shown in Figure 10.6 and is suggestive as a precursor to the formation of a channeled mush. Chiareli and Worster (1995) verified that the more general model gives qualitatively the same instability.

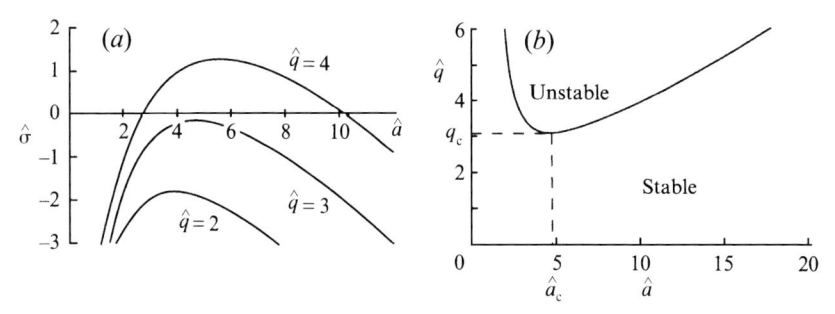

Figure 10.5. (a) Dispersion relation $\hat{\sigma}$ versus $\hat{a}$ for the "near-eutectic" limit. It can be seen that the system is completely stable for small values of $\hat{q}$, whereas, for sufficiently large values of $\hat{q}$, there is a range of unstable wave numbers $\hat{a}$ bounded away from zero and infinity. (b) A marginal stability curve $\hat{q}$ versus $\hat{a}$ showing the critical value $\hat{q}_c$, which is the minimum value of the permeability variation parameter for which the system is unstable. These curves were calculated with $\rho = 0.5$. From Chiareli and Worster (1995).

10.5 Mushy Zones with Buoyancy-Driven Convective Instability

In principle, the rejection of solute and the release of latent heat will allow convective motions in the mush and in the liquid. In order to describe these one must set up equations that describe the two-layer system. For simplicity, consider only the case $\rho = 1$.

In the liquid there are the Navier–Stokes equations, subject to the Boussinesq approximation, and the equation of continuity:

$$\rho_0^\ell(\mathbf{v}_t + \mathbf{v} \cdot \nabla \mathbf{v}) = -\nabla p + \mu \nabla^2 \mathbf{v} + \rho_0^\ell \left[\alpha^*(T - T_0) + \beta^*(C - C_0)\right] \mathbf{g} \quad (10.21\text{a})$$

and

$$\nabla \cdot \mathbf{v} = 0. \quad (10.21\text{b})$$

Further, there are the heat and solute balances:

$$\rho^\ell c_p^\ell (T_t + \mathbf{v} \cdot \nabla T) = k_T^\ell \nabla^2 T \quad (10.21\text{c})$$

and

$$C_t + \mathbf{v} \cdot \nabla C = D \nabla^2 C. \quad (10.21\text{d})$$

In the mush, one must use Eqs. (10.7c,f,i) in which

$$c_p^m = \phi c_p^s + (1 - \phi) c_p^\ell \quad (10.21\text{e})$$

$$k_T^m = \phi k^s + (1 - \phi) k^\ell, \quad (10.21\text{f})$$

and, for the one-sided model

$$D^m = (1 - \phi) D^\ell. \quad (10.21\text{g})$$

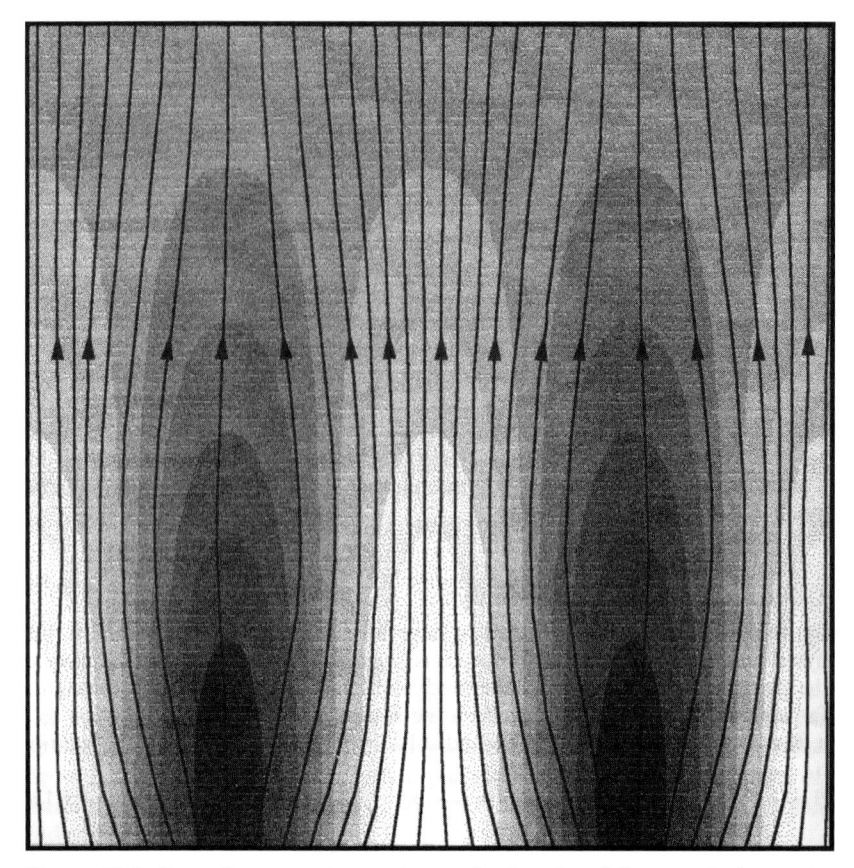

Figure 10.6. Streamlines superimposed on a density plot of the perturbation to the solid fraction of the mushy layer. Light regions correspond to negative perturbations that represent local melting of the dendrites. Darker regions show where solidification is enhanced. The nondimensional parameter values used in the calculation of these marginal eigenfunctions were $q = 42$, $S_1 = C = \theta_\infty = 1$, and $\rho = 0.9$. From Chiareli and Worster (1995).

The Darcy velocity $\mathbf{V}$ is related to the point liquid velocity $\mathbf{v}$ by

$$\mathbf{V} = (1 - \phi)\mathbf{v}. \tag{10.21h}$$

As argued in Section 10.4, the temperature and concentration in the mush are taken to be on the liquidus, for example Eq. (10.9a) (see Figure 10.1 for $C > C_E$). The fluid flow is given by Darcy's law

$$\frac{\mu \mathbf{V}}{\Pi} = -\nabla p + \rho_0^\ell \beta (C - C_0) \mathbf{g}, \tag{10.22}$$

and ρ_0 is a reference value of density. The permeability Π depends on the local solid fraction ϕ, where $d\Pi/d\phi < 0$.

10.5 Mushy zones with buoyancy-driven convective instability

Notice that condition (10.22) implies that, in the mush,

$$\beta = \beta^* + m\alpha^* \tag{10.23}$$

and $\beta > 0$, usually.

Introduce the following scaled quantities. In the mush $z < h(x, y, t)$,

$$\theta = \frac{T - T_\mathrm{L}(C_0)}{\Delta T}, \quad \Theta = \frac{C - C_0}{\Delta C}, \tag{10.24}$$

where $\Delta T = m\Delta C = T_\mathrm{L}(C_0) - T_\mathrm{E}$. Scale speeds on V, lengths on δ_T, time on δ_T/V, and pressure on $\beta \Delta C \rho_0^\ell g \kappa / V$, where V is the constant rate of solidification.

The nondimensional governing system (Worster 1992b) is then as follows: liquid ($z > h$):

$$\theta_t - \theta_z + \mathbf{v} \cdot \nabla\theta = \nabla^2 \theta \tag{10.25a}$$

$$\Theta_t - \Theta_z + \mathbf{v} \cdot \nabla\Theta = \epsilon \nabla^2 \Theta \tag{10.25b}$$

$$Pr^{-1}(\mathbf{v}_t - u_z + \mathbf{v} \cdot \nabla \mathbf{v}) = \nabla^2 u + R^\mathrm{T} \theta \mathbf{k} - R^\mathrm{C}\left[\Theta \mathbf{k} + \frac{\beta}{\beta^*}\nabla p\right] \tag{10.25c}$$

mush ($z < h$):

$$\left(\frac{\partial}{\partial t} - \frac{\partial}{\partial z}\right)\left[\theta + S\frac{C - \xi}{C - \theta}\right] + \mathbf{V} \cdot \nabla\theta = \nabla^2 \theta \tag{10.25d}$$

$$\left(\frac{\partial}{\partial t} - \frac{\partial}{\partial z}\right)\xi + \mathbf{V} \cdot \nabla\theta = 0 \tag{10.25e}$$

$$\frac{\mathbf{V}}{\Pi} = -R_\mathrm{m}[\nabla p + \theta \mathbf{k}] \tag{10.25f}$$

$$\nabla \cdot \mathbf{V} = 0 \tag{10.25g}$$

Interface ($z = h$):

$$\theta = \Theta \tag{10.25h}$$

$$\mathbf{n} \cdot \nabla\theta = \mathbf{n} \cdot \nabla\Theta \tag{10.25i}$$

$$[\mathbf{n} \cdot \mathbf{V}] = 0 \tag{10.25j}$$

$$[\mathbf{t} \cdot \mathbf{V}] = 0 \tag{10.25k}$$

Infinity ($z \to \infty$):

$$\theta = \theta_\infty \tag{10.25l}$$

$$\Theta \to 0 \tag{10.25m}$$

$$\mathbf{V} \to 0. \tag{10.25n}$$

In the preceding, the velocity is written in terms of $\mathbf{V} = (1 - \phi)\mathbf{v}$, and thus in the liquid region $\phi = 0$ and $\mathbf{V} = \mathbf{v}$.

Here

$$\begin{aligned}
R^{\mathrm{T}} &= \alpha^* \Delta T g \delta_T^3 / \kappa \nu \\
R^{\mathrm{C}} &= \beta^* \Delta C g \delta_T^3 / \kappa \nu \\
R_m &= \beta \Delta C g \Pi^* \delta_T / \kappa \nu \\
S &= L_\nu / c_p \Delta T \\
\mathcal{C} &= (C^s - C_0) / \Delta C \\
\epsilon &= D / \kappa \\
\xi &= (1 - \phi)\Theta + \phi \mathcal{C}
\end{aligned} \tag{10.26}$$

In the coordinate system moving with the interface, there is a static basic state. In the liquid,

$$\bar{\theta} = \theta_\infty + \left(\theta^{\mathrm{I}} - \theta_\infty\right) e^{-(z-\bar{h})} \tag{10.27a}$$

$$\bar{\theta} = \theta^{\mathrm{I}} e^{-(z-\bar{h})/\epsilon} \tag{10.27b}$$

with

$$\theta^{\mathrm{I}} = -\frac{\epsilon}{1-\epsilon} \theta_\infty. \tag{10.27c}$$

In the mush,

$$z = \frac{\alpha - \mathcal{C}}{\alpha - \beta} \ln\left(\frac{\alpha + 1}{\alpha - \bar{\theta}}\right) + \frac{\mathcal{C} - \beta}{\alpha - \beta} \ln\left(\frac{\beta + 1}{\beta - \bar{\theta}}\right), \tag{10.27d}$$

where

$$\alpha = A + B, \ \beta = A - B, \ A = 1/2\,(\mathcal{C} + \theta_\infty + S), \ B^2 = A^2 - \mathcal{C}\theta_\infty - S\theta^{\mathrm{I}} \tag{10.27e}$$

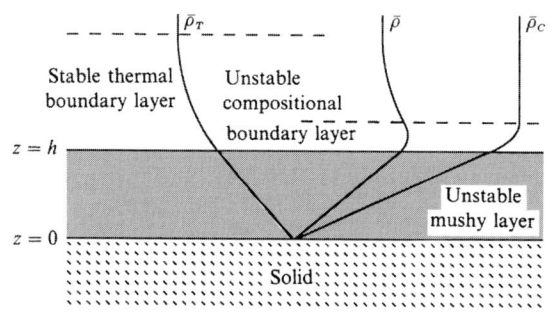

10.5 Mushy zones with buoyancy-driven convective instability

Figure 10.7. The density distributions caused by the temperature field $\bar{\rho}_T$, the solute field $\bar{\rho}_C$, and the total density of the liquid $\bar{\rho}$. The density field in the mushy layer is statically unstable in a gravitational field directed vertically downwards, as is the density in the compositional boundary layer just above the mush–liquid interface. From Worster (1991).

and

$$1 - \bar{\phi} = \frac{C - \theta^I}{C - \bar{\theta}}, \qquad (10.27\text{f})$$

The parameter $\bar{h}$ is defined from Eq. (10.27d) with $\bar{\theta} = 0$.

Figure 10.7 shows the density distribution of the basic state (10.27). The density $\bar{\rho}$ is composed of thermal, $\bar{\rho}_T$, and solutal, $\bar{\rho}_c$, components. As expected, the system is thermally stably stratified, $d\bar{\rho}_T/dz < 0$ everywhere. The system is solutally unstably stratified in the mush and in the solutal boundary layer at the interface; it is neutrally stable above this. The composite distribution is gravitationally unstable in both the mush and the boundary layer; hence, in principle there should be two modes of instability for the system (Worster 1991).

Further, this theory delivers a prediction of the solid mass fraction $\bar{\phi}$ as a function of z. Figure 10.8 shows this for various values of the concentration ratio C. When C is large, $\bar{\phi}$ is linear in z and uniformly small (Figure 10.8(a)). When C is small, then $\bar{\phi}$ is unity everywhere except near the interface. When C is arbitrary but θ_∞ is small, then a relatively large region near the interface has $\bar{\phi}$ small, and Worster (1991) calls the appearance of the mush "feathery." He attributes this prediction to the loss of thermal control, in which case the present model becomes inaccurate and other physical effects become unimportant.

Worster (1992b) considered the linear instability of solution (10.27) using normal modes $\exp(\sigma t + iax)$ and found a neutral stability curve of R_m versus a, as shown in Figure 10.9. This bimodal curve shows the existence of two "independent" instabilities, one scaling on the depth of the mush and the other on the thickness of the solutal boundary layer. The eigenfunctions, shown in

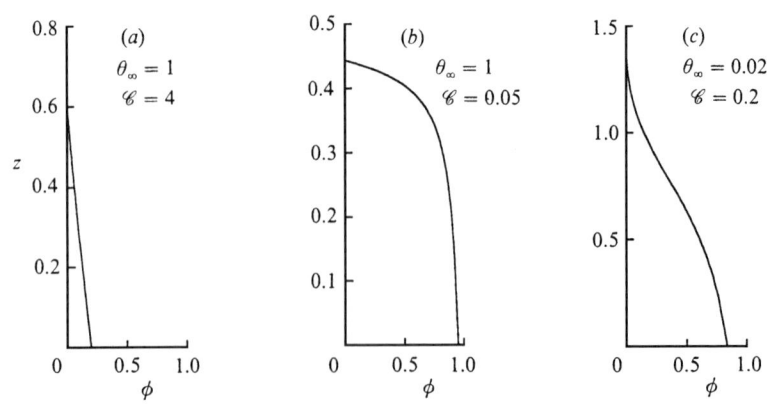

Figure 10.8. Various profiles of solid fraction ϕ as a function of height z in a mushy layer for different values of the concentration ratio $\mathcal{C}$ for $S = 1$. (a) When $\mathcal{C}$ is large, the solid fraction is small throughout the layer. (b) When $\mathcal{C}$ is small, ϕ can be almost equal to unity through much of the layer. (c) For any value of $\mathcal{C}$, letting $\theta_0 \to 0$ produces a "feathery" top to the layer, where the solid is very small. From Worster (1991).

Figure 10.10, are consistent with this view. Figure 10.10(a) shows the mushy-layer mode in which fluid is transported on the large scale linking the pure liquid with the mush. Figure 10.10(b) shows the boundary-layer mode of much smaller scale barely penetrating the mush. The modes are analogous to those first found by Chen and Chen (1988) for a two-layer system of pure material with unstable thermal stratification.

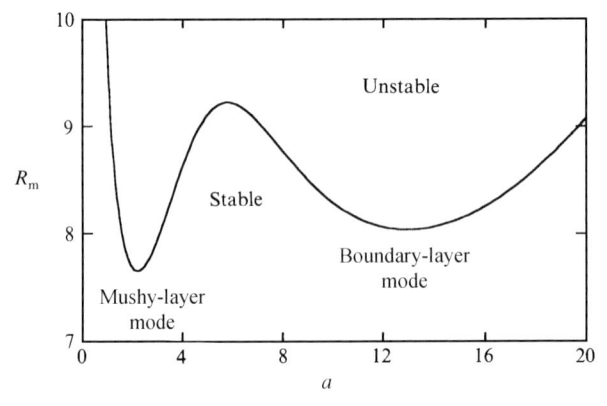

Figure 10.9. The marginal stability curve for the onset of compositional convection. The two minima correspond to distinct modes of convection. This graph was calculated for the diffusivity ratio $D/\kappa = 0.025$ and the ratio $\delta_T^2/\Pi^* = 10^5$. Either mode can be the more unstable, depending on the values of these parameters. From Worster (1992b).

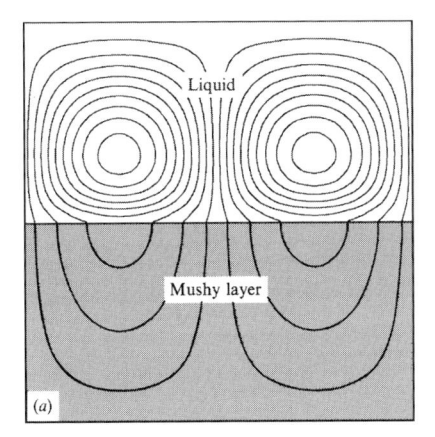

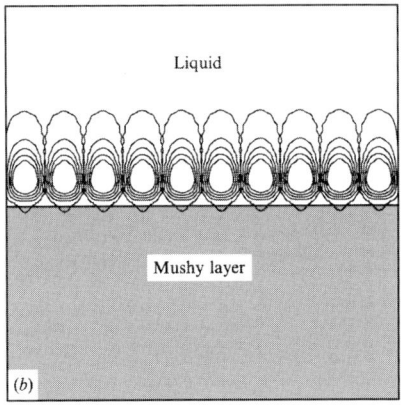

10.5 Mushy zones with buoyancy-driven convective instability

Figure 10.10. Streamlines for the marginally stable modes corresponding to the two minima in Figure 10.9. (a) The mushy-layer mode has a wavelength comparable to the depth of the mushy layer, and the flow penetrates the layer. (b) The boundary-layer mode has a wavelength comparable to the thickness of the compositional boundary layer above the mush–liquid interface. The flow in this mode barely influences the underlying mushy layer. From Worster (1992b).

Worster introduced new scales appropriate to the mush by writing $\mathcal{H} = \delta_T^2/\Pi^*$ and $\mathcal{A} = ma^*/\beta$ in place of $R^{\mathrm{T}} = \mathcal{A}\mathcal{H}R_m$, and $R^{\mathrm{C}} = (1 + \mathcal{A})\mathcal{H}R_m$, and thus the characteristic equation has the form $F(\sigma, a; R_m, \epsilon, \mathcal{H}, \mathcal{A}, Pr, S, \mathcal{C}, \theta_\infty, \Pi) = 0$, and on the neutral curve, $\sigma = 0, R_m = R_m(\sigma; a)$, with all other parameters fixed. When $\mathcal{A} = 0$, thermal buoyancy is negligible and double-diffusive effects are absent. For the moment take $\mathcal{A} = 0$ and let the permeability be constant, $\Pi \equiv 1$, that is $q = 0$. Normally $\epsilon \equiv 10^{-2}$ and $\mathcal{H}$ becomes a crucial parameter because it measures the square of the ratio of the boundary-layer thickness δ_T to an appropriate length scale based on Π^* in the mush. One expects that when $\mathcal{H}$ is small enough, the Rayleigh numbers in the boundary layer are relatively small and the mushy-layer mode will be destabilized first as R_m increases; see Figure 10.9. When $\mathcal{H}$ is large enough, the boundary-layer mode will be destabilized first. The parameter ϵ measures the ratio of the solutal boundary layer thickness by the thickness of the mush. As $\epsilon \to 0$, the boundary-layer mode becomes more stable, whereas the mushy-layer mode barely changes (Chen and Chen 1988). As the Prandtl number Pr decreases to values appropriate to metals, $Pr \ll 1$, the basic state is stabilized, as shown in Figure 10.11.

The boundary-layer mode has been studied by Hurle, Jakeman, and Wheeler (1983) for $\mathcal{A} = 0$, and if the boundary-layer scales are used (Worster 1992b), $R_{\delta c} = \epsilon^3 \mathcal{H} R_m$, a critical value 10–15 is obtained.

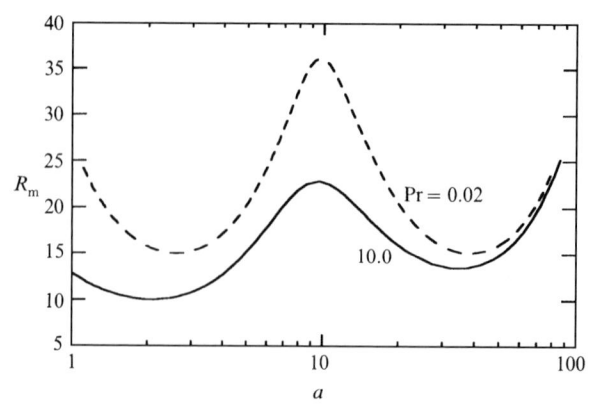

Figure 10.11. Marginal stability curves for two different values of the Prandtl number Pr, with $\epsilon = 10^{-2}$ and $\mathcal{H} = 10^6$; $Pr = 10$ is an appropriate magnitude for many aqueous solutions, whereas $Pr = 0.02$ is appropriate for metallic alloy. From Worster (1992b).

The properties of the mushy-layer mode can be studied for $\epsilon \to 0$, but now with a permeability Π depending on ϕ. Worster (1992b) used

$$\Pi = (1 - \phi)^3 \tag{10.28}$$

and varied θ_∞, S, and $\mathcal{C}$.

As θ_∞ increases, the mush thickness decreases, the mush becomes more stable, and the critical wave number a_c increases. As S increases, the mush decreases in thickness but becomes more unstable, and a_c increases. Both of these effects are relatively small.

When $\mathcal{C}$ increases, the mush increases in thickness and the permeability increases. Thus, the mush is strongly destabilized; R_{m_c} decreased by 10^3 as $\mathcal{C}$ increased from 10^{-1} to 10; a_c decreased by a factor of 6 or so in this range. Figure 10.8 shows how $\bar{\phi}$ of the basic state changes with $\mathcal{C}$. The resulting eigenfunctions in the stability problem show that the convective flow penetrates fully in Case (a) (Figure 10.8) and only within the variable $\bar{\phi}$ region in Case (b) (Figure 10.8).

Experiments (Hellawell 1987, Huppert 1990, Chen and Chen 1991) with aqueous solutions show various forms of convection. In some cases small-scale convection is seen near the interfaces (the boundary-layer mode?). In other cases, such as ammonium chloride, this convection is present but is superseded by large-scale convection (the mushy-layer mode?) that couples the mush to the pure liquid. Experiments by Tait and Jaupert (1992) on aqueous solutions of ammonium chloride seem reasonably well explained by the present theory, as seen in Figure 10.12. The critical condition has been given further support from the experiments of Wettlaufer, Worster, and Huppert (1997).

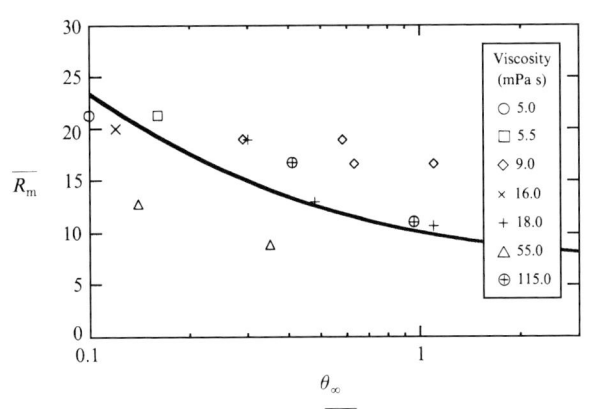

Figure 10.12. The critical Rayleigh number $\overline{R_m}$ as a function of the superheat θ_z for parameter values $S = 5$ and $C = 20$. The solid line shows the prediction of the linear-stability theory; the data are from experiments by Tait and Jaupart (1992). Different symbols are used to distinguish the different viscosities of the melt utilized in the experiments. From Worster (1992b).

It is sometimes useful to examine the mush unconnected to the liquid above it and thus address a one-layer problem. Consider a *planar* boundary S between a liquid and a mushy layer. A typical "pillbox" argument, using the fact that $[\mathbf{V}] = 0$, suggests that the normal stress is continuous across S. On the liquid side there is pressure and viscous normal stress τ_N. The latter is systematically smaller than the pressure, and thus to good approximation $[p] = 0$ across S. To see this, in the mush $\Pi p/L \sim \mu |\mathbf{V}|$, where L is a macroscopic dimension. In the liquid region $\tau_N \sim \mu |\mathbf{V}| \ell^{-1}$, where ℓ measures a shear length normal to S, say $\ell = \sqrt{\Pi}$ and thus ℓ is comparable to a pore width. Then $p/\tau_N \sim L/\sqrt{\Pi}$, which is huge. Thus, only the pressure "survives" (Worster 1999 private communication). In addition to

$$[p] = 0 \tag{10.29a}$$

the no-slip condition

$$[\mathbf{V} \cdot \mathbf{t}] = 0 \tag{10.29b}$$

holds. This is a special case of the Beavers and Joseph (1967) relation with the slip coefficient set to zero. This seems to give results here that fit experiment reasonably well.

10.6 An Oscillatory Mode of Convective Instability

The stability results of Section 10.5 show that the boundary-layer mode is only weakly dependent on the presence of the mush and that the mushy-layer mode penetrates deeply into the liquid.

When δ_T is much larger than the average spacing between the dendrites, Emms and Fowler (1994) showed that the interfacial condition $[p] = 0$ reduces to p-is-constant for the mush. In the present analysis a model system is used in which the liquid–mush interface is a fixed plane and p-is-constant is replaced with $W = 0$. This model, introduced by Amberg and Homsy (1993), permits analysis of the dynamics within the mush, free from coupling with the liquid.

This model further restricts the mush thickness, now called d, by letting $\delta = d/\delta_T \ll 1$. Anderson and Worster (1995) equated the heat fluxes at the interface $\Delta T/d \sim [T_\infty - T_L(C_0)]/\delta_T$ and found that

$$\theta_\infty \equiv \frac{T_\infty - T_L(C_0)}{\Delta T} \sim \frac{1}{\delta}. \tag{10.30a}$$

Amberg and Homsy (1993) likewise showed that, when δ is small, $\mathcal{C}$ is large, namely,

$$\mathcal{C} = \frac{\bar{\mathcal{C}}}{\delta} \tag{10.30b}$$

where $\bar{\mathcal{C}} = O(1)$ as $\delta \to 0$, which corresponds to the near-eutectic approximation used by Fowler (1985).

Emms and Fowler (1994) further assumed that the Stefan number is large, that is,

$$S = \frac{\bar{S}}{\delta}, \tag{10.30c}$$

where $\bar{S} = O(1)$ as $\delta \to 0$.

Through this model, a term in Eq. (10.25d) can be simplified,

$$\frac{\mathcal{C} - \xi}{\mathcal{C} - \theta} = \frac{\mathcal{C} - (1-\phi)\theta - \mathcal{C}\phi}{\mathcal{C} - \theta} = 1 - \phi, \tag{10.31}$$

and in Eq. (10.27d), the constant, unity, is annihilated by the derivatives. The model system related to Eq. (10.28) then becomes

$$\left(\frac{\partial}{\partial t} - \frac{\partial}{\partial z}\right)(\theta - S\phi) + \mathbf{V} \cdot \nabla\theta = \nabla^2\theta \tag{10.32a}$$

$$\left(\frac{\partial}{\partial t} - \frac{\partial}{\partial z}\right)\{(1-\phi)\theta + \mathcal{C}\phi\} + \mathbf{V} \cdot \nabla\theta = 0 \tag{10.32b}$$

$$\Pi^{-1}(\phi)\mathbf{V} = -\nabla p - R_m \theta \mathbf{k} \tag{10.32c}$$

$$\nabla \cdot \mathbf{V} = 0 \tag{10.32d}$$

10.6 An oscillatory mode of convective instability

with the boundary conditions

$$\theta = -1, \quad W = 0 \qquad \text{on} \quad z = 0 \qquad (10.33\text{a,b})$$

$$\theta = 0, \quad W = 0, \quad \phi = 0 \qquad \text{on} \quad z = \delta. \qquad (10.33\text{c,d,e})$$

Introduce the rescalings

$$(X, Y, Z) = \frac{(x, y, z)}{\delta}, \quad T = t/\delta^2, \quad R^2 = \delta R_m \qquad (10.34)$$

$$p_1 = p/R, \quad \mathbf{V}_1 = \frac{\delta}{R}\mathbf{V}$$

as suggested by Amberg and Homsy (1993). The subscript unity is now dropped. The basic state is steady and horizontally uniform and satisfies

$$-\delta \left[\bar{\theta} - \frac{\bar{S}}{\delta}\bar{\phi} \right]' = \bar{\theta}'' \qquad (10.35\text{a})$$

$$-\delta \left[(1-\bar{\phi})\bar{\theta} + \frac{\bar{C}}{\delta}\bar{\phi} \right]' = 0 \qquad (10.35\text{b})$$

$$-\bar{p}' - R\bar{\theta} = 0 \qquad (10.35\text{c})$$

with

$$\bar{\theta} = -1 \quad \text{on} \quad Z = 0 \qquad (10.36\text{a})$$

and

$$\bar{\theta} = \bar{\phi} = 0 \quad \text{on} \quad Z = 1, \qquad (10.36\text{b,c})$$

where a prime denotes d/dZ.

Approximate solutions to systems (10.35) and (10.36) can be obtained in powers of δ, namely.

$$\bar{\theta} = Z - 1 - \frac{1}{2}\delta\Omega\left(Z^2 - Z\right) + \cdots \qquad (10.37\text{a})$$

$$\bar{\phi} = \delta\tilde{\phi} = -\delta\frac{Z-1}{\bar{C}} + \delta^2 \left[-\frac{(Z-1)^2}{\bar{C}^2} + \frac{\Omega}{2\bar{C}}(Z^2 - Z) \right] + \cdots, \qquad (10.37\text{b})$$

where

$$\Omega = 1 + \bar{S}/\bar{C}. \qquad (10.37\text{c})$$

Note that $\bar{\phi}$ is small, and $\bar{\theta}$ is linear to leading order, which is directly related to convection in a passive porous layer as considered by Palm, Weber, and

Kvernvold (1972). It is the higher-order term, $O(\delta)$ that models the "mushiness." Anderson and Worster (1996) examined the linear instabilities of the basic state using normal modes $\exp(\sigma T + iaX)$. They wrote

$$\begin{aligned}
\Pi^{-1}(\delta\tilde{\phi}) &= 1 + \delta\tilde{\phi}_0 K_1 + \cdots \\
\sigma &= \sigma_0 + \delta\sigma_1 + \cdots \\
R &= R_0 + \delta R_1 + \cdots \\
\hat{\theta} &= \theta_0 + \delta\theta_1 + \cdots \\
\hat{\phi} &= \phi_0 + \delta\phi_1 + \cdots \\
\hat{\mathbf{V}} &= \mathbf{V}_0 + \delta\mathbf{V}_1 + \cdots,
\end{aligned} \tag{10.38}$$

where carets denote normal-mode amplitudes, and from Eq. (10.37b)

$$\tilde{\phi}_0 = (1 - Z)/\bar{C}. \tag{10.39}$$

The constant K_1 characterizes variations in permeability with solid fraction:

At $O(\delta)$ an orthogonality condition determines $\sigma_1 = \sigma_{1R} + i\omega_1$ and R_1. When conditions are neutral, $\sigma_{1R} = 0$, the orthogonality condition has a real part

$$\frac{R_1}{R_0} = \frac{1}{4}\frac{K_1}{\bar{C}} + \frac{\bar{S}}{\bar{C}^2\Omega}\left[\frac{1}{4} + \frac{\pi^2(1 + \cos\omega_1)}{\left(\pi^2 - \omega_1^2\right)^2}\right] \tag{10.40a}$$

and an imaginary part

$$\omega_1\left\{1 + \frac{\bar{S}}{\bar{C}^2\Omega^2}\frac{\pi^2 + a^2}{\pi^2 - \omega_1^2}\left[1 - \frac{2\pi^2}{\pi^2 - \omega_1^2}\frac{\sin\omega_1}{\omega_1}\right]\right\} = 0 \tag{10.40b}$$

There is the steady mode with $\omega_1 = 0$ and

$$R = R_0\left[1 + \delta\frac{R_1}{R_0}\right] + O(\delta^2)$$

$$= \frac{a^2 + \pi^2}{a\Omega^{1/2}}\left\{1 + \delta\left[\frac{K_1}{4\bar{C}} + \left(\frac{1}{4} + \frac{2}{\pi^2}\right)\frac{\bar{S}}{\bar{C}^2\Omega}\right]\right\} + O(\delta^2), \tag{10.41}$$

where $R_0^2 = (a^2 + \pi^2)^2/\Omega a^2$ has been used. Result (10.41) agrees with that of Amberg and Homsy (1993) when proper regard is taken of the different scalings as well as the results in Section 10.5.

When conditions allow oscillations, $\omega_1 \neq 0$, (Anderson and Worster 1996). Figure 10.13 shows ω_1 versus a/π for fixed values of $\bar{S}/\bar{C}^2\Omega^2 = S/(C+S)^2\delta$. When $\bar{S}/\bar{C}^2\Omega^2 < 1$, the oscillatory mode exists only for a large enough. When $\bar{S}/\bar{C}^2\Omega^2 \geq 1$, it exists for all $a \geq 0$.

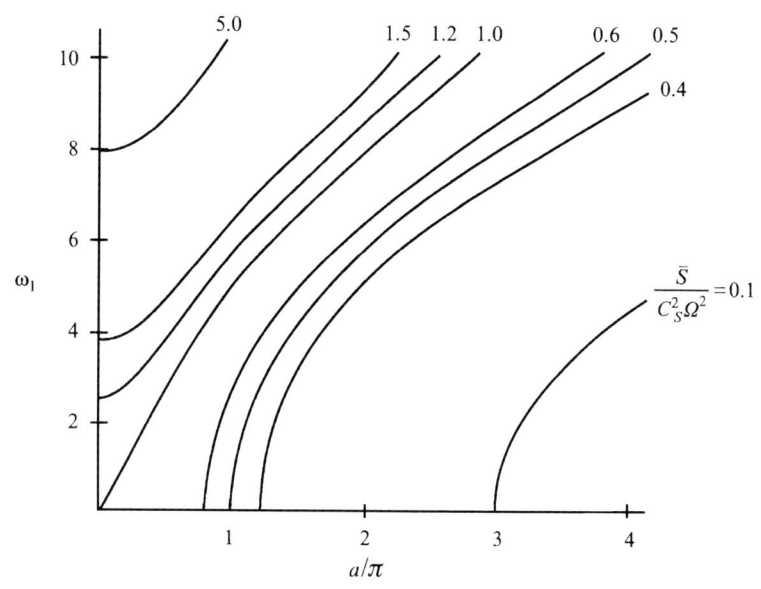

10.6 An oscillatory mode of convective instability

Figure 10.13. The angular frequency ω_1 is shown as a function of the perturbation wave number a for different values of the parameter combination $\bar{S}/(\bar{C}^2\Omega^2)$. Nonzero values of ω_1 indicate the presence of an oscillatory instability. For $\bar{S}/(\bar{C}^2\Omega^2) < 1$, nonzero ω_1 exist only beyond a nonzero wave number whose value depends on $\bar{S}/(\bar{C}^2\Omega^2)$. For $\bar{S}/(\bar{C}^2\Omega^2) > 1$, nonzero ω_1 exist for all wave numbers, including zero wave number. The first appearance of an oscillatory instability at wave number π occurs for $\bar{S}/(\bar{C}^2\Omega^2) = 0.5$. From Anderson and Worster (1996).

In general, when $\sigma_{1R} \neq 0$, one can calculate growth rates and neutral curves. Figure 10.14(a) shows the steady mode (solid), the oscillatory mode (dashed), and the curve separating oscillatory from monotone behavior (dotted curve). When $\bar{S}/\bar{C}^2\Omega^2 < 1$, Figure 10.14(a) shows that the steady mode occurs first as R increases from zero. As $\bar{S}/\bar{C}^2\Omega^2$ increases, the intersection point moves around the steady branch and ultimately completely bounds it from below.

Anderson and Worster (1996) also study how R_c varies with parameters. It decreases with S and increases with C and δ. Finally, they examined the eigenfunctions of the modes to see what signatures such modes would leave behind in the solid. A traveling mode is shown in Figure 10.15.

The oscillatory mode discussed is intrinsic to the mush, for only a one-layer model has been considered. It is intrinsic to the interaction of melting–freezing process with convection. Because the thermal buoyancy has been neglected, double-diffusive effects are absent. One should note that double-diffusive effects can give rise to oscillations in the mush through a mechanism different from the one discussed here (Chen, Lu, and Yang 1994).

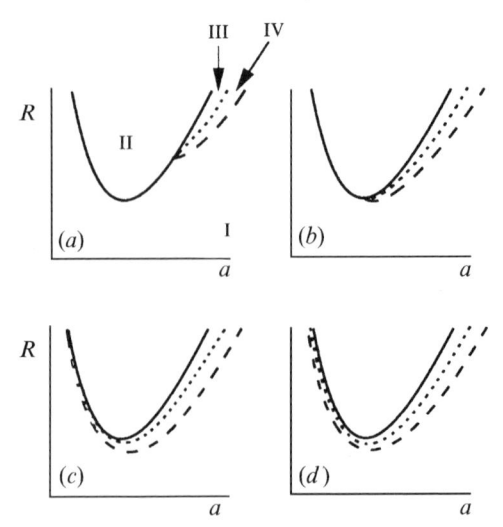

Figure 10.14. The neutral stability curves for the real mode (solid curve), the oscillatory mode (long-dashed curve), and the transition boundary marking the transition between real and imaginary growth rates (short-dashed curve) for representative values (a) $\bar{S}/(\bar{C}^2\Omega^2) < 0.5$, (b) $\bar{S}/(\bar{C}^2\Omega^2) = 0.5$, (c) $\bar{S}/(\bar{C}^2\Omega^2) = 1$, and (d) $\bar{S}/(\bar{C}^2\Omega^2) > 1$. The domains marked in (a) indicate the regions where (I) no growing modes exist, (II) a single growing real mode exists, (III) two growing real modes exist, and (IV) two growing oscillatory modes exist. As the value of $\bar{S}/(\bar{C}^2\Omega^2)$ increases, the point at which the oscillatory branch attaches to the real branch moves in the direction of smaller wave numbers. In (c) the real and oscillatory branches intersect at $a = 0$, and in (d) they do not intersect. From Anderson and Worster (1996).

Figure 10.15. Neutrally stable eigenfunctions. The mushy layer advances with speed V, and the structure of the solutions for three different values of the frequency ω_1 are illustrated in a laboratory frame of reference. (a) $\omega_1 = 0$, which corresponds to the real mode of convection. Here the solid-fraction channels in the mushy layer and the compositional stripes in the solid are vertically oriented. (b) Typical results for the oscillatory mode of convection when $0 < \omega_1 < 3\pi$. At any instant in time, the convection rolls and thermal fields are, to leading order, vertically oriented. However, as a result of the interaction between the solid fraction and the thermal and flow fields, the entire pattern translates horizontally as the mushy layer advances. In this case, the solid fraction channels have a slope that increases monotonically from $-k/\omega_1$ at the mush–solid interface to $-3k/\omega_1$ at the mush–liquid interface. The slope of the compositional stripes in the solid, $-k/\omega_1$, is determined by the vertical growth velocity of the mushy layer and the horizontal translation speed of the pattern. (c) Typical results for the oscillatory mode of convection when $5\pi < \omega_1 < 7\pi$. The notable difference between this case and that in (b) is that the slope of the solidfraction channels in the mushy layer now is nonmonotonic. The values of the slope at the bottom and top of the mushy layer are still $-k/\omega_1$ and $-3k/\omega_1$, respectively. The slope of the compositional stripes in the solid is again $-k/\omega_1$. From Anderson and Worster (1996).

(a)

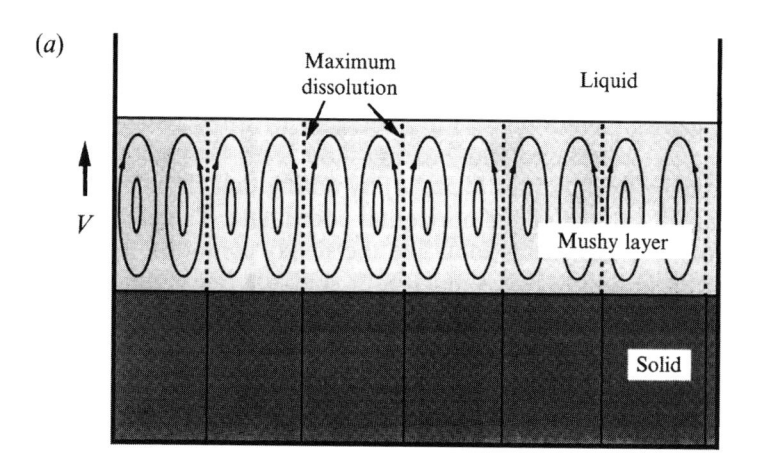

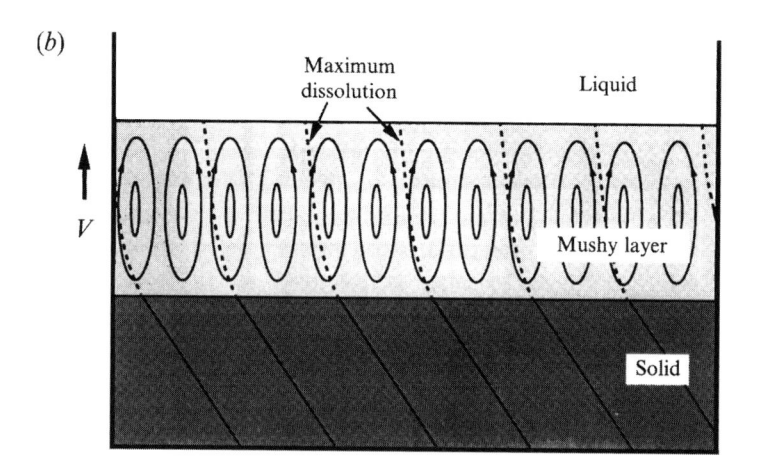

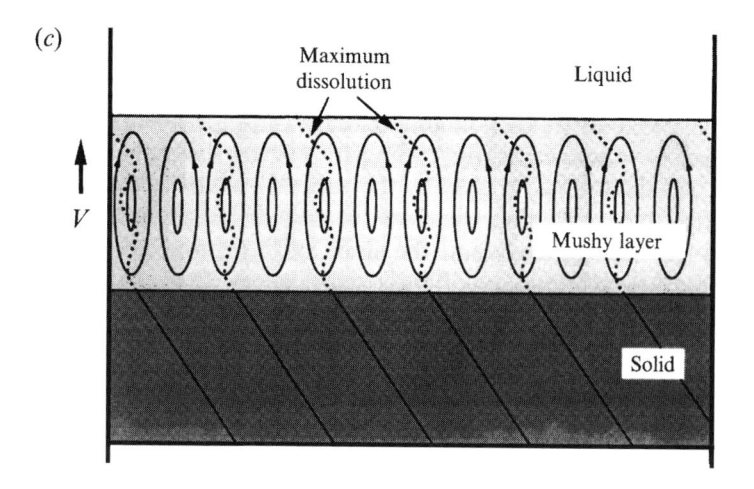

(b)

(c)

10.7 Weakly Nonlinear Convection

In Section 10.6 a model appropriate to thin mushy layers, $\delta \to 0$, was examined, particularly for $\mathcal{C} \sim \delta^{-1}$, $\mathcal{S} \sim \delta^{-1}$ as $\delta \to 0$. The governing system for this model is Eq. (10.32) with rescaling (10.34) for X, Y, Z, T, and R, and this will be used to examine pattern selection for the steady mode in the one-layer case.

Anderson and Worster (1995) used a double expansion in δ and ϵ, the disturbance amplitude for the limiting case $0 \leq \epsilon \ll \delta \ll 1$. The basic state is given by the analysis of Section 10.6 and is of the form $(\theta, \phi, \mathbf{V}, p) = [\bar{\theta}(z), \bar{\phi}(z), \mathbf{0}, R\bar{p}(z)]$ where each element depends on δ. Because $\bar{\phi} \sim \delta$, one can represent the permeability as follows:

$$\Pi^{-1}(\phi) = 1 + K_1\phi + K_2\phi^2 + K_3\phi^3 + \cdots \tag{10.42}$$

with $K_1 > 0$ and $K_2, K_3, \ldots$ characterizing variations in permeability with solid fraction.

The disturbed fields are written

$$\begin{aligned}
\theta &= \bar{\theta} + \epsilon\hat{\theta}(X, Y, Z, T) \\
\phi &= \bar{\phi} + \epsilon\hat{\phi}(X, Y, Z, T) \\
\mathbf{V} &= 0 + \frac{\epsilon R}{\delta}\hat{\mathbf{V}}(X, Y, Z, T) \\
p &= R\bar{p} + \epsilon R\hat{p}(X, Y, Z, T)
\end{aligned} \tag{10.43}$$

Note that the asymptotic representation for $\mathbf{V} \sim \epsilon/\delta$ is well defined. At $O(1)$ in ϵ, the basic state is recovered. At $O(\epsilon)$, the linear theory of Section 10.6 is recovered. At $O(\epsilon^2)$, one obtains an integrable system, and at $O(\epsilon^3)$, an orthogonality condition that produces a set of amplitude equations is enforced.

Anderson and Worster (1995) investigated the competition between rolls and hexagons, for $a = a_c$, where the linear theory eigenfunctions are proportional to $\eta(X, Y)$,

$$\eta = \sum_{j=1}^{3} A_j(\tau)e^{i\mathbf{a}_j \cdot \mathbf{r}} + A_j^*(\tau)e^{-i\mathbf{a}_j \cdot \mathbf{r}} \tag{10.44}$$

where $\tau = \epsilon^2 T$ and

$$\mathbf{a}_j = \left\{ (0, \pi), \left(\frac{1}{2}\pi\sqrt{3}, \frac{1}{2}\pi\right), \left(\frac{1}{2}\pi\sqrt{3}, -\frac{1}{2}\pi\right) \right\}, \quad \mathbf{r} = (X, Y).$$

This gives rise to amplitude equations of the form (4.18). Most important, the coefficient b_0 of the quadratic terms determines the character of the hexagonal pattern that is stable. Here, in unscaled variables

$$-\epsilon b_0 = \frac{6\pi^3}{\Omega}\left\{\frac{K_1}{C\delta} + \frac{2\delta}{3\pi^2}\left[\frac{4}{3}\Omega^2 + \left(\frac{44}{9} + \frac{3\pi^2}{2}\right)\frac{K_2}{C^2\delta^2} - \frac{26}{7}\bar{C}\right.\right.$$

(5)

and all terms here are $O(\delta)$.

Figure 4.11 shows the bifurcation diagrams for $\epsilon b_0 < 0$ and $\epsilon b_0 > 0$ (downflow at the center), respectively. As R increases, there is a jump transition to hexagons, and as R is further increased, rolls become stable as well. For R sufficiently large, the hexagons become unstable, and only rolls persist.

The first analysis of this system was that of Amberg and Homsy (1993), who considered the case $S = O(1)$, and took into account nonlinear perturbations in $\Pi(\phi)$. By taking $S/C \sim 1$ as $\delta \to 0$, Anderson and Worster were able to account for additional physical effects such as nonlinear variations in $\bar{\theta}(z)$, higher-order effects in $\Pi(\phi)$, and interactions between the temperature and ϕ. It is this last effect that gives rise to the last term in Eq. (10.45) and allows the possibility that ϵb_0 can be positive. When $S \sim 1$, this term disappears. Further, Amberg and Homsy found that rolls bifurcate supercritically only when $K_1/\bar{C} < 0.226$. In the present analysis, $K_1/\bar{C} = O(\delta)$, and thus the rolls are as pictured in Figure 4.11.

Amberg and Homsy found that $\epsilon b_0 < 0$, and thus only upflow hexagons can be stable. The inclusion of further physical effects by Anderson and Worster shows that upflow or downflow hexagons are physically possible. Figure 10.16 for $K_2 = 0$ shows where one expects either up or down hexagons at onset and where the oscillatory mode, discussed in Section 10.7, is present. The weakly nonlinear analysis here describes only the steady bifurcations. Stable down hexagons can be found only for $K_2 < 0.131$.

These results can be compared qualitatively with the observations of Tait, Jahrling, and Jaupart (1992) on ammonium chloride solutions in a tank cooled from below. They saw arrays with the hexagonal pattern of chimneys (see Section 10.8) with downflow at the centers consistent with the preceding theory for $K_2 < 0.131$. Without accounting for temperature–ϕ interactions, only upflow at centers would be predicted (Amberg and Homsy 1993).

10.8 Chimneys

When alloys are directionally solidified at speeds well within the region where dendrites are present, certain defects can occur. One of these is the "freckle," which is a long solid-free channel parallel to the pulling direction. The

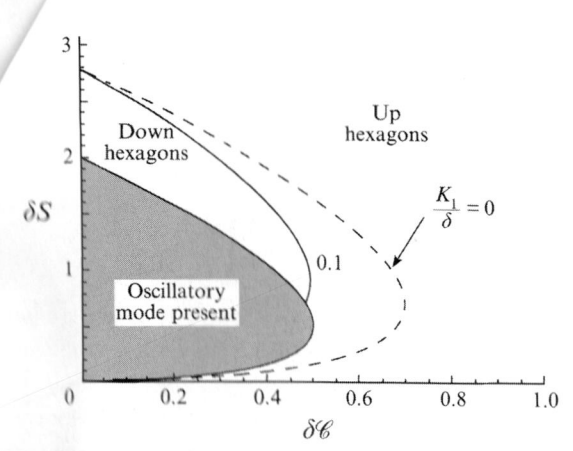

Figure 10.16. Parameter regimes for δS versus δC with fixed $K_1/\delta = 0.1$ and $K_2 = 0$, where various features of convection in the mushy layer can be identified. The shaded portion corresponds to those parameter values at which our weakly nonlinear analysis breaks down owing to the interaction of an oscillatory instability with the steady convective mode. Outside this boundary, we can identify the parameter values for which stable down hexagons or stable up hexagons are predicted. The solid curve (corresponding here to $K_1/\delta = 0.1$) separates the regions where down and up hexagons are predicted. This solid curve moves to the left (right) when K_1/δ increases (decreases). The dashed curve indicates the outermost position that this separating curve can attain (corresponding to $K_1/\delta = 0$). As a result, only up hexagons are possible for parameter values outside the dashed boundary. From Anderson and Worster (1995).

occurrence of freckles is associated with effects of convection (Hellawell 1987) in which vertical convective flows of solute-depleted liquid melt away the dendritic structure and create the solid-free channels.

One can imagine the rejection of low-density solute on an upflow. This flow is solute rich and so depresses the melting temperature in its vicinity, dissolving some dendrites and increasing the permeability of the upflow region. The decreased flow resistance there will further increase the flow rate, and a focusing of the flow may result.

Tait, Jahrling, and Jaupart (1992) observed convection in near-eutectic ammonium chloride solutions heated from below. At first under such conditions double-diffusive finger convection is observed in the liquid that seemingly does not affect the mush; see Chen (1995). At later times (higher ΔT), regions of upflow through the mush, Figure 10.17, are seen in a roughly hexagonal pattern. There is a reduced ϕ along the edges. Later, the upflow occurs mainly at the nodes of the hexagons, whereas the edges between the nodes close off. The result is a planform of chimneys with upflow at the edges and downflow in the centers similar to the weakly nonlinear patterns predicted in Section 10.7.

10.8 Chimneys

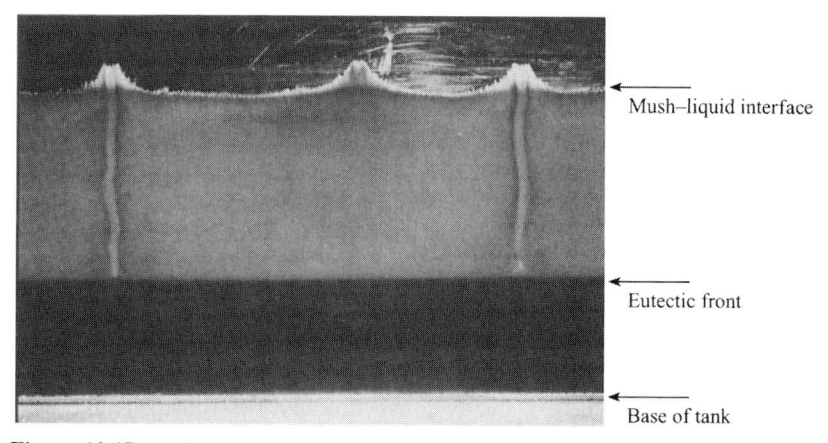

Figure 10.17. A photograph of mushy-layer chimneys during an experiment with an ammonium–chloride solution. In this system, pure ammonium–chloride crystals are formed when the solution is cooled below its freezing temperature, leaving behind a diluted solution with a density lower than that of the bulk fluid. In the present case, the mushy layer is growing away from a fixed cold base, which is at a temperature below the eutectic point; thus, the solid–mush and mush–liquid interfaces are advancing at a decreasing rate. At the time the photograph was taken, the distance between the base of the tank and the eutectic front was about 3 cm. Notice that the chimney walls and the mush–liquid interface are flat to a good first approximation. From Schulze and Worster (1998).

Given this scenario, one would expect the weakly nonlinear theory to govern the first appearance of the incipient chimneys, but fully developed chimneys should be strongly nonlinear (in a chimney, ϕ goes from $O(1)$ to zero).

Further, the weakly nonlinear analyses considered only a one-layer model with an impermeable lid over the mush. In an effort to generalize the model to a two-layer system and to investigate the strongly nonlinear regime, Schulze and Worster (1999) considered directional solidification having Stokes flow in the liquid, a mush, and a deformable mush–liquid interface. They considered equal thermal properties and densities in the two phases, negligible thermal buoyancy, zero solute diffusivity, and zero distribution coefficient. The whole system is two-dimensional. This model is similar to that of Emms and Fowler except that they took the permeability to be constant.

Schulze and Worster (1999) found steady solutions that bifurcate subcritically consistent with the results of Amberg and Homsy (1993). Now the branch extends deeply below the critical Rayleigh number, and the upper branch is so far distant that $\phi > 0$ is no longer possible everywhere. They computed steady solutions from perturbed positions along the (unstable) branch and found that there is a point locally at which W equals V, and later, when $W > V$, a region

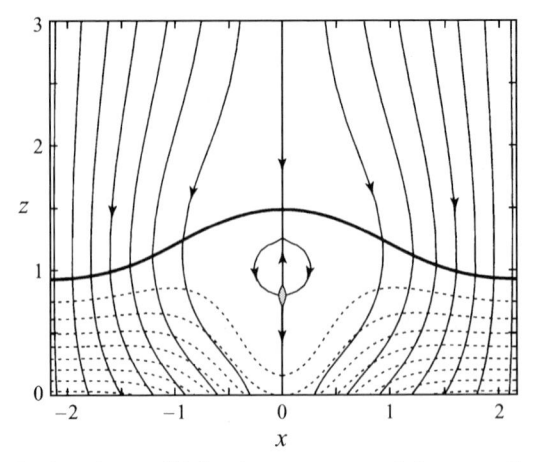

Figure 10.18. The interface, solid fraction contours, and the streamlines for the total volume flux of solutions just beyond the point at which the solid fraction has become negative; notice the closed zero-solid-fraction contour in the center of the second figure. The solid fraction inside this contour is just below zero, indicating that the mushy layer in this region should be replaced by a fluid inclusion. The solid fraction contours range from 0.0 to 0.08. From Schulze and Worster (1999).

in which ϕ is negative. Clearly, the model breaks down when $\phi < 0$, but the presence of a contour, $\phi = 0$, suggests the generation of a liquid-inclusion interior to the mush; see Figure 10.18. The existence of a nodal point near the birth of an inclusion supports suggestions by Fowler (1985) and Tait and Jaupart (1992). In addition to the appearance of an inclusion of initially interior to the mush, the analysis shows that, above a rising current, the interface is elevated and the thermal gradient above it (in the liquid) decreases.

Presumably, a liquid inclusion will elongate in the growth direction as the Rayleigh number is increased, push through the interface, and at sufficiently large R_m, form a steady, nearly cylindrical channel. (Emms and Fowler 1994 argued that at the mush–solid interface, the channel must close at the base.) Schulze and Worster (1998) let the liquid have infinite depth and examined the asymptotic scaling in the limit $R_m \to \infty$, $Da \to 0$, and $C \sim R_m$ so that the mush was thin. The Darcy number $Da = \Pi^*/\delta_T^2$. Here $\Pi^* = \Pi(0)$, and $\Pi(\phi) = (1-\phi)^3$. They found the structure shown in Figure 10.19. In nondimensional terms $r \sim Da^{1/3} R_m^{-2/9}$. There is a thermal boundary layer outside the chimney of thickness $R^{-1/3}$ and a liquid boundary layer of thickness $R_m^{-1/3}$ below an isothermal core.

To study such steady channels, Schulze and Worster (1998) simplified the preceding model by making the mush–liquid interface a rigid, porous plane (having $p = 0$) and studying the one-layer case with a two-dimensional,

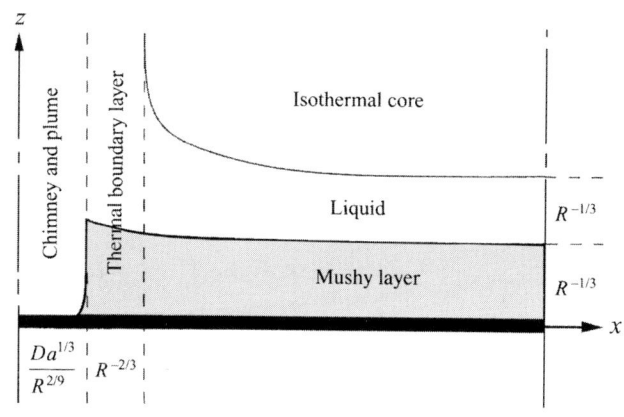

Figure 10.19. A diagram of the boundary layer structure in the mush–chimney system for the limit $R_m \gg 1$, $Da \ll 1$ and $\mathcal{C} \sim R_m$. In this limit, the mass flux is large and the solid fraction is small. The chimney and thermal boundary layer scale like $Da^{1/3}/R_m^{2/9}$ and $R_m^{-2/3}$, respectively. The height of the mushy layer and the thickness of the boundary layer in the liquid above the mushy layer both scale like $R_m^{-1/3}$. From Schulze and Worster (1998).

spatially periodic array of pores of widths 2ℓ and separations $2L$. They performed numerical simulations for the sides of the channel, straight and vertical, with ℓ left as a parameter.

The numerical results are consistent with the boundary-layer structure estimated – even to quite small R_m. The steady-state computations give two distinct solutions corresponding to different amplitudes of motion consistent with subcritical instabilities. The lower branch, presumably unstable, is a continuation from a no-chimney-like region. The upper branch, presumably stable, has a strong chimney-like structure in which the solid fraction is nonmonotonic laterally, as shown in Figure 10.20. It is "zero" at the chimney wall and, as the distance increases, rises to a maximum in the thermal layer and then decreases afterward. Figure 10.21 shows a bifurcation diagram for the scaled chimney scale r versus Rm for various spacings L. As L increases, the flat mush becomes more unstable, and the upper branch moves to higher amplitude.

In summary, then, near R_c one can predict a scale of instability from linear theory and the persistence of hexagonal convection. If the scale is fixed, one can predict the appearance of liquid inclusions and the steady structure of individual channels in two dimensions. It is not yet possible to predict the strongly nonlinear preferred pattern nor the preferred value of the channel spacings. Worster (1997) gives a clear description of the models and the literature.

The model of the mushy zone discussed gives two steady convective solutions in two dimensions, one with strong and one with weak channels. Experiments

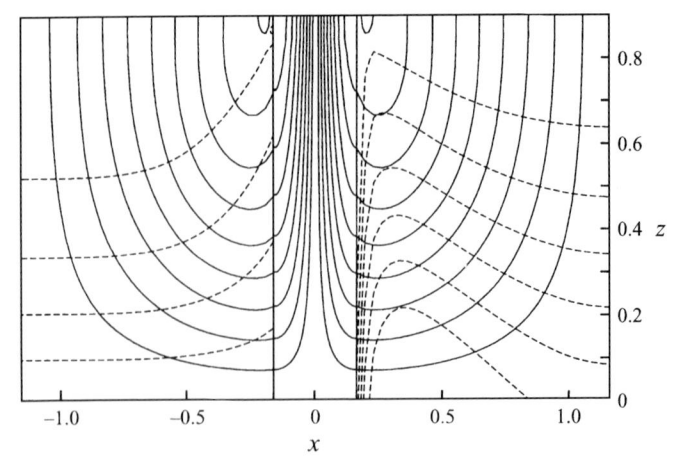

Figure 10.20. Streamlines in both the chimney and mushy regions along with contours for temperature and solid fraction in the mushy region for $R_m = 30$, upper branch. Here $C = 10$, $S = 1$, $\theta_\infty = 0.1$, $Da = 0.001$, and $L = 1.0$. The contour values for the right-hand-side streamlines are $\psi = 0.1$ to $\psi = 1.5$ by 0.2 increments, starting from the outside. The temperature contours (dashed lines) are shown on the left-hand side of the mushy region only, and their values are $\theta = -1.0$ to $\theta = 0$ by 0.2 increments. Solid fraction contours (dashed curves on right-hand side of the mushy region) have values in the range $\phi = 0$ to $\phi = 0.15$ by 0.03 increments, starting from the top. From Schulze and Worster (1998).

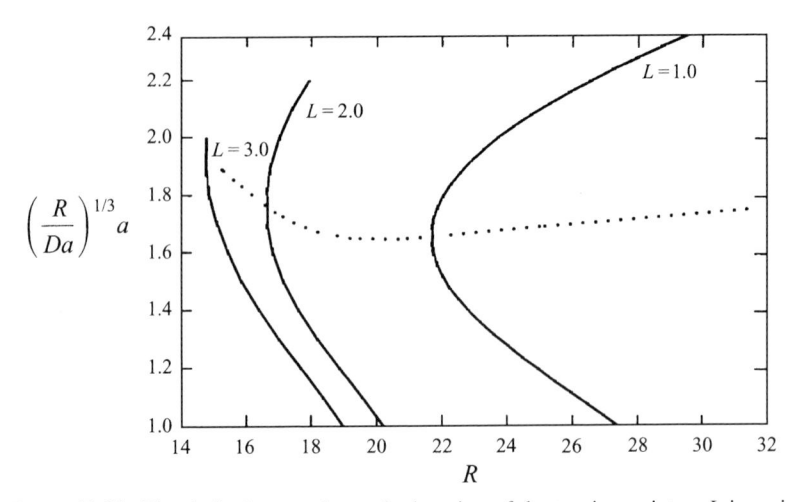

Figure 10.21. The dashed curve shows the location of the turning point as L is varied from 0.5 to 2.7 with the parameters C, θ_∞, and S held fixed at the values used in the previous figures. The three solid curves represent the solution as R_m is varied for L fixed at 1.0, 2.0, and 3.0. The figure demonstrates that the point where the numerical method fails moves closer to the turning point as L is increased. From Schulze and Worster (1998).

with ammonium chloride solutions exhibit channels, though experiments on other salt solutions (Huppert 1990) often do not exhibit channels. Huppert, Hallworth, and Lipson (1993) obtained insight into this discrepancy by doping NH_4Cl with $CuSO_4$. When the dopant had concentrations over 0.3 wt % $CuSO_4$, chimneys did not form. The addition of $CuSO_4$ makes the crystals more faceted (Mellor 1981) and hence increases the importance of the kinetic undercooling. Worster and Kerr (1994) measured μ in the relation $v_n = \mu[T_L(C_0) - T^1]^2$ and found that μ is a strong function of the concentration of $CuSO_4$ and that its increase strengthens the boundary-layer convective mode, which retards the growth of the mush, increases its solid reaction, and decreases ΔC across it. All of these reduce R and may prevent R from rising to values necessary for the onset of channels (Worster 1997).

10.9 Remarks

Mushy zones have been modeled as porous media whose matrices can grow or shrink by phase transformation. The model, coupled with convection, has been used as a basis of instability theory that yields three distinct modes: a steady boundary-layer mode in the liquid, a steady mushy-layer mode, and a time-periodic mushy-layer mode. The three-dimensional, weakly nonlinear theory for the steady mushy-layer mode shows a preference for hexagonal convection near the onset of motion.

The mushy-layer models yield predictions for mushy-layer thickness and planform preference in good agreement with observation, though all the experiments have been done with various salt solutions. What remains to be done are experiments on metallic alloys, which have low Prandtl numbers and negligible kinetic undercoolings, in order to see whether the mushy-layer theory for preferred microstructure compares well with observation.

References

Amberg, G., and Homsy, G. M. (1993). Nonlinear analysis of buoyant convection in binary solidification with application to channel formation, *J. Fluid Mech.* **252**, 79–98.

Anderson, D. M., and Worster, M. G. (1995). Weakly nonlinear analysis of convection in mushy layers during the solidification of binary alloys, *J. Fluid Mech.* **302**, 307–331.

_____. (1996). A new oscillatory instability in a mushy layer during the solidification of binary alloys, *J. Fluid Mech.* **307**, 245–267.

Beavers, G. S., and Joseph, D. D. (1967). Boundary conditions at a naturally permeable wall, *J. Fluid Mech.* **30**, 197–207.

Chen, C. F. (1995). Experimental study of convection in a mushy layer during directional solidification, *J. Fluid Mech.* **293**, 81–98.

Chen, F., and Chen, C. F. (1988). Onset of finger convection in a horizontal porous layer underlying a fluid layer, *Trans. ASME C: Heat Transfer* **110**, 403–407.

———. (1991). Experimental study of directional solidification of aqueous ammonium chloride solution, *J. Fluid Mech.* **227**, 567–586.

Chen, F., Lu, J. W., and Yang, T. L. (1994). Convective instability in ammonium chloride solution directionally solidified from below, *J. Fluid Mech.* **276**, 163–187.

Chiareli, A. O. P., Huppert, H. E., and Worster, M. G. (1994). Segregation and flow during the solidification of alloys, *J. Crystal Growth* **139**, 134–146.

Chiareli, A. O. P., and Worster, M. G. (1995). Flow focusing instability in a solidifying mush, *J. Fluid Mech.* **297**, 293–306.

Davis, S. H., Müller, U., and Dietsche, C. (1984). Pattern selection in single-component systems coupling Bénard convection and solidification, *J. Fluid Mech.* **144**, 133–151.

Emms, P. W., and Fowler, A. C. (1994). Compositional convection in the solidification of binary alloys *J. Fluid Mech.* **262**, 111–139.

Fowler, A. C. (1985). The formation of freckles in binary alloys, *IMA J. Appl. Math.* **35**, 159–174.

Hellawell, A. (1987). Local convective flows in partially solidified systems, in D. E. Loper, editor, *Structure and Dynamics of Partially Solidified Systems*, pp. 5–22, Martinus Nijhoff, Dordrecht.

Hills, R. N., Loper, D. E., and Roberts, P. H. (1983). A thermodynamically consistent model of a mushy zone, *Q. J. Appl. Math.* **36**, 505–539.

Huppert, H. E. (1990). The fluid dynamics of solidification, *J. Fluid Mech.* **212**, 209–240.

———. (1993). Bulk models of solidification, in D. T. J. Hurle, editor, *Handbook of Crystal Growth*, Vol. 1b, pp. 741–784, North-Holland, Amsterdam.

Huppert, H. E., Hallworth, M. A., and Lipson, S. G. (1993). Solidification of NH_4Cl and NH_4Br from aqueous solutions contaminated by $CuSO_4$: The extinction of chimneys, *J. Crystal Growth* **130**, 495–506.

Huppert, H. E., and Worster, M. G. (1985). Dynamic solidification of a binary melt, *Nature* **314**, 703–707.

———. (1991). Vigorous motions in magma chambers and lava lakes, in D. A. Yuen, editor, *Chaotic Processes in the Geological Sciences, IMA Series* **41**, p. 41.

Hurle, D. T. J., Jakeman, E., and Wheeler, A. A. (1983). Hydrodynamic stability of the melt during the solidification of a binary alloy, *Phys. Fluids* **26**, 624–626.

Kerr, R. C., Woods, A. W., Worster, M. G., and Huppert, H. E. (1990). Solidification of an alloy cooled from above. Part 1. Equilibrium growth, *J. Fluid Mech.* **216**, 323–342.

Mellor, J. W. (1981). *A Comprehensive Treatise on Inorganic and Theoretical Chemistry*, pp. 383–389, Longmans, London.

Palm, E., Weber, J. E., and Kvernvold, O. (1972). On steady convection in a porous medium, *J. Fluid Mech.* **54**, 153–161.

Schulze, T. P., and Worster, M. G. (1998). A numerical investigation of steady convection in mushy layers during the directional solidification of binary alloys, *J. Fluid Mech.* **356**, 199–202.

———. (1999). Weak convection, liquid inclusions and the formation of chimneys in mushy layers, *J. Fluid Mech.* **388**, 197–215.

Shirtcliffe, T. G. L., Huppert, H. E., and Worster, M. G. (1991). Measurement of solid fraction in the crystallization of a binary melt, *J. Crystal Growth* **113**, 566–574.

Tait, S., Jahrling, K., and Jaupart, C. (1992). The planform of composition convection and chimney formation in a mushy layer, *Nature* **359**, 406–408.

Tait, S., and Jaupart, C. (1992). Compositional convection in a reactive crystalline mush and melt differentiation, *J. Geophys. Res.* **97**, 6735–6756.

Turner, J. S. (1979). *Buoyancy Effects in Fluids*, Cambridge University Press, New York.
Turner, J. S., Huppert, H. E., and Sparks, R. S. J. (1986). Komatiites II: Experimental and theoretical investigations of post-emplacement cooling and crystallization, *J. Petrol.* **27**, 397–437.
Wettlaufer, J. S., Worster, M. G., and Huppert, H. E. (1997). Natural convection during solidification of an alloy from above with application to the evolution of sea ice, *J. Fluid Mech.* **344**, 291–316.
Woods, A. W., and Huppert, H. E. (1989). The growth of compositionally stratified solid above a horizontal boundary, *J. Fluid Mech.* **199**, 29–53.
Worster, M. G. (1986). Solidification of an alloy from a cooled boundary, *J. Fluid Mech.* **167**, 481–502.
———. (1990). Structure of a convecting mushy layer, *Appl. Mech. Rev.* **43**, S59–S62.
———. (1991). Natural convection in a mushy layer, *J. Fluid Mech.* **224**, 335–339.
———. (1992a). The dynamics of mushy layers, in S. H. Davis, H. E. Huppert, U. Müller, and M. G. Worster, editors, *Interactive Dynamics of Convection and Solidification*, pp. 113–138, NATO ASO **E219**, Dordrecht, Kluwer.
———. (1992b). Instabilities of the liquid and mushy regions during solidification of alloys, *J. Fluid Mech.* **237**, 649–669.
———. (1997). Convection in mushy layers, *Ann. Rev. Fluid Mech.* **29**, 91–122.
Worster, M. G., and Kerr, R. C. (1994). The transient behavior of alloys solidified from below prior to the formation of chimneys, *J. Fluid Mech.* **269**, 23–44.

11

Phase-field models

In all the systems discussed heretofore the solidification front was considered to be a mathematical interface of zero thickness endowed with surface properties deemed appropriate to the physics. In this chapter, another approach is taken. The front is allowed to be diffuse, and the fields of interest, such as T and C, are supposed to have well-defined bulk behaviors away from the interfacial region and rapid, though continuous, variations within it. Minimally, one would wish the model to satisfy the laws of thermodynamics, appropriately extended into the nonequilibrium regions, and regain the interfacial properties and jump conditions appropriate to the thin-interface limit when the interfacial thickness approaches zero.

On the one hand one would anticipate that an infinite number of such models likely exists. On the other hand one would anticipate that the thin-interface limit might be taken a number of ways, each giving distinct properties to the front. Nonetheless, conceptually there are two possible virtues of the diffuse-interface approach. If the models are well chosen on the basis of some underlying framework, then there would be a *systematic* means of generalizing the models to systems such as rapid solidification. In Chapter 6 high rates of solidification were modeled by appending to the standard model variations $k = k(V_n), m = m(V_n)$ with, for example, the equilibrium Gibbs–Thomson undercooling. A systematic generalization could indicate how the relationships for k and m emerge, and what other alterations to the model should simultaneously be included. Call this "model building."

The second virtue is the practical one of being able to compute complex microstructures. Given the sharp interface model, one can track the front when dendrites begin to appear, but as a dendrite develops secondary and tertiary sidearms, front tracking becomes impossible. The "diffuse" interface model treats all the phases as one with spatially varying properties, and thus it **may** be possible to simulate more complex structures numerically for longer times. Call this "simulation building."

Consider a system in which an interfacial region of thickness W is included between two bulk phases, solid and liquid. To identify the phases one defines a "marker function" ϕ, called the phase field, which takes on two different constant values (say ± 1) in the bulks but varies smoothly between them. The task is to find a system of continuum equations that govern C, T, and ϕ. One takes the convention that the interface lies at $\phi = 0$. A simulation of the interface then involves the location of the zeros of ϕ and not the solution of a free-boundary problem.

The discussion will begin with a model system for a simple pure-material system. This will be generalized to a deduced system, and then the results of numerical computation will be discussed.

11.1 Pure Materials – A Model System

Consider first a model solid–liquid system in which the thermal properties of the two phases are identical. For the nondimensional temperature θ and the phase field ϕ in planar growth write

$$\alpha p^2 \phi_t = p^2 \phi_{zz} - f'(\phi) - \lambda g'(\phi)\theta \tag{11.1a}$$

$$\theta_t = \theta_{zz} + \frac{1}{2}\phi_t, \tag{11.1b}$$

where

$$\begin{aligned} f(\phi) &= -\left[\tfrac{1}{2}\phi^2 - \tfrac{1}{4}\phi^4\right] \\ g(\phi) &= \phi - \tfrac{2}{3}\phi^3 + \tfrac{1}{5}\phi^5. \end{aligned} \tag{11.2}$$

Here, a Peclét number p represents the ratio of the thickness W of the interfacial region to $\delta_T = \kappa/v$ and hence is considered to be small; α and the linking parameter λ are constants. Here v is some estimate of the freezing rate.

In this model $\phi = +1$ in the solid and -1 in the liquid, and it varies continuously in between them. One then sees that the temperature θ in each bulk phase satisfies the heat-conduction equation (11.1b) in which $(1/2)\phi_t$ represents a source of heat localized to the interfacial region and is related to the liberation of latent heat. The sources $f'(\phi)$ and $g'(\phi)$ in Eq. (11.1a) reflect the phase change alterations as the temperature field is varied. When $\lambda = 0$, the phase-field decouples from the temperature and develops independently, though the temperature continues to depend on ϕ. As the interface thickness $W \to 0$, $p \to 0$, and the thin-interface limit should be approached though, as is clear, $p \to 0$ is a singular perturbation.

One would expect that if system (11.1) represents the standard solidification equations, it should do so in a limit that involves $p \to 0$, or $\lambda \to 0$, or both. Two of the possibilities have been used extensively in the literature and will be discussed now.

Consider first the case $p \to 0$ with $\lambda = O(1)$, which was posed by Fife and Penrose (1995) and elaborated by Karma and Rappel (1998).

In the outer region, far from the interface, write

$$\theta = \theta_0 + p\theta_1 + \cdots ; \qquad (11.3)$$

one finds that at all orders

$$\theta_{it} = \theta_{izz}, \; i = 0, 1, 2, \ldots \qquad (11.4)$$

as expected; the temperature satisfies a heat-conduction equation.

To analyze the inner region it is convenient to introduce a coordinate $\hat{z}$ moving with the front whose speed is $v(t)$,

$$\hat{z} = z - \int_0^t v(t')\,dt'. \qquad (11.5)$$

One can then introduce the inner variable η,

$$\eta = \hat{z}/p. \qquad (11.6)$$

In terms of η, the system (11.1) and (11.2) becomes

$$\alpha p^2 \left(\phi_t - \frac{v}{p}\phi_\eta\right) = \phi_{\eta\eta} + \phi(1-\phi^2) + \lambda(1-\phi^2)^2\theta \qquad (11.7a)$$

$$p^2 \left(\theta_t - \frac{v}{p}\theta_\eta\right) = \theta_{\eta\eta} + \frac{1}{2}p^2\left(\phi_t - \frac{v}{p}\phi_\eta\right). \qquad (11.7b)$$

Write an asymptotic expansion in the inner region

$$(\theta, \phi) = (\Theta_0, \Phi_0) + p(\Theta_1, \Phi_1) + \cdots. \qquad (11.8)$$

When this is substituted into system (11.7), one obtains at leading order in p

$$\Phi_{0\eta\eta} + \Phi_0(1-\Phi_0^2) + \lambda(1-\Phi_0^2)^2\Theta_0 = 0 \qquad (11.9a)$$

$$\Theta_{0\eta\eta} = 0. \qquad (11.9b)$$

It is immediately evident that Θ_0 must be constant in order to match to the outer fields, and hence one can take

$$\Theta_0 \equiv 0. \qquad (11.10)$$

11.1 Pure materials – A model system

Equation (11.9a) can then be solved, giving

$$\Phi_{0\eta} = -\tanh\frac{\eta}{\sqrt{2}}, \quad (11.11)$$

which gives the principal behavior of the phase-field structure. At $O(p)$ one has, using result (11.10), that

$$\Phi_{1\eta\eta} + (1 - 3\Phi_0^2)\Phi_1 = -\alpha v \Phi_{0\eta} - \lambda(1 - \Phi_0^2)\Theta_1 \quad (11.12a)$$

$$\Theta_{1\eta\eta} - \frac{1}{2}v\Phi_{0\eta} = 0. \quad (11.12b)$$

Equation (11.12b) can be integrated directly twice, giving

$$\Theta_{1\eta} - \frac{1}{2}v\Phi_0 = A \quad (11.13a)$$

and

$$\Theta_1 = \bar{\theta}_1 + A\eta + \frac{1}{2}v\int_0^\eta \Phi_0(\eta')\,d\eta', \quad (11.13b)$$

where A and $\bar{\theta}_1$ are constants of integration. Notice that at $\eta = 0$, $\Theta_1 = \bar{\theta}_1$.

The linear operator on the left-hand side of Eq. (11.12a) is self-adjoint and is the Frechet derivative of that of Eq. (11.9a). Hence, a complementary solution of Eq. (11.12a) is $\Phi_{0\eta}$, and its right-hand side must be orthogonal to this for solutions to exist. This Fredholm condition gives

$$\int_{-\infty}^{\infty} \left[\alpha v \Phi_{0\eta} + \lambda(1 - \Phi_0^2)\Theta_1\right]\Phi_{0\eta}\,d\eta = 0. \quad (11.14)$$

The condition (11.14) determines the constant $\bar{\theta}_1$. To see this, eliminate Θ_1 using Eq. (11.13b) to obtain

$$\alpha v \int_{-\infty}^{\infty} \Phi_{0\eta}^2\,d\eta + \lambda \bar{\theta}_1 \int_{-\infty}^{\infty} (1 - \Phi_0^2)^2 \Phi_{0\eta}\,d\eta$$

$$+ \frac{1}{2}\lambda v \int_{-\infty}^{\infty} (1 - \Phi_0^2)^2 \Phi_{0\eta} \left[\int_0^\eta \Phi_0(\eta')\,d\eta'\right] d\eta = 0. \quad (11.15)$$

Notice that the constant A is not present because it multiplies the integral of an odd function.

Now let us turn to matching. The heat fluxes match if

$$\lim_{\eta \to \pm\infty} \Theta_{1\eta} = \lim_{z \to 0^\pm} \theta_{0z} \equiv \theta_{0z}^\pm. \quad (11.16)$$

Using solution (11.13a) on each side of the region gives, in the solid,

$$\theta_{0z}^+ - \frac{1}{2}v = A, \quad (11.17a)$$

and in the liquid,

$$\theta^-_{0z} + \frac{1}{2}v = A \qquad (11.17b)$$

By eliminating A between these, one obtains the first condition satisfied by the outer field,

$$v = \theta^+_{0z} - \theta^-_{0z}, \qquad (11.18a)$$

which is the scaled thin-interface heat balance.

One can also solve for the integration constant A,

$$A = \theta^\pm_{0z} \mp \frac{1}{2}v. \qquad (11.18b)$$

Now consider the temperature itself. Because $\Theta_0 \equiv 0$, one can write near $\eta = 0$, that

$$\theta_{\text{outer}} \sim p\theta^\pm_1 + \theta^\pm_{0z} z \qquad (11.19)$$

and using Eq. (11.13b), one obtains

$$\theta_{\text{inner}} \sim p\left[\bar{\theta}_1 + A\eta + \frac{1}{2}v\int_0^\eta \Phi_0(\eta')\,d\eta'\right]$$

$$= p\left[\bar{\theta}_1 + \frac{1}{2}v\int_0^\eta \Phi_0(\eta')\,d\eta'\right] + Az. \qquad (11.20)$$

Now, from relation (11.18b), one obtains

$$\theta_{\text{inner}} \sim p\left\{\bar{\theta}_1 + \frac{1}{2}v\int_0^\eta [\Phi_0(\eta') \pm 1]\,d\eta' + \theta^\pm_{0z} z\right\}. \qquad (11.21)$$

When relations (11.19) and (11.21) are matched, one finds that

$$\theta^\pm_1 = \bar{\theta}_1 + \frac{1}{2}v + I, \qquad (11.22)$$

where

$$I = \int_0^\infty [\Phi_0(\eta') + 1]\,d\eta' = \int_0^{-\infty} [\Phi_0(\eta') - 1]\,d\eta', \qquad (11.23)$$

and thus from relation (11.19) one finds

$$\theta_{\text{outer}} \sim p\left(\bar{\theta}_1 + \frac{1}{2}I\right) + \theta^\pm_{0z} z. \qquad (11.24)$$

The temperature θ^I at the interface, $\theta_{\text{outer}}|_{z=0}$, is thus continuous. Given the values of $\bar{\theta}_1$ from Eq. (11.22), the interfacial temperature through $O(p)$ becomes

$$\theta^I \sim -\frac{a_1\alpha}{\lambda}\left[1 - a_4\frac{\lambda}{\alpha}\right]pv, \qquad (11.25)$$

where the a_i are positive constants obtained by definite integrations and

$$a_4 = 1 + \frac{1}{2}\frac{a_2}{a_1}. \tag{11.26}$$

Equation (11.25) shows that as $p \to 0$, the interface temperature recovers the form

$$\theta^I = -\mu^{-1}vp, \tag{11.27}$$

and μ is the nondimensional kinetic coefficient appropriate to the planar interface.

Karma and Rappel (1998) considered the more general case of a curved interface and found that Eqs. (11.25) and (11.27) become, respectively,

$$\theta^I \sim \left\{\frac{2Ha_1}{\lambda} - \frac{a_1\alpha}{\lambda}\left[1 - a_4\frac{\lambda}{\alpha}\right]v\right\}p \tag{11.28}$$

and

$$\theta_I = 2H\Gamma - \mu^{-1}v, \tag{11.29}$$

where Γ denotes nondimensional surface energy.

Thus, at $O(p)$ one recovers the appropriate jump conditions. Karma and Rappel (1998) noted that μ^{-1} can be controlled by the choice of λ because so far λ is arbitrary. For example, to examine solidification with instantaneous kinetics appropriate to small undercooling, one merely chooses $\lambda = \alpha/a_4$.

The second limit appropriate to thin interfaces dates back to the work of Caginalp (1989) and is related to a vanishingly *small linking parameter* λ. In the present notation, $p \to 0$, $\lambda \to 0$ such that $p/\lambda = O(1)$. The asymptotics follow as before, and the result for a curved front is condition (11.25) but with the term $a_4\lambda/\alpha$ neglected compared with unity. Now μ^{-1} is no longer controllable; it cannot be chosen to be zero.

Karma and Rappel (1998) discussed, on dimensional grounds, the implications for numerical computation. In the small-linking-parameter limit, the ratio of the thickness W to the capillary length $\delta_{cap} = \gamma T_m c_p/L_v^2$ must be very small, whereas this ratio need only be of unit order in the small-Peclét number limit, which is a huge advantage given that δ_{cap} is typically of molecular dimensions.

Note: The equations involving the phase-field approach look startlingly similar to those of the mushy zone, as discussed in Chapter 10. A contrast is between phase-field models in which one takes $p \to 0$ and mushy models in which one allows the equivalant of p to remain of unit order.

11.2 Pure Materials – A Deduced System

The system examined in Section 11.1 was asserted with only verbal justification. In this section, a more elaborate system will be deduced using the ideas of thermodynamics for systems out of equilibrium.

Phase-field theory for the solidification can be traced to 1978 when Langer extracted from analysis of critical phenomena a model for the freezing of an undercooled melt (later published in Langer 1986). Caginalp (1985), Collins and Levine (1985), and Umantsev and Roitburd (1988) did so as well. The issue of "thermodynamic self-consistency" was treated by Penrose and Fife (1990) and Wang et al. (1993).

The necessity of including anisotropy follows in part from the MST for dendrites. Caginalp and Fife (1986) and Caginalp (1986) replaced the magnitude of the gradient by anisotropic quadratic forms. Langer (1986) added higher-order gradients. Kobayashi (1993) and Wheeler, Murray, and Schaefer (1993) allowed the coefficient of the gradient energy to depend on orientation through $\mathbf{n} = \nabla\phi/|\nabla\phi|$.

Consider

$$\mathcal{F}(\theta, \phi) = \int_\Omega \left[\frac{1}{2} W^2 |\nabla\phi|^2 + f(\phi) + \frac{1}{2} b\lambda\theta^2 \right] dV, \quad (11.30)$$

where W is the interfacial thickness, $\theta = (T - T_m)c_p/L$ is the scaled temperature, ϕ is the phase field, Ω is the volume occupied by the material, and $\partial\Omega$ is its zero-flux boundary. Here

$$f(\phi) = -\frac{1}{2}\phi^2 + \frac{1}{4}\phi^4, \quad (11.31)$$

which has a local maximum at $\phi = 0$ and local minima at $\phi = \pm 1$.

One *supposes* that θ and ϕ evolve by the rules

$$\tau \phi_t = -\frac{\delta \mathcal{F}}{\delta \phi} \quad (11.32a)$$

and

$$b\lambda \hat{\Theta}_t = \kappa \nabla^2 \frac{\delta \mathcal{F}}{\delta \hat{\Theta}}, \quad (11.32b)$$

where δ denotes the variational derivative. The mobility parameter τ^{-1} and b, λ, and κ are positive. Here the enthalpy $\hat{\Theta}$ is related to the temperature by

$$\hat{\Theta} = \theta - \frac{1}{2} h(\phi) \quad (11.33)$$

and $\frac{1}{2}[h(1) + h(-1)] = 1$ for reasons of normalization.

11.2 Pure materials – A deduced system

This formulation guarantees that the free energy decreases locally and that the enthalpy balance holds (Warren and Boettinger 1995; Wheeler, Boettinger, and McFadden 1992)

When

$$g(\phi) = \frac{1}{2}bh(\phi), \tag{11.34}$$

Eqs. (11.32) reduce to

$$\tau \phi_t = W^2 \nabla^2 \phi - \frac{\partial}{\partial \phi} F(\phi, \lambda \theta) \tag{11.35a}$$

$$\theta_t = \kappa \nabla^2 \theta + \frac{1}{2}[h(\phi)]_t \tag{11.35b}$$

and

$$F(\phi, \lambda \theta) = f(\phi) + \lambda g(\phi)\theta. \tag{11.35c}$$

The models in the literature differ by their choices of f, g, and h, and reasonably wide ranges of choices give consistent results for $p \to 0$. Caginalp and Chen (1992) and Wang et al. (1993) took

$$g(\phi) = \phi - \frac{2}{3}\phi^3 + \frac{1}{5}\phi^5 \tag{11.36}$$

When τ, λ, κ, and W are constants, the system describes an isotropic material. Following Karma and Rappel (1998), scale lengths on $\delta_T = \kappa/V$, time on δ_T/V, and system (11.35) becomes

$$\alpha p^2 \phi_t = p^2 \nabla^2 \phi + \phi - \phi^3 - \lambda(1 - \phi^2)^2 \theta \tag{11.37a}$$

$$\theta_t = \nabla^2 \theta + \frac{1}{2}[h(\phi)]_t, \tag{11.37b}$$

where either

$$h = \frac{15}{8}\left(\phi - \frac{2}{3}\phi^3 + \frac{1}{5}\phi^5\right), \quad \left(b = \frac{15}{16}\right) \tag{11.37c}$$

or

$$h = \phi. \tag{11.37d}$$

The latter is not a special case of Eq. (11.34). Here p is the interfacial Peclét number,

$$p = \frac{WV}{\kappa}, \tag{11.38}$$

which is presumed to be very small.

This system can be seen to be related directly to Eq. (11.1) and the limit $p \to 0$, $\lambda = O(1)$, one regains the first thin-interface limit.

When the material is anisotropic, and τ and W depend on the local unit normal $\mathbf{n}$, then, with choice (11.37d), system (11.37) becomes

$$A^2(\mathbf{n})\phi_t = \nabla \cdot [A^2(\mathbf{n})\nabla\phi] + [\phi - \lambda\theta(1-\phi^2)](1-\phi^2) \quad (11.39a)$$

$$+ \left[|\nabla\phi|^2 A(\mathbf{n})\frac{\partial}{\partial \phi_x} A(\mathbf{n})\right]_x + \left[|\nabla\phi|^2 A(\mathbf{n})\frac{\partial}{\partial \phi_y} A(\mathbf{n})\right]_y$$

$$\theta_t = D\nabla^2\theta + \frac{1}{2}\phi_t \quad (11.39b)$$

and

$$D = \kappa\tau_0/W_0^2. \quad (11.39c)$$

The anisotropy is defined by $W(\mathbf{n}) = W_0 A(\mathbf{n})$, where W_0 is a scale for the interface width and $\tau(\mathbf{n}) \equiv \tau_0 A^2(\mathbf{n})$, A is in the range $[0, 1]$, and

$$A(\mathbf{n}) = (1 - 3\alpha)\left[1 + \frac{4\alpha}{1-3\alpha}\frac{\phi_x^4 + \phi_y^4}{|\nabla\phi|^4}\right]. \quad (11.40)$$

Equation (11.40) defines α, which is zero in the isotropic case and is related to α_4 in the case of fourfold symmetry.

11.3 Pure Materials – Computations

The systems (11.37) and (11.39) can be integrated numerically in a variety of ways given that a free boundary need not be tracked. The earliest numerical simulations derive from Fix (1983), from whom the questions of adequate resolution and sufficiently small W arose. Langer (1986) and Caginalp (1989) obtained expressions for $\lambda \to 0$ of the form

$$d_0 = a_1\frac{W}{\lambda}, \quad \mu^{-1} = a_1\frac{\tau}{\lambda W} \quad (11.41)$$

that relate the sharp-interface parameters d_0 and μ to the limiting forms of the phase field. Kobayashi (1993) and McFadden et al. (1993) extended these to systems with anisotropy. Following Karma and Rappel (1998), d_0 and μ^{-1} must scale W/λ and $\tau/W\lambda$, respectively. Relations (11.41) only fix W/λ and $\tau/W\lambda$ and not W and τ independently. Hence, numerical convergence requires decreasing $W \sim \lambda$ and $\tau \sim \lambda^2$ until the results become W independent. This requires, $W \to 0$, $\tau \to 0$, $\lambda \to 0$ such that d_0 and μ remain fixed. In the limit, one wishes the variation $\delta\theta$ across the interfacial region to vanish, as seen from

11.3 Pure materials – computations

the outer region. Here $\delta\theta \sim W/\delta_T$, and if the computation is done at unit-order undercooling, then necessarily $\mu^{-1}V \gg d_0(2H)$; the kinetic undercooling is dominant. Now, if W/δ_T is small compared with the interfacial temperature, one wishes that $W/\delta_T \ll \mu^{-1}V$, and thus, using Eq. (11.41) one finds that the limitation is

$$\frac{W}{d_0} \ll \frac{\kappa\tau}{W^2} \tag{11.42}$$

and, because $\kappa\tau/W^2$ is of unit order, W must be much smaller than the capillary length, which is exactly the same conclusion derived by Brattkus, Meiron, and Spencer (1993) and Braun, McFadden, and Coriell (1994).

The asymptotics of Section 11.2 take the linking parameter of unit order and allow the interfacial Péclet number $p \to 0$. Fife and Penrose (1995) first addressed this limit, but it was Karma and Rappel (1996) who extended it appropriately to numerical computation. Karma and Rappel found that, corresponding to relations (11.41), the new relations become

$$d_0 = a_1 \frac{W}{\lambda}, \quad \mu^{-1} = a_1\left(\frac{\tau}{\lambda W} - a_2 \frac{W}{D}\right) \tag{11.43}$$

Constraint (11.42) no longer applies because the inequality $W/\kappa \ll \tau/\lambda W$ necessarily applies as well; if one chooses $\lambda = \tau\kappa/W^2 a_2$, then $\mu^{-1} = 0$. No longer must $W/\delta_T \ll \mu^{-1}V$, but now $W/\delta_T \ll d_0$, and thus $W/d_0 \ll \delta_T$, which is a much weaker requirement.

The most recent and accurate computations entail the use of a finite-element method using adaptive grids. Provatas, Goldenfeld, and Dantzig (1998) considered in two dimensions the growth and instability of an initially circular nucleus. Figure 11.1 shows their typical result for the case of fourfold symmetry with $\alpha_4 = 0.05$ and undercooling $S^{-1} = c_p \Delta T/L = 0.55$. Notice that sidearms have appeared and that the thermal boundary layer is thin because S^{-1} is large. In order to get numerical convergence, the ratios of the sizes of the largest to the smallest elements are made 2^n. For $S^{-1} = 0.25$ they can use $n = 12$, whereas for $S^{-1} = 0.10$ they require $n = 17$. For $S^{-1} \geq 0.30$ the computations give equilibration to a steady state consistent with MST. At lower S^{-1}, such approach is inconsistent with the theory for an isolated finger because the thermal boundary layer overlaps neighboring arms; the MST result is not recovered. Kim et al. (1999) examined the phase-field solutions for large S^{-1}, but for various choices of $g(\phi)$ and $h(\phi)$. They found that the *same* long-time asymptotic state is obtained for all cases, verifying numerically that, for sufficient resolution, the MST prediction is obtained for large enough S^{-1}.

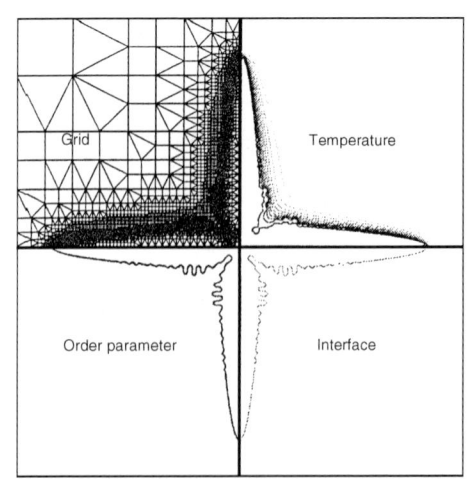

Figure 11.1. A dendrite grown using the adaptive-grid method for $S^{-1} = 0.55$. Clockwise, beginning at the upper right, the figures show contours of the T-field, the contour $\phi = 0$, contours of the ϕ-field, and the current m. From Provatas et al. (1998).

11.4 Remarks

"Model building" using phase-field methods gives a systematic process for creating systems out of equilibrium using analogies to equilibrium thermodynamics. When the interface is taken to be small asymptotically, one can in principle obtain expressions for capillary, kinetic, and constitutional undercoolings that can give insight into the alterations necessary for the sharp-interface models under extreme conditions of processing. These models are, of course, not unique, nor are the sharp-interface asymptotic limits. Therefore, the results should be one factor out of several in assessing the appropriateness of a given sharp-interface model.

"Simulation building" using phase-field methods provides an attractive bypass to the free-boundary nature of the sharp-interface model. The appearance of the numerical results is exactly what one would expect of dendritic growth. However, appearances are deceptive because, before quantitative validity is given to the simulations, one must be assured that the diffusive interface is thin enough and the system boundaries are sufficiently distant.

There has been a start at examining directional solidification. Charach, Chen, and Fife (1999), Wheeler, Boettinger, and McFadden (1993), and Charach and Fife (1998) have found in specialized models that, as the pulling speed increases, there is evidence of solute trapping that in some cases verifies the models discussed in Chapter 6. The phase-field approach is in its infancy, and many new ideas are still to be expected.

Clearly, carefully controlled use of phase-field simulations can be extraordinarily useful in assessing and predicting complex microstructures.

References

Brattkus, K., Meiron, D. I., and Spencer, B. J. (1993). Consistent nucleation in diffusive phase-field models, unpublished.

Braun, R. J., McFadden, G. B., and Coriell, S. R. (1994). Morphological instability in phase-field models of solidification, *Phys. Rev. E* **49**, 4336–4352.

Caginalp, G. (1985). Surface tension and supercooling in solidification theory, in L. Garrido, editor, *Applications of Field Theory to Statistical Mechanics*, pages 216–226, Notes in Physics 216, Springer-Verlag, Berlin.

———. (1986). An analysis of a phase field model of a free boundary, *Arch. Rat. Mech. Anal.* **92**, 205–245.

———. (1989). Stefan and Hele–Shaw type models as asymptotic limits of the phase-field equations, *Phys. Rev. A* **39**, 5887–5896.

Caginalp, G., and Chen (1992). Phase field equations in the singular limit of sharp interface problems, in M. E. Gwinn and G. B. McFadden, editors, *On the Evolution of Phase Boundaries,* pages 1–27, IMA Volumes in Mathematics and Its Applications **43**, Springer Verlag, Berlin.

Caginalp, G., and Fife, P. C. (1986). Phase field methods of interfacial boundaries, *Phys. Rev. B* **33**, 7792–7794.

Charach, C., Chen, C. K., and Fife, P. C. (1999). Developments in phase-field modeling of thermoelastic and two-component materials, *J. Stat. Phys.* **95**, 1141–1164.

Charach, C., and Fife, P. C. (1998). Solidification fronts and solute trapping in a binary alloy, *SIAM J. Appl. Math.* **58**, 1826–1851.

Collins, J. B., and Levine, H. (1985). Diffuse interface model of diffusion-limited cyrstal growth, *Phys. Rev. B* **31**, 6119–6122.

Fife, P. C., and Penrose, O. (1995). Interfacial dynamics for thermodynamically consistent phase-field models with non-conserved order parameter, *Electronic J. Diff. Eqns.* **16**, 1–19.

Fix, G. J. (1983). Phase field methods for free boundary problems, in A. Fasano and M. Primicerio, editors *Free boundary Problems: Theory and Applications.*

Kim, Y.-T., Provatos, N., Goldenfeld, N., and Dantzig, J. A. (1999). Universal dynamics of phase-field models for dendritic growth, *Phys. Rev. E* **59**, R2546–R2549.

Karma A., and Rappel, W.-J. (1996). Numerical simulation of three-dimensional dendritic growth, *Phys. Rev. Lett.* **77**, 4050–4053.

———. (1998). Quantitative phase-field modeling of dendritic growth in two and three dimensions, *Phys. Rev. E* **57**, 4323–4339.

Kobayashi, R. (1993). Modeling and numerical simulations of dendritic crystal growth, *Physica D* **63**, 410–423.

Langer, J. S. (1986). Models of pattern formation in first-order phase transitions, in *Directions in Condensed Matter*, World Scientific, Singapore, 165–186.

McFadden, G. B., Wheeler, A. A., Braun, R. J., and Coriell, S. R., and Sekerka, R. F. (1993). Phase-field models for anisotropic interfaces, *Phys. Rev. E* **48**, 2016–2024.

Penrose, O., and Fife, P. C. (1990). Thermodynamically consistent model of phase-field type for the kinetics of phase transitions, *Physica D* **43**, 44–62.

Provatas, N., Goldenfeld, N., and Dantzig, J. A. (1998). Efficient computation of dendritic microstructures using adaptive mesh refinement, *Phys. Rev. Lett.* **80**, 3308–3311.

Umantsev, A. R., and Roitburd, A. L. (1988). Non-isothermal relaxation in a nonlocal medium, *Sov. Phys. Solid State* **30**, 594–598.

Wang, S.-L., Sekerka, R. F., Wheeler, A. A., Murray, B. T., Coriell, S. R., Braun, R. J., and McFadden, G. B. (1993). Thermodynamically consistent phase-field models for solidification, *Physica D* **69**, 189–200.

Warren, J. A., and Boettinger, W. J. (1995). Prediction of dendritic growth and microsegregation on patterns in a binary alloy using the phase-field model, *Acta Metall. Mater.* **43**, 689–703.

Wheeler, A. A., Boettinger, W. J., and McFadden, G. B. (1992). Phase-field model for isothermal phase transitions in binary alloys, *Phys. Rev. A* **45**, 7424–7439.

———. (1993). A phase-field model of solute trapping during solidification, *Phys. Rev. E* **47**, 1893–1909.

Wheeler, A. A., Murray, B. T., and Schaefer, R. J. (1993). Computation of dendrites using a phase field model, *Physica D* **66**, 243–262.

Index

A

Amberg–Homsy model, 350, 359
Anderson–Worster model, 352–357
Anisotropy, 86–161
    attachment kinetics, 92, 98, 102–103, 106–110, 124, 228–229
    Brattkus–Davis equation, 94, 103, 105
    Burgers equation, 137
    capillarity and, 100
    coarsening and, 152–156
    coefficients of, 94, 102, 117–119
    corners and, 121, 125,130
    dendrites and, 228
    faceting and, 139–152
    fingers and, 228
    fourfold, 68, 89–94, 106–108
    Kuramoto–Sivashinsky equation, 119–120
    large, 89, 121–139, 229
    microscopic solvability theory (MST) and, 229–230, 238–239, 372
    needles and, 229–230
    phase-field, theory and, 372–374
    Sivashinsky equation, 92, 100
    small, 89, 91–121
    step-wise growth, 97–105, 121–124
    surface energy, 86–91, 97, 105, 110–122, 138–146, 230, 236–238
    symmetry and, 82–83, 131, 145, 238
    thin-interface limit and, 374–375
    travelling waves and, 234–237
    unconstrained growth and, 105–139
    See also specific models, parameters
Antikinks, 126, 128–129
Approximate selection of dendrites, 221–229
    of eutectics, 261–267
Aqueous solutions, 325–326
Asymptotic suction profile, 293–294

Attachment kinetics, 18, 164, 181–183, 196
    anisotropy and. *See* Anisotropy
    fingers and, 228
    growth and, 138
    high speeds and, 203
    Ivantsov solution, 251–252
    step density and, 97
    See also specific model, parameters

B

Bands, 162–163, 171, 177, 204, 205, 208
Bénard convection, 72, 74, 280–283, 301
    pulsations, 190–203
Bifurcation, 62, 94, 265
    anisotropy on, 94–97
    coupling and, 288–289
    degenerate, 206
    dimension and, 78–89
    directional solidification and, 62–85
    free boundary problems and, 76
    interfacial shape and, 101
    Landau coefficient and, 70, 73, 101, 181–183
    nonlinear theory, 62–85
    subcritical, 64, 76, 199, 265, 283
    three-dimensional theory, 72–75
    tilting cells and, 266
    tip-splitting and, 265
    transition point, 64
    two-dimensional theory, 62–71, 181–183
Binary substances, 42–61
    approximate models, 46–48
    directional solidification, 45–56
    linear instability, 48–56
    model for, 42–45
    morphological instability, 56–57
    See also specific alloys, effects
Binodal points, 126–127

380 Index

Boundary conditions, 21–26
Boundary-integral methods, 71
Boundary-layer effects, 285, 293, 346.
    See also specific effects, parameters
Boundary-layer mode, 346
Boussinesq approximation, 276–277, 304, 341
Brattkus-Davis equation, 80–82, 94, 303
    anisotropy, 103, 105
Buoyancy, 331
    convective effects, 162, 275, 287, 341–349
    directional solidification and, 287–292
    instabilities, 280, 341–349
    mushy zones and, 341–349
    solutal, 288, 289

C

Cahn-Hilliard equation, 126, 131, 134, 144, 157
Capillarity, anisotropy and, 100
Capillary number (C), 36
Capillary undercooling, 24, 57
Cellular-dendritic transition, 245
Cellular modes, in FTA, 181–183
Chaotic states, 82, 188, 272, 295
Chemical potential, 123
Chiareli–Worster limit, 340
Chimneys, 5, 357–363
Coarsening, 134, 144, 152–156
Cold-boundary condition, 171
Compatibility conditions, 142, 145, 192, 195, 199–200
Composite structures, 311
Concentration boundary-layer thickness, 47
Constitutional undercooling, 44, 51, 54, 57, 59, 287
Contact angle, 21, 82–83, 259
Continuity, equation of, 341
Continuity of temperature, 39, 59
Continuous-growth model, 166–167
Convection
    Bénard, 280–283
    binodal points, 127
    buoyancy and, 3, 275, 287–292, 341–349
    cellular, 304–311
    coarsening and, 134–135, 154
    directional solidification and, 283–287
    free, 279–280
    mesoscale fluid flow, 341–357
    oscillatory instability, 341–356
    thermal boundary layer and, 289–291
    thermocapillary, 275
    volume-change, 275, 283–287
    weakly nonlinear, 356–357
Corners, 121, 125, 130
Couette flow, 283, 293
Coupling effects, 205, 288–270, 205, 288

Critical nucleation radius, 30, 34, 215–221
Cross-gap curvatures, 83
Curved interfaces, 21–39
    boundary conditions, 21–26
    undercooling and, 26–39

D

Darcy's law, 334–335, 338, 342
Darcy number, 360
Dendrites, 215–254
    approximate selection, 221–229
    banded transition, 179
    critical radius, 215–221
    crystals, 217–221
    definition of, 215
    mushy zones and, 330–334, 350–352
    needles. *See* Needles
    prediction of, 215
    selection theories, 221–237
    See specific types, conditions, parameters
Diffusion
    diffusivity ratios, 172
    double-diffusive effects, 353
    interdiffusion coefficient, 257
    modified problem of, 296
    thermodynamics and, 1
Directional solidification, 45–46, 275
    anisotropy and, 91–105
    binary substances and, 45–46, 48–56
    buoyancy-driven convection, 287–292
    cellular convection and, 304–311
    forced flows and, 292–304
    microscale fluid flow and, 283–311
    nonlinear theory for, 62–85
    volume-change convection, 283–287
Disequilibrium, 156, 190, 199, 162–214
    bifurcation theory, 181–189
    cellular modes, in FTA, 181–183
    linear stability theory, 167–181
    mode coupling, 204–208
    nonlinear pulsations, 189–204
    oscillatory–cellular interactions, 204–206
    oscillatory modes, in FTA, 183–189
    oscillatory–pulsatile interactions, 206–208
    parameter, 181
    phenomenological models, 208–211
    pulsatile–cellular interactions, 204–205
    rapid solidification and, 164–167
    thermal effects, 171–181
    undercooling and, 166
Distribution coefficient. *See* segregation coefficient
Double-diffusive effects, 353
Double-well free energy, 126–127
Drop-shaped roots, 71
Droplets, 280
Dufour effect, 221

E

Eckhaus instability, 67–68, 72, 187–190, 267–268
Emms-Fowler model, 359
End effects, 298
Epitaxial growth, 156
Equilibrium distribution coefficient, 165
Equilibrium shape, 88, 89
Euler–Lagrange equation, 23, 87
Eutectics, 255–273
    approximate theories for, 261–267
    eutectic point, 255
    hypereutectic, 272, 326–328
    hypoeutectic, 272, 326–328
    lamellar, 255
    rod, 255
    selection models, 261–267
    See also specific effects
Evaporation–condensation model, 138, 152, 156
Evolution equations, 14, 62, 77, 82, 113, 124
    anisotropic, 92, 101, 113–114, 119, 125–126, 137–139, 144, 147
    See
        Kardar–Parisi–Zhang
        Korteweg–de Vries
        Cahn–Hilliard
        Brattkus–Davis
        Riley–Davis
        Sivashinsky
        Kuramoto–Sivashinsky

F

Faceting, 138
    anisotropy and, 139–152
    constant driving force, 139–152
    line tension and, 127
    spinodal decomposition and, 127, 130
Fickian law, 42
Finger growth, 25–26, 106, 228. *See also* Needles
Fisher approximation, 220
Floquet theory, 69, 81, 188, 247, 299, 306
Flow-induced instabilities, 274, 296
Forced flows, 274, 292–304
Fourfold symmetry, 68, 89–91, 108
Fourier's law, 8
Fowler model, 340
Freckles, 3, 357
Free convection, 279–280
Free energy, 127
Frozen temperature approximation (FTA), 47, 55–57, 76, 92, 181–189, 287–288
FTA. *See* Frozen temperature approximation

G

G-jitter effect, 274
Gaussian curvature, 139
Gibbs free energy, 23
Gibbs-Thomson equation, 24–25, 27, 29, 40, 44, 59
    anisotropy and, 88, 157
    disequilibrium, 166
Ginzburg–Landau equation
    bifurcation and, 66
    Eckhaus limit for, 190
Grashof number, 279, 314
Greengard–Strain method, 203
Groove structures, 148
Growth models, 30–39, 166–167, 261–267.
    See specific models

H

Heat balance, 26, 39, 44, 59
Heat equation, 8–10, 37, 42
Hele–Shaw cell, 46, 62, 72, 82–83, 237
Helical growth, 189
Helmholtz free energy, 21
Hemispherical-needle approximation, 220
Herring model, 88
Hexagonal structures, 73–75, 79–82, 95–97, 282, 356–358
Hole solutions, 189
Hong–Langer model, 238
Hopf bifurcations, 206–208
Hydrodynamic effects, 134
Hypercooling, 20, 35, 106–161, 138–139

I

Imposed cellular convection, 304
Imposed flows, 275
Instabilities
    absolute-stability limit, 52–58, 80–82, 102–104, 169, 177, 294, 303
    buoyancy-driven, 280, 292
    chaotic states, 82, 188, 272, 295
    coarsening and, 152–156
    convective, 291
    critical radius for, 34
    Eckhaus, 67–68, 72, 187–190, 267–268
    eutectics, 267–269
    flow-induced, 274
    latent-heat effects, 57–59, 172–183
    long-scale theories, 76–82
    mechanism of, 56–57
    morphological, 46–49, 55–60, 271, 292, 302–303
    mushy zones and, 340–347
    shrinkage and, 285
    soluto-convective, 288

382 Index

Instabilities (*Cont.*)
  subharmonic, 251
  See also Linear stability theory
Interdiffusion coefficient, 257
Interfaces
  anisotropy. *See* Anisotropy
  boundary conditions, 37–40, 59, 88, 164–166
  attachment coefficient, 97
  heat balance at, 25, 58
  diffuse, 366–378
  no-slip condition, 277, 285, 349
  stretching, 25
  thin-interface limit, 366–367
  See also specific parameters, effects
Interfacial wave theory (IWT), 228, 236–238, 252
Isolated needles, 217–221, 228
IWT. *See* Interfacial wave theory
Ivantsov solutions, 251–252
  needles and, 210, 246–247, 311–319
  Peclet number and, 231, 314
  Xu basic state and, 232

J

Jackson–Hunt model, 261, 264–267

K

Kardar–Parisi–Zhang (KPZ) equation, 137
KPZ. *See* Kardar–Parisi–Zhang equation
KdV. *See* Korteweg–de Vries operator
Kinetic coefficient, 17, 371
Kinks, 126, 128–130
  antikinks and, 129
  coarsening and, 134–135
  facets and, 127
  phase separation and, 131
  symmetry and, 131
Korteweg–deVries (KdV) equation, 113, 120
Kuramoto–Sivashinsky equation, 113, 114, 118–119, 295
Kurz–Fisher system, 219
Kuzmak–Luke procedure, 193

L

Lamellar eutectics, 255
Landau coefficient, 63, 70–73, 101, 181–183
Landau equations, 251, 304
  bifurcation types, 62–64
  coupled, 205
  regular planforms and, 72
Langer–Muller–Krumbhaar model, 224, 228, 231
Large anisotropy, 89, 121–139, 229
Latent heat effects, 7, 57–58, 171
  pulling speed and, 210

Lateral transport effects, 274, 286
  convective effects, 288
Leading-order problem, 340
Line tension, 121, 127
Linear stability theory, 36, 38, 108, 282
  basic state, 167–171
  curved interfaces and, 32–35
  directional solidification, 48–56
  disequilibrium, 167–181
  fingers and, 223–225
  Floquet theory and, 301
  Jackson–Hunt model and, 267–268
  marginal stability, 224
  Warren–Langer model, 250
  See also Instabilities
Liquid phase epitaxy, 140
Long-scale theories
  See Evolution equations, 82

M

Maximum velocity hypothesis, 222
Mesoscale fluid flow, 324–365
  buoyancy effects, 341–349
  chimneys, 357–363
  convective instability, 341–356
  formulation, 325–326
  mushy-zone models, 331–349
  nonlinear convection, 356–357
  oscillatory mode, 349–356
  volume-change convection, 336–341
Microscale fluid flow, 274–323
  Bénard convection, 280–283
  buoyancy-driven convection, 287–292
  cellular convection, 304–311
  directional solidification, 283–311
  equations for, 276–279
  forced flows, 292–304
  free convection, 279–280
  Ivantsov needles, 311–319
  prototype flows, 276–279
  volume-change convection, 283–287
Microscopic solvability theory (MST), 234–238, 249
Mobility parameter, 126, 372
Mode coupling
  disequilibrium, 204–208
  oscillatory–cellular interactions, 204–206
  oscillatory–pulsatile interactions, 206–208
  pulsatile-cellular interactions, 204–205
Models, 70
  binary substances, 42–48, 57–59
  disequilibrium, 208–211
  phase-field, 366–376
  phenomenological, 208–211
  See also specific effects, models
Modulational stability, 187–189
Morphological instability. *See* Instabilities

Morphological number, 49, 260
MST. *See* Microscopic solvability theory
Mullins–Sekerka theory, 66, 103, 162, 223, 232, 296
Mushy zones, 311, 324
    buoyancy and, 341–349
    convective instability, 341–349
    mesoscale fluid flow, 336–349
    models for, 331–336
    with volume-change convection, 336–341

N

Natural parameters, 53
Needles, 210, 216, 228, 231
    arrays of, 237–251
    dendrites, 216–221, 237–251
    hemispherical-needle approximation, 220
    Ivantsov needles, 220, 311–319
    kinetics and, 228–230
    See Isolated needles
    sidebranching, 225–227, 233, 249
    Stokes flow, 317
    tilting, 317–318
    See also specific effects, systems
Newell–Whitehead-Segel equation, 62, 72, 295
Nucleate growth, 7–40, 215
    critical nucleation radius, 30, 34, 215–221
    linearized instability, 32–35
Numerical simulation methods, 69–70, 190–203, 367, 375

O

Oldfield model, 224, 231
Optical mode, 82
Order parameter, 126, 127, 134
    phase separation and, 130
Oscillatory behavior, 167–173, 177, 182, 299, 353
    bifurcation and, 183–189
    cellular interactions, 204–206
    convective instability, 341–356
    disequilibrium, 183–189, 204–208
    FTA and, 183–189
    mesoscale fluid flow, 349–356
    mode coupling, 204–208
    pulsatile interactions, 206–208
Overgrowth instability, 251

P

Partial solute trapping, 165
Peclet number, 223, 227, 241
    interfacial, 373
    Ivantsov solution, 312–314
    selection criteria and, 263
    undercooling and, 218–219

Permeability, 338–3399
Phase-evolution equation, 67, 196
Phase–field models, 366–376
Phase separation, 130
Prandtl number, 275, 279, 280, 293, 347
Prototype flows, 275
Pulsatile instabilities, 169
Pulsatile modes, 184–186, 204–208
Pulsatile–oscillatory interactions, 206–208
Pulsations, 190–203
Pyramids, 140–143
    cones and, 156
    magic slopes, 144
    rhombic, 148, 150–154

Q

Quasi-periodic motion, 188
Quasi-steady linear stability analysis, 250
Quenched interfaces, 74

R

Rapid solidification, 164, 189–214
Rayleigh numbers, 275, 278, 282, 288, 347
Rejection, of solute, 163
Reynolds number, 278
Riley–Davis equation, 78–80
Rhombic pyramids, 150–154
Rod eutectics, 255
Roots, 71
Rounded corners, 129–130

S

Saffman–Taylor problem, 237, 238
Sausage structures, 119
Scheil approximation, 246–247
Scheil region, 240
Schmidt number, 278, 295
Schulze–Davis model, 299
Schulze–Worster model, 360
SCN. *See* Succinonitrile
Segregation coefficient, 42, 45, 59, 197, 198, 286, 289
Semiconductors, 165, 256, 292, 316
Shocklike solutions, 189
Shortwave cutoff, 121, 137, 167, 169
Shrinkage, instability and, 285
Sideband disturbances, 189
Sidebranches, 225–227, 233, 249
Similarity solutions, 10, 12, 15–16, 221
Simulation methods, 69–70, 190–203, 367, 375
Sivashinsky equation, 77–78, 92, 100
Slender-body theory, 240
Snowflake formation, 2, 34
Solitary waves, 189
Solute balance, 44

384

Solute bands, 171
Solute rejection, 44, 163–165
Solute trapping, 165
Soluto-convective instability, 288
Soret effect, 221, 283
Spencer–Huppert model, 243–248, 249
Spinodal decomposition, 126–127, 130–131
Spiral modes, 280
Square patterns, 283
Square pyramids, 145
Squares, 73, 94
Stability
  absolute-stability limit, 52–58, 80–82, 102–104, 164, 177, 294
  Datye–Langer theory, 271
  modulational, 189
  nearly absolute, 80–82
  window of, 171
  See Instabilities
Stagnation-point flow, 295–296, 298
Stationary cells, 109, 111, 120
Steady cellular mode, 173
Steady growth, 261–267
Stefan number, 11, 28, 47, 111, 112, 119, 127, 314, 350
  anisotropy and, 112
Stefan problem, 333, 336
Step density, 97, 121
Stepwise growth, 97–105
Stretching, interfacial, 25
Subharmonic instability, 251
Succinonitrile (SCN), 29, 222
  acetone alloys, 55, 66
  ethanol and, 289
Surface energy, 34, 109
  anisotropy and, 86–91, 97, 100, 105, 106, 111–112, 117, 122, 230, 236
  corners and, 121
  curvature-dependent, 125
  diffusion and, 154
  homogenization and, 288
  Ivantsov function, 252
  kinetics and, 86–91
  parameter, 28, 36, 49, 260
  segregation coefficient and, 78–80
  Spencer–Huppert theory and, 248
  undercooling and, 44
  weighted, 90, 106
  WKB approximation, 231
  Xu basic state and, 232
  zero value, 151
Swirl, 275
Symmetry, 42, 82–83, 268
  aspect ratio and, 145
  Cahn–Hilliard equation and, 131
  kinks and, 131

surface energy and, 237–238
symmetry-breaking, 238
See also Anisotropy

T

Tertiary arms, 249
Thermal boundary layer, 155
Thermal conductivity ratio, 28, 171
Thermal-diffusivity ratio, 28
Thermal length, 336
Thermodynamically instability, 137, 138
Thermosoluto-capillary convection, 275
Thin-interface limit, 367
Tilted cells, 94, 97, 116
Tip splitting
  anisotropy and, 105
  bifurcation and, 265
  chaotic states and, 272
  instabilities and, 249
  Jackson–Hunt theory, 264
  steady growth and, 264
  supercriticality and, 69–70
  surface tension and, 225
  symmetry and, 69–70
Total grand potential, 24
Transition point, 64
Trapping, of solute, 165
Travelling-wave mode, 20, 93, 97, 109–120, 234–237
Triangular pyramids, 145
Trijunction conditions, 259
Trivedi–Kurz parameter, 244

U

Ultraviolet forcing, 251
Unconstrained growth, 26–34, 105–161
  large anisotropy, 121–139, 161
Undercooling, 13, 17
  anisotropy and, 92, 98, 102–103, 106, 109, 124
  capillary, 57
  constitutional, 44, 51, 59, 287
  curved interfaces and, 26–32, 35–39
  equilibrium and, 166
  Gibbs–Thomson relation and, 25, 35, 44, 218
  instability and, 35
  Ivantsov analysis and, 237
  kinetic, 18–20, 25, 35, 92, 124
  linearized instability, 35–39
  nucleus growth, 26–32
  Peclet number, 218
  solute rejection, 163
  surface energy and, 17, 44, 92
  unit undercooling, 17
  See specific models

V

Volume-change convection, 275, 286
  mushy-zones and, 334–341

W

Warren–Langer model, 60, 249–250
Weakly nonlinear systems, 62, 77, 124, 356–357
WKB approximation, 231, 232

Wollkind–Segel model, 64
Wulff construction, 89–90, 121

X

Xu model, 231–237

Y

Young–Laplace equation, 24, 88